A Literature Guide for the Identification of Plant Pathogenic Fungi

Amy Y. Rossman, Mary E. Palm,
and Linda J. Spielman

In cooperation with David F. Farr

APS PRESS
The American Phytopathological Society
St. Paul, Minnesota

Cover: Line drawing courtesy of G. Morgan-Jones. Previously published in Morgan-Jones, G., et al, 1972, *Icones Genera Coelomycetarum II*, University of Waterloo (Canada) Biology Series No. 4, p. 31.

This publication is No. 1 in the series Contributions from the U.S. National Fungus Collections. Information on other publications in this series is available from the Systematic Botany, Mycology, and Nematology Laboratory, USDA/ARS, Beltsville, MD 20705.

This book has been reproduced directly from typewritten copy submitted in final form to APS Press by the authors of the volume. No editing or proofreading has been done by the Press.

Library of Congress Catalog Card Number: 87-070764
International Standard Book Number: 0-89054-080-2

Published 1987 by The American Phytopathological Society
Second Printing, 1990

Printed in the United States of America

The American Phytopathological Society
3340 Pilot Knob Road
St. Paul, Minnesota 55121, USA

TABLE OF CONTENTS

1   Introduction

5   General References

5   General

7   Aphyllophorales

9   Ascomycotina

11   Deuteromycotina

12   Erysiphales

13   Mucorales

13   Peronosporales

14   Uredinales

15   Ustilaginales

17   Literature By Genus

237 Index to Authors

250 Index to Genera

INTRODUCTION

Plant pathologists and mycologists are frequently called upon to identify unknown specimens of plant pathogenic fungi, either in the course of their own research or as a service to other scientists and practitioners. In order to do this, the pathologist or mycologist must have access to keys and descriptions suitable for identifying the fungus at hand. The ability to identify plant pathogenic fungi is often limited by the difficulty of locating the vast amount of taxonomic literature which has been amassed over the years. This literature is scattered throughout a variety of journals and books, and few groups are covered by comprehensive monographs. Thus, it is difficult to keep abreast of the literature for the identification of even a small group of plant pathogenic fungi and almost impossible to be in command of it. The purpose of this guide is to bridge the gap between the worldwide mycological literature and those who wish to use it.

In compiling this book we have assumed that its users will have some knowledge of fungi and the techniques used to study them, either from a general course in mycology or through practical experience. This guide is not intended to teach mycology nor does it describe genera or species in detail, rather it will direct the user to the sources of such information.

The initial stage in the identification of plant pathogenic fungi consists of the proper placement of the unknown fungus in class, order, family, and genus. A number of books that are helpful at this stage in the identification are listed in the **General References** section. This is divided into subsections according to subject matter, i.e. General or major taxonomic group. Those references marked with an asterisk (*) are considered essential resources which should be a part of the library of anyone who is involved in the identification of fungi. For definitions of basic mycological terms, the works of Snell and Dick (1971) and Hawksworth et al. (1983) are useful, in addition to the glossaries included in many works dealing with specific groups of fungi. Also listed in the General subsection are sources of information concerning the isolation of fungi and the preparation of fungal specimens for microscopic examination.

Following the initial identification at the higher taxonomic levels, the investigator should have in mind one genus or a few genera to which the specimen at hand might belong. The **Literature by Genus** section of this guide should then be used to select the literature needed for further consultation. The guide is organized alphabetically by genus and includes 607 genera, primarily those containing plant pathogenic species.

The first two lines of each genus entry provide the following information:

1. **Author citation.** This consists of the author(s) who originally introduced the name, and, where appropriate, that author(s) in parentheses is followed by the author(s) who recognized the name at the generic rank. Author citations were updated in accordance with the most recent version of the International Code of Botanical Nomenclature (Voss et al., 1983). For example, genera that have been sanctioned by Fries or Persoon are indicated using ":Fr." or ":Pers.", respectively. Author names are abbreviated according to Stafleu and Cowan (1976-1986); if not included in that work, author names are spelled out.

2a. **Classification above the genus level.** For all fungi except the Deuteromycotina (Fungi Imperfecti), the order is given, e.g. Aphyllophorales. For the Deuteromycotina the class is given, either Hyphomycetes, Coelomycetes, or Agonomycetes (Mycelia Sterilia).

2b. **Number of described species.** In genera with few species this number is based on the latest monographic account. For large genera that have not been monographed this number is intended to indicate the general size of the genus.

Hawksworth et al. (1983) served as the basis for the above information although modifications were made based on more recent data.

Following the factual data for each genus, the literature references are presented in the **Literature by Genus** section. In compiling these lists, emphasis was placed on recent references considered most useful in identification. The criteria used to determine usefulness included the presence of keys, descriptions and illustrations, a taxonomic discussion, the inclusion of more than one species, and the availability of the publication. Descriptive comments are given after each citation in order to give the user an idea of the applicability of the publication.

A short discussion of the biology of the genus follows the references. When applicable, information is given about the teleomorph, the sexual form, or the anamorph, the asexual form. LITERATURE LISTED UNDER BOTH FORMS SHOULD BE CONSULTED. A cross reference is also given if literature listed for a related genus is relevant. Publications found in the **General References** are cited where they might be particularly helpful. The **Index to Authors** can be used to locate the subsection, e.g. Aphyllophorales, of the **General References** in which a particular article is listed.

Some of the important plant pathogenic species in the genus are listed, particularly those species included in the Commonwealth Mycological Institute Descriptions of Pathogenic Fungi (CMI) or in Fungi Canadenses (F. Can.). These descriptions are listed according to the currently accepted name of the species. After the CMI or F. Can. number, the term "under" is used to designate a species that is treated under a different morph. For example, the anamorph *Curvularia lunata* is included in the CMI Descriptions under the teleomorph name and is listed here as "CMI 474 under *Cochliobolus lunatus*." Species that are treated as the synonym of a currently accepted name are cited after the CMI or F. Can. number with an "as." For example, *Bipolaris incurvata* is listed as "CMI 342 as *Drechslera incurvata*." Reference is often given to synonyms well-established in the literature.

The **Index to Authors** lists all authors included in this book, whether a first author or subsequent author. The author's publications are listed alphabetically by genus or by subsection of the **General References** followed by the date. An **Index to Genera** is provided.

ACKNOWLEDGMENTS

We would like to express our sincere appreciation to David F. Farr for his patience and the many hours spent on all the computer-related activities of this endeavor. We gratefully acknowledge Flora G. Pollack and Alice J. Watson, the previous mycologists in the Animal and Plant Health Inspection Service, who started and maintained the literature files on which this book was based. Finally, we thank Gerald F. Bills and George P. Chamuris for ongoing discussions and Larry F. Grand and Clark T. Rogerson for reviewing the manuscript.

Amy Y. Rossman
United States Department of Agriculture
Agricultural Research Service

Mary E. Palm
United States Department of Agriculture
Animal and Plant Health Inspection Service

Systematic Botany, Mycology and Nematology Laboratory
Biosystematics and Beneficial Insects Institute
Beltsville Agricultural Research Center
Beltsville, Maryland 20705 USA

Linda J. Spielman
Plant Pathology Department
Cornell University
Ithaca, New York 14853 USA

GENERAL REFERENCES

GENERAL

*Ainsworth, G. C., Sparrow, F. K., and Sussman, A. S., eds. 1973. The Fungi. An Advanced Treatise. Vol. IVA. A Taxonomic Review with Keys: Ascomycetes and Fungi Imperfecti. Academic Press, New York. 621 pp. (taxonomic treatment of major groups of Ascomycetes and Fungi Imperfecti, keys to genera within each group)

*Ainsworth, G. C., Sparrow, F. K., and Sussman, A. S., eds. 1973. The Fungi. An Advanced Treatise. Vol. IVB. A Taxonomic Review with Keys: Basidiomycetes and Lower Fungi. Academic Press, New York. 504 pp. (taxonomic treatment of major groups of Basidiomycetes and "lower fungi", keys to genera within each group)

Arx, J. A. von 1981. The Genera of Fungi Sporulating in Pure Culture. Third Edition. J. Cramer, Vaduz. 424 pp. (keys to genera, teleomorph and anamorph information, references for each genus, sparse illustrations)

Booth, C., ed. 1971. Methods in Microbiology. Vol. 4. Academic Press, New York. 795 pp. (manual of methods used in mycological studies with 24 chapters covering general methodology, specialized techniques for specific groups of fungi, and methods used to study physical, biochemical, and genetic aspects)

Brandenburger, W. 1985. Parasitische Pilze an Gafaerrspflanzen in Europe. Gustav Fischer Verlag, New York. 1248 pp. (lists by host, brief species descriptions, generic descriptions, illustrations)

*Commonwealth Mycological Institute. 1964-1985. Descriptions of Pathogenic Fungi and Bacteria. Numbers 1-840. By various authors. Commonwealth Mycological Institute, Kew, Surrey, England. (useful at the species level, descriptions and illustrations, numerous references, see comments after each genus for species treated; Descriptions 841- are also published in Mycopathologia beginning with Vol. 94, 1986)

Commonwealth Mycological Institute. 1983. Plant Pathologist's Pocketbook. Second Edition. Commonwealth Mycological Institute, Kew, Surrey, England. 439 pp. (a compilation of information covering many aspects of plant pathology, includes both methods and general information about plant diseases, lists, references)

*Domsch, K. H., Gams, W., and Anderson, T-H. 1980. Compendium of Soil Fungi. Vol. 1, Vol. 2. Academic Press, New York. 859 pp. (Vol. 1, text), 405 pp. (Vol. 2, bibliography). (keys to genera and species, detailed information with numerous references, descriptions and illustrations)

Ellis, M. B., and Ellis, J. P. 1985. Microfungi on Land Plants. An Identification Handbook. Macmillan Publishing, New York. 818 pp. (organized into sections based on host biology, e.g.

GENERAL - cont.
plurivorous wood and bark fungi, plurivorous leaf-litter fungi, etc.; keys to some genera and species, brief descriptions, illustrations)

Funk, A. 1981. Parasitic Microfungi of Western Trees. Canadian Forestry Service, Pacific Forest Research Centre, Victoria, B.C. 190 pp. (keys to genera, descriptions and illustrations, host index, glossary)

Funk, A. 1985. Foliar Fungi of Western Trees. Canadian Forestry Service, Pacific Forest Research Centre, Victoria, B.C. 159 pp. (keys to genera, descriptions and illustrations, host index, glossary)

Hawksworth, D. L. 1974. Mycologist's Handbook. An Introduction to the Principles of Taxonomy and Nomenclature in the Fungi and Lichens. Commonwealth Mycological Institute, Kew, Surrey, England. 231 pp. (information on: collection and preservation; naming, describing, and publishing; nomenclature, including a copy of the 1972 International Code of Botanical Nomenclature; herbarium locations and abbreviations; and a glossary of taxonomic and nomenclatural terms)

*Hawksworth, D. L., Sutton, B. C., and Ainsworth, G. C. 1983. Ainsworth & Bisby's Dictionary of the Fungi. Seventh Edition. Commonwealth Mycological Institute, Kew, Surrey, England. 445 pp. (lists all genera of fungi with nomenclatural and taxonomic information, references, distribution, etc.; includes some mycological and lichenological terms)

Malloch, D. 1981. Moulds - Their Isolation, Cultivation, and Identification. University of Toronto Press, Toronto. 97 pp. (general information on moulds, methods for their isolation and cultivation, dichotomous key to common genera, descriptions and illustrations)

Snell, W. H., and Dick, E. A. 1971. A Glossary of Mycology. Harvard University Press, Cambridge, MA. 181 pp. (a glossary with some illustrations, 15 plates of fungal structures with descriptive terms, color term equivalents)

Stafleu, F. A., and Cowan, R. S. 1976-1985. Taxonomic Literature. A Selective Guide to Botanical Publications and Collections with Dates, Commentaries and Types. Second Edition. Bohn, Scheltema, and Holkema, Utrecht. 1136 pp. (Vol. I), 991 pp. (Vol. II), 980 pp. (Vol. III), 1214 pp. (Vol. IV), 1066 pp. (Vol. V). (bibliographic information on botanical authors and collectors, basis for authority citations in this reference book)

Stevens, R. B., ed. 1981. Mycology Guidebook. University of Washington Press, Seattle. 712 pp. (useful compilation of information on methods of dealing with fungi for research and as teaching tools)

Tuite, J. 1969. Plant Pathological Methods. Fungi and Bacteria. Burgess Publishing, Minneapolis, MN. 239 pp. (general methods

GENERAL - cont.
for working with pathogens and studying diseases)
Voss, E. G., et al. 1983. International Code of Botanical
Nomenclature adopted by the Thirteenth International Botanical
Congress, Sydney, Australia, August 1981. Regnum Vegetabile
Series #111. Bohn, Scheltema, and Holkema, Utrecht. 472 pp. (in
English, French, and German; principles, rules, and
recommendations concerning botanical nomenclature, lists of
conserved and rejected family and generic names; guide for
determining types)

APHYLLOPHORALES

Bakshi, B. K. 1971. Indian Polyporaceae. Indian Council of Agric.
Res., New Delhi. 246 pp. (key to 105 species, descriptions and
illustrations, recognizes broad generic concepts)
Bondartsev, A. S. 1953. The Polyporaceae of the European USSR and
Caucasia. Translated from Russian 1971. U.S. Dep. Comm., Nat.
Tech. Info. Serv., Springfield, VA. 896 pp. (keys to genera and
species of poroid Aphyllophorales, descriptions and sparse
illustrations)
Breitenbach, J., and Kraenzlin, F. 1986. Fungi of Switzerland. Vol.
2. Heterobasidiomycetes, Aphyllophorales, Gasteromycetes. Verlag
Mykologia, Lucerne. 412 pp. (covers 528 species many of which
are present in the U.S., useful for persons not familiar with
modern generic concepts, descriptions and excellent
illustrations, key to species by group based on fruiting body
characters)
Burt, E. A. 1966. The Thelephoraceae of North America I-XV. Reprint
of original Ann. Missouri Bot. Gard. Papers, 1914-1926. Hafner
Publishing, New York. 354 pp. (nomenclature outdated but useful
for identification, keys to genera and species, descriptions and
sparse illustrations)
Domanski, S. 1965. Fungi II. Polyporaceae I (resupinatae)
Mucronoporaceae I (resupinatae). Transl. from Polish 1972. U.S.
Dep. Comm., Nat. Tech. Info. Serv., Springfield, VA. 234 pp.
(treats resupinate Polyporaceae and Mucronoporaceae, keys to
genera and species, descriptions and illustrations)
Domanski, S., Orlos, H., and Skirgiello, A. 1967. Fungi III.
Polyporaceae II (pileatae) Mucronoporaceae II (pileatae),
Ganodermataceae, Bondarzewiaceae, Boletopsidaceae, Fistulinaceae.
Transl. from Polish 1973. U.S. Dep. Comm., Nat. Tech. Info.
Serv., Springfield, VA. 330 pp. (treats poroid fungi with pilei,
keys to genera and species, descriptions and illustrations)
Donk, M. A. 1974. Checklist of European Polypores. Verh.
Koninklijke Nederlandse Akad. van Wetenschappen, afd. Natuurkunde
62:1-469. (an extensive list with a discussion of each genus)
Eriksson, J., Hjortstam, K., and Ryvarden, L. 1978. The Corticiaceae

APHYLLOPHORALES - cont.
of North Europe. Vol. 5. *Mycoaciella-Phanerochaete*. Fungiflora,
Oslo. Pp. 889-1407. (see comments under Eriksson and Ryvarden,
1973)
Eriksson, J., Hjortstam, K., and Ryvarden, L. 1981. The Corticiaceae
of North Europe. Vol. 6. *Phlebia-Sarcodontia*. Fungiflora, Oslo.
Pp. 1051-1276. (see comments under Eriksson and Ryvarden, 1973)
Eriksson, J., Hjortstam, K., and Ryvarden, L. 1984. The Corticiaceae
of North Europe. Vol. 7. *Schizopora-Suillosporium*. Fungiflora,
Oslo. Pp. 1282-1449. (see comments under Eriksson and Ryvarden,
1973)
Eriksson, J., and Ryvarden, L. 1973. The Corticiaceae of North
Europe. Vol. 2. *Aleurodiscus-Confertobasidium*. Fungiflora,
Oslo. Pp. 60-261. (keys to species, descriptions and
illustrations, keys to genera to be published in forthcoming Vol.
1)
Eriksson, J., and Ryvarden, L. 1975. The Corticiaceae of North
Europe. Vol. 3. *Coronicium-Hyphoderma*. Fungiflora, Oslo. Pp.
289-546. (see comments under Eriksson and Ryvarden, 1973)
Eriksson, J., and Ryvarden, L. 1976. The Corticiaceae of North
Europe. Vol. 4. *Hyphodermella-Mycoacia*. Fungiflora, Oslo. Pp.
549-886. (see comments under Eriksson and Ryvarden, 1973)
Gilbertson, R. L. 1971. Fungi that Decay Ponderosa Pine. University
of Arizona Press, Tucson. 197 pp. (checklist by family, keys to
orders, families, genera, and species, descriptions and
illustrations)
*Gilbertson, R. L., and Ryvarden, L. 1986. North American Polypores.
Vol. 1. *Abortiporus-Lindtneria*. Fungiflora, Oslo. 433 pp.
(treats all polypores up to *Lindtneria*, keys to families, genera,
and species, descriptions and illustrations)
Harrison, K. A. 1973. Aphyllophorales III: Hydnaceae and
Echinodontiaceae. Pages 369-395 in: The Fungi. Vol. IVB. G. C.
Ainsworth, F. K. Sparrow, and A. S. Sussman, eds. Academic
Press, New York. (overview chapter, key to genera of Hydnaceae,
discussion of each genus)
*Juelich, W. 1984. Die Nichtblaetterpilze, Gallertpilze und
Bauchpilze. Aphyllophorales, Heterobasidiomycetes,
Gasteromycetes. Gustav Fischer Verlag, New York. 626 pp. (in
German, keys to genera and species, descriptions and
illustrations)
Juelich, W., and Stalpers, J. A. 1980. The Resupinate Non-poroid
Aphyllophorales of the Temperate Northern Hemisphere.
North-Holland Publishing, Amsterdam. 335 pp. (comprehensive
reference for identification of corticioid fungi of temperate
northern hemisphere, keys to genera and species, descriptions,
references, no illustrations)
Lindsey, J. P., and Gilbertson, R. L. 1978. Basidiomycetes that
Decay Aspen in North America. J. Cramer, Vaduz. 406 pp.

APHYLLOPHORALES - cont.
(checklist by family, keys to orders, families, genera, and species, descriptions and illustrations)
Overholts, L. O. 1953. The Polyporaceae of the United States, Alaska and Canada. University of Michigan Press, Ann Arbor. 466 pp. (nomenclature outdated but useful for identification of non-resupinate polypores, keys, descriptions and illustrations)
Pegler, D. N. 1973. Aphyllophorales IV: Poroid families. Pages 397-420 in: The Fungi. Vol. IVB. G. C. Ainsworth, F. K. Sparrow, and A. S. Sussman, eds. Academic Press, New York. (overview chapter, keys to families and genera)
Ryvarden, L. 1976. The Polyporaceae of North Europe. Vol. 1. *Albatrellus-Incrustoporia*. Fungiflora, Oslo. Pp. 1-214. (keys to genera and species, descriptions and illustrations)
Ryvarden, L. 1978. The Polyporaceae of North Europe. Vol. 2. *Inonotus-Tyromyces*. Fungiflora, Oslo. Pp. 219-507. (keys to genera and species, descriptions and illustrations)
Ryvarden, L., and Johansen, I. 1980. A Preliminary Polypore Flora of East Africa. Fungiflora, Oslo. 636 pp. (keys to genera and species, descriptions and illustrations)
Talbot, P. H. B. 1973. Aphyllophorales I: General characteristics; thelephoroid and cupuloid families. Pages 327-349 in: The Fungi. Vol. IVB. G. C. Ainsworth, F. K. Sparrow, and A. S. Sussman, eds. Academic Press, New York. (overview chapter, keys to families and genera)

ASCOMYCOTINA

Arx, J. A. von, and Mueller, E. 1954. Die Gattungen der Amerosporen Pyrenomyceten. Beitr. Kryptogamenflora Schweiz 11(1):1-434. (in German, keys to genera, translated into English by Butterfill, 1969, descriptions and illustrations)
Arx, J. A. von, and Mueller, E. 1975. A re-evaluation of the bitunicate Ascomycetes with keys to families and genera. Stud. Mycol. 9:1-159. (keys to families and genera, descriptions of families, brief information on genera, references, illustrations)
Barr, M. E. 1972. Preliminary studies on the Dothideales in temperate North America. Contrib. Univ. Mich. Herb. 9:523-638. (keys to genera, subgenera, and sections, descriptions and illustrations)
Barr, M. E. 1978. The Diaporthales in North America, with Emphasis on *Gnomonia* and its Segregates. Mycol. Mem. 7:1-232. (keys to genera and species, descriptions and illustrations, host index, list of excluded genera)
Barr, M. E. 1979. A classification of Loculoascomycetes. Mycologia 71:935-957. (dichotomous keys to 58 families in eight orders, genera listed)
Breitenbach, J., and Kraenzlin, F. 1983. Fungi of Switzerland. Vol.

ASCOMYCOTINA - cont.
1. Ascomycetes. Edition Mykologia, Lucerne. 310 pp. (originally in German and French, now also in English, key to species, descriptions, excellent color photographs)
Butterfill, G. B. 1969. Keys to the Genera of Amerospored and Didymospored Pyrenomycetes. Commonwealth Mycological Institute, Kew, Surrey, England. 53 pp. (translation into English of keys included in Arx and Mueller, 1954 and Mueller and Arx, 1962)
Cannon, P. F., Hawksworth, D. L., and Sherwood-Pike, M. A. 1985. The British Ascomycotina. An Annotated Checklist. Commonwealth Mycological Institute, Kew, Surrey, England. 302 pp. (lists 5,100 species with anamorph information and references)
Dennis, R. W. G. 1978. British Ascomycetes. J. Cramer, Vaduz. 585 pp. (keys to genera, descriptions and illustrations)
Holm, L. 1975. Nomenclatural notes on Pyrenomycetes. Taxon 24:475-488. (provides nomenclatural status and information on many generic names)
Kobayashi, T. 1970. Taxonomic Studies of Japanese Diaporthaceae with Special Reference to their Life-histories. Bull. Gov. For. Exp. Stn. (Jpn.) 226:1-242. (key to genera, species descriptions and illustrations)
Korf, R. P. 1972. Synoptic key to the genera of the Pezizales. Mycologia 64:937-994. (discussion of synoptic keys, key to 93 genera, annotated list of genera)
Korf, R. P. 1973. Discomycetes and Tuberales. Pages 249-319 in: The Fungi. Vol. IVA. G. C. Ainsworth, F. K. Sparrow, and A. S. Sussman, eds. Academic Press, New York. (overview chapter, keys to orders, families, and genera)
Luttrell, E. S. 1973. Loculoascomycetes. Pages 135-219 in: The Fungi. Vol. IVA. G. C. Ainsworth, F. K. Sparrow, and A. S. Sussman, eds. Academic Press, New York. (overview chapter; discussion of taxonomic characters; synoptic guide to orders, families, and genera; keys to orders, families, and genera; illustrations)
Mueller, E., and Arx, J. A. von 1962. Die Gattungen der Didymosporen Pyrenomyceten. Beitr. Kryptogamenflora Schweiz 11(2):1-922. (in German, keys to genera, translated into English by Butterfill, 1969, descriptions and illustrations)
Mueller, E., and Arx, J. A. von 1973. Pyrenomycetes: Meliolales, Coronophorales, Sphaeriales. Pages 87-132 in: The Fungi. Vol. IVA. G. C. Ainsworth, F. K. Sparrow, and A. S. Sussman, eds. Academic Press, New York. (keys to orders, families of Sphaeriales, and genera)
Munk, A. 1957. Danish Pyrenomycetes. A Preliminary Flora. Einar Munksgaard, Copenhagen. 491 pp. (key to genera, descriptions and illustrations)
Pfister, D. H. 1982. A Nomenclatural Revision of F. J. Seaver's North American Cup-fungi (Operculates). Occasional Pap. Farlow

ASCOMYCOTINA - cont.
Herb. No. 117. 32 pp. (updated names for Seaver, 1928)
Seaver, F. J. 1928. The North American Cup-fungi. (Operculates). Originally published by the author, reprinted 1978 by Lubrecht and Cramer, Monticello, NY. 377 pp. (see Pfister, 1982 for an update of the names used by Seaver, keys, descriptions and illustrations)
Seaver, F. J. 1951. The North American Cup-fungi. (Inoperculates). Originally published by the author, reprinted 1978 by Lubrecht and Cramer, Monticello, NY. 428 pp. (keys, descriptions, illustrations, nomenclature outdated but still useful for identification)
*Sivanesan, A. 1984. The Bitunicate Ascomycetes and their Anamorphs. J. Cramer, Vaduz. 701 pp. (keys to species based on teleomorph and anamorph, limited to bitunicate Ascomycetes with known anamorphs, descriptions and illustrations)

DEUTEROMYCOTINA

Barnett, H. L., and Hunter, B. B. 1972. Illustrated Genera of Imperfect Fungi. Burgess Publishing, Minneapolis, MN. 241 pp. (keys to genera, brief generic descriptions and illustrations, broad generic concepts outdated)
Barron, G. L. 1968. The Genera of Hyphomycetes from Soil. The Williams and Wilkins Co., Baltimore, MD. 364 pp. (keys to genera, generic descriptions, discussion, illustrations)
*Carmichael, J. W., Kendrick, W. B., Conners, I. L., and Sigler, L. 1980. Genera of Hyphomycetes. University of Alberta Press, Edmonton. 386 pp. (synoptic keys to genera, illustrations, nomenclatural information and references for each genus)
*Ellis, M. B. 1971. Dematiaceous Hyphomycetes. Commonwealth Mycological Institute, Kew, Surrey, England. 608 pp. (keys to genera and species, descriptions and illustrations)
*Ellis, M. B. 1976. More Dematiaceous Hyphomycetes. Commonwealth Mycological Institute, Kew, Surrey, England. 507 pp. (a companion volume to Ellis, 1971, keys to genera and species, descriptions and illustrations)
Grove, W. B. 1935, 1937. British Stem and Leaf Fungi. Vol. 1, Vol. 2. Cambridge University Press, Cambridge. 488 pp., 407 pp. (nomenclature outdated but useful for identification, keys to genera and species, descriptions, sparse illustrations, host index, most species treated by host)
Hughes, S. J. 1958. Revisiones hyphomycetum aliquot cum appendice de nominibus rejiciendis. Can. J. Bot. 36:727-836. (lists nearly 400 generic and 1000 specific names of Hyphomycetes and indicates taxonomic placement)
Kendrick, W. B., and Carmichael, J. W. 1973. Hyphomycetes. Pages 323-509 in: The Fungi. Vol. IVA. G. C. Ainsworth, F. K.

DEUTEROMYCOTINA - cont.
Sparrow, and A. S. Sussman, eds. Academic Press, New York. (lists generic names of Hyphomycetes with information on taxonomy, nomenclature, and morphological characters; references and illustrations; synoptic key to genera; replaced by Carmichael et al., 1980)

Matsushima, T. 1971. Microfungi of the Solomon Islands and Papua-New Guinea. Published by the author, Kobe. 78 pp. (in Latin, descriptions, high quality illustrations)

Matsushima, T. 1975. Icones Microfungorum a Matsushima Lectorum. Published by the author, Kobe. 209 pp. (in Latin, descriptions, high quality illustrations)

Michaelides, J., Hunter, L., Kendrick, B., and Nag Raj, T. R. 1979. Icones Generum Coelomycetum, Supplement. Synoptic Key to 200 Genera of Coelomycetes. University of Waterloo Biology Series. University of Waterloo, Ontario. 41 pp. (synoptic key to the 200 genera treated in the Icones Generum Coelomycetum series, some illustrations)

Sprague, R. 1950. Diseases of Cereals and Grasses in North America. Ronald Press, New York. 538 pp. (outdated but useful, key to genera, descriptions and illustrations)

Sutton, B. C. 1973. Coelomycetes. Pages 513-582 in: The Fungi. Vol. IVA. G. C. Ainsworth, F. K. Sparrow, and A. S. Sussman, eds. Academic Press, New York. (overview chapter, keys to genera, some illustrations)

Sutton, B. C. 1977. Coelomycetes VI. Nomenclature of generic names proposed for Coelomycetes. Mycol. Pap. 141:1-253. (provides nomenclatural status and history for a large number of coelomycete genera)

*Sutton, B. C. 1980. The Coelomycetes. Fungi Imperfecti with Pycnidia, Acervuli and Stromata. Commonwealth Mycological Institute, Kew, Surrey, England. 696 pp. (keys to genera and species, descriptions, illustrations, host index)

Vassiljevsky, N. I., and Karakulin, B. P. 1950. [Fungi Imperfecti Parasitici II. Melanconiales.] Academiae Scientiarum, Moscow. 680 pp. (in Russian, host index)

ERYSIPHALES

Junell, L. 1967. Erysiphaceae of Sweden. Symb. Bot. Ups. 19:1-117. (keys to genera and species, descriptions and illustrations, host list)

Parmelee, J. A. 1977. The fungi of Ontario. II. Erysiphaceae (mildews). Can. J. Bot. 55:1940-1983. (keys to six genera and 28 species, descriptions and illustrations, review of taxonomy of *Erysiphe polygoni*, host list, useful for northeastern North America)

Spencer, D. M., ed. 1978. The Powdery Mildews. Academic Press, New

ERYSIPHALES - cont.
York. 565 pp. (20 chapters, generally according to host, one chapter on taxonomy with keys to genera and to some species based on conidial state characters)
Yarwood, C. E. 1973. Pyrenomycetes: Erysiphales. Pages 71-86 in: The Fungi. Vol. IVA. G. C. Ainsworth, F. K. Sparrow, and A. S. Sussman, eds. Academic Press, New York. (overview chapter, keys to genera based on teleomorph and to some species based on conidial state)
Zheng, R.-y. 1985. Genera of the Erysiphaceae. Mycotaxon 22:209-263. (synoptic and diagnostic keys, recognizes 19 teleomorph and four anamorph genera)

MUCORALES

Hesseltine, C. W., and Ellis, J. J. 1973. Mucorales. Pages 187-217 in: The Fungi. Vol. IVB. G. C. Ainsworth, F. K. Sparrow, and A. S. Sussman, eds. Academic Press, New York. (overview chapter, keys to families and genera, illustrations)
O'Donnell, K. L. 1979. Zygomycetes in Culture. University of Georgia Press, Athens. 257 pp. (keys to families and 66 genera, descriptions and illustrations)
Zycha, H., Siepmann, R., and Linnemann, G. 1969. Mucorales, eine Beschreibung aller Gattungen und arten dieser Pilzgruppe mit einem Beitrag zur Gattung *Mortierella*. J. Cramer, Lehre. 355 pp. (in German, key to genera and species, descriptions and illustrations)

PERONOSPORALES

Constantinescu, O., and Negrean, G. 1983. Checklist of Romanian Peronosporales. Mycotaxon 16:537-556. (lists 112 species in eight genera according to fungus and host)
Kenneth, R. G., and Palti, J. 1984. The distribution of downy and powdery mildews and of rusts over tribes of Compositae (Asteraceae). Mycologia 76:705-718. (lists fungal genera according to host genus)
Lucas, M. T., and Dias, M. R. de Sousa 1976. Peronosporaceae Lusitaniae. Agron. Lusit. 37:281-299. (lists 19 species in three genera with brief descriptions)
Lucas, M. T., Dias, M. R. de Sousa, and Lopes, M. C. 1982. Peronosporaceae Lusitaniae II. Agron. Lusit. 41:165-174. (lists 15 species in three genera with brief descriptions)
Spencer, D. M., ed. 1981. The Downy Mildews. Academic Press, New York. 636 pp. (28 chapters, generally according to host)
Waterhouse, G. M. 1973. Peronosporales. Pages 165-183 in: The Fungi. Vol. IVB. G. C. Ainsworth, F. K. Sparrow, and A. S. Sussman, eds. Academic Press, New York. (keys to families and

PERONOSPORALES - cont.
genera, discussion, sparse illustrations)

UREDINALES

*Arthur, J. C. 1934. Manual of the Rusts in United States and Canada. Purdue Research Foundation, Lafayette, IN. 438 pp. (outdated but useful for hosts not included in Cummins 1971, 1978, descriptions and illustrations)
Bushnell, W. R., and Roelfs, A. P., eds. 1984. The Cereal Rusts Vol. 1. Origins, Specificity, Structure, and Physiology. Academic Press, New York. 546 pp. (one chapter on taxonomy, keys to many cereal rusts, descriptions, other chapters on formae speciales and races)
Cummins, G. B. 1962. Supplement to Arthur's Manual of the Rusts in United States and Canada. Hafner Publishing, New York. 24 pp. (updates taxonomy and nomenclature of Arthur, 1934)
*Cummins, G. B. 1971. The Rust Fungi of Cereals, Grasses and Bamboos. Springer-Verlag, New York. 570 pp. (key to genera, keys to species by host genus and by rust genus, descriptions and illustrations)
*Cummins, G. B. 1978. Rust Fungi on Legumes and Composites in North America. University of Arizona Press, Tucson. 424 pp. (keys to genera and species, descriptions and illustrations)
*Cummins, G. B., and Hiratsuka, Y. 1983. Illustrated Genera of Rust Fungi. Revised Edition. American Phytopathological Society, St. Paul, MN. 152 pp. (key to 105 genera, descriptions and illustrations of each genus with additional information and many references, a useful reference text for rust fungi, replaces *Illustrated Genera of Rust Fungi* by Cummins, 1959)
Eboh, D. O. 1986. A taxonomic survey of Nigerian rust fungi: Uredinales nigerianensis. IV. Mycologia 78:577-586. (descriptions and illustrations of 17 taxa, references to previous papers in this series)
Gaeumann, E. 1959. Die Rostpilze Mitteleuropas. Buechler et Cie, Berne. 1407 pp. (in German, keys to rust fungi of central Europe, especially Switzerland, descriptions and illustrations)
Gallegos, H. L., and Cummins, G. B. 1981. Uredinales (Royas) de Mexico. Vol. 1, Vol. 2. Secretaria de Agricultura y Recursos Hidraulicos, Sinaloa, Mexico. 440 pp., 492 pp. (in Spanish, keys to genera and species, descriptions and illustrations)
Gjaerum, H. B. 1986. East African rusts (Uredinales), mainly from Uganda 5. On families belonging to Gamopetalae. Mycotaxon 27:507-550. (reports 71 taxa, descriptions and illustrations, provides host list for previously published records of east African rusts, references)
Laundon, G. F. 1973. Uredinales. Pages 247-279 in: The Fungi. Vol. IVB. G. C. Ainsworth, F. K. Sparrow, and A. S. Sussman, eds.

UREDINALES - cont.
Academic Press, New York. (overview chapter, key to anamorph and teleomorph genera)
Littlefield, L. J., and Heath, M. C. 1979. Ultrastructure of Rust Fungi. Academic Press, New York. 277 pp. (review of rust ultrastructure and host-parasite relations, electron micrographs)
Savulescu, T. 1953. Monografia Uredinalelor Republica Populara Romana. Vol. 1, Vol. 2. Editura Academiei Republicii Populare Romane. 1166 pp. (in Rumanian, rusts of Rumania, Vol. 1 provides a general overview, Vol. 2 is taxonomic with keys, descriptions, and illustrations)
Sydow, H., and Sydow, P. 1904-1924. Monographia Uredinarum. (reprint 1971). Bibl. Mycol. 33:1-2794. (German text, Latin descriptions, comprehensive to 1924)
Wilson, M., and Henderson, D. M. 1966. British Rust Fungi. Cambridge University Press, Cambridge. 384 pp. (keys to genera, descriptions and illustrations, replaces *British Rust Fungi* by Grove, 1913)
Ziller, W. G. 1974. The Tree Rusts of Western Canada. Canadian Forestry Service, Dept. of the Environment, Victoria, B.C. 272 pp. (keys to species based on host genus, descriptions and color illustrations)

USTILAGINALES

Duran, R. 1973. Ustilaginales. Pages 281-300 in: The Fungi. Vol. IVB. G. C. Ainsworth, F. K. Sparrow, and A. S. Sussman, eds. Academic Press, New York. (overview chapter, key to genera)
Duran, R., and Fischer, G. W. 1961. The Genus *Tilletia*. Washington State University, Pullman. 138 pp. (key to species based on host, descriptions and illustrations)
*Fischer, G. W. 1953. Manual of the North American Smut Fungi. Ronald Press, New York. 343 pp. (nomenclature outdated but useful for identification, keys to species based on host, descriptions and illustrations)
Harada, Y. 1983. Material for the smut flora of Japan I. Trans. Mycol. Soc. Jpn. 24:299-306. (reports ten species in six genera, illustrations)
Hirschhorn, E. 1986. Las Ustilaginales de la Flora Argentina. Publicacion Especial, Comision de Investigaciones Cientificas, La Plata, Argentina. 530 pp. (in Portuguese, descriptions and illustrations of 109 species in 14 genera)
Kakishima, M. 1982. A taxonomic study on the Ustilaginales in Japan. Mem. Inst. Agr. & For. Univ. Tsukuba 1:1-124. (in Japanese with English summary, keys to 141 species in 14 genera, classification on basis of teliospore surface morphology, host list)
Mordue, J. E. M., and Ainsworth, G. C. 1984. The Ustilaginales of the British Isles. Mycol. Pap. 154:1-96. (replaces Ainsworth and

USTILAGINALES - cont.
Sampson's *British Smut Fungi*, 1950, keys to families, genera, and species, key to species based on host, descriptions and illustrations)

Savulescu, T. 1957. Ustilaginalele din Republica Populara Romina. Vol. 1, Vol. 2. Editura Academiei Republicii Populare Romine. 1168 pp. (in Rumanian, smuts of Rumania, Vol. 1 provides a general overview, Vol. 2 is taxonomic with keys, descriptions and illustrations)

Vanky, K. 1985. Carpathian Ustilaginales. Symb. Bot. Ups. 24:2:1-309. (keys to genera and 299 species, descriptions, light and scanning electron photomicrographs, host list)

Zundel, G. L. 1953. The Ustilaginales of the World. Dept. of Botany, Pennsylvania State College, Contrib. No. 176. 410 pp. (key to genera, descriptions, host index)

LITERATURE BY GENUS

ACANTHARIA Theissen & Sydow
  Pleosporales   6 spp.
Arx, J. A. von, and Mueller, E. 1984. Notes on some Ascomycetes. Sydowia 37:6-10. (disagrees with Sivanesan's placement of *A. sinensis* in *Gibbera*)
Bose, S. K., and Mueller, E. 1965. Central Himalayan Fungi-II. Indian Phytopathol. 18:340-355. (describes and illustrates a new species)
Sivanesan, A. 1984. *Acantharia, Gibbera* and their anamorphs. Trans. Br. Mycol. Soc. 82:507-529. (describes and illustrates two *Acantharia* spp. and their anamorphs)
Acantharia species occur on living leaves of Fagaceae. The genus has been monographed by Mueller and Arx (1962). Anamorphs have been placed in *Fusicladium* and *Stigmina*.

ACREMONIUM Link:Fr.
  Hyphomycetes   105 spp.
Gams, W. 1971. *Cephalosporium*-artige Schimmelpilze (Hyphomycetes). Gustav Fischer Verlag, New York. 262 pp. (in German with summary, glossary, and keys in English, treats 82 *Acremonium* spp. and nearly 50 spp. in related genera)
Hinton, D. M., and Bacon, C. W. 1985. The distribution and ultrastructure of the endophyte of toxic tall fescue. Can. J. Bot. 63:36-42. (illustrates fungus within the host)
Latch, G. C. M., Christensen, M. J., and Samuels, G. J. 1984. Five endophytes of *Lolium* and *Festuca* in New Zealand. Mycotaxon 20:535-550. (includes *A. coenophialum* and other *Acremonium* spp.)
Morgan-Jones, G., and Gams, W. 1982. Notes on Hyphomycetes. XLI. An endophyte of *Festuca arundinacea* and the anamorph of *Epichloe typhina*, new taxa in one of two new sections of *Acremonium*. Mycotaxon 15:311-318. (description of *A. coenophialum*, the endophyte frequently misidentified as *Epichloe typhina*, description of *A. typhinum*, anamorph of *E. typhina*)
Samuels, G. J. 1976. Perfect states of *Acremonium*. The genera *Nectria, Actiniopsis, Ijuhya, Neohenningsia, Ophiodictyon,* and *Peristomialis*. N. Z. J. Bot. 14:231-260. (key to 12 *Nectria* spp. with *Acremonium* anamorphs, descriptions and illustrations)
Siegel, M. R., Latch, G. C. M., and Johnson, M. C. 1985. *Acremonium* fungal endophytes of tall fescue and perennial ryegrass: significance and control. Plant Dis. 69:179-183. (review article, useful reference for detection and identification using ELISA or direct staining of host)
White, J. F., Jr., and Cole, G. T. 1985. Endophyte-host associations in forage grasses. II. Taxonomic observations on the endophyte of *Festuca arundinacea*. Mycologia 77:483-486. (description of *A. coenophialum*, discusses taxonomy of this species and its

ACREMONIUM - cont.
relations to *Epichloe typhina*)
White, J. F., Jr., and Morgan-Jones, G. 1987. Endophyte-host associations in forage grasses. VII. *Acremonium chisosum*, a new species isolated from *Stipa eminens* in Texas. Mycotaxon 28:179-189. (description and illustrations, reports endophytes from other *Stipa* spp.)
Most *Acremonium* species have previously been placed in *Cephalosporium*. Teleomorphs belong to *Epichloe, Nectria*, and *Neocosmospora*. Domsch et al. (1980) is useful in the identification of *Acremonium* species from soil. *Acremonium coenophialum* Morgan-Jones & Gams, *A. typhinum* Morgan-Jones & Gams, and other *Acremonium* spp. are endophytic inhabitants of tall fescue and perennial ryegrass that cause summer toxicosis in cattle and ryegrass staggers in sheep (Morgan-Jones and Gams, 1982). *Acremonium kiliense* Grutz is a soil fungus attacking humans and cattle but may attack maize (CMI 741); *A. recifei* (Leao & Lobo) Gams attacks humans and a number of plants (CMI 742); and *A. zonatum* (Sawada) Gams causes fig zonate leaf spot and zonal leaf spot of coffee (CMI 502).

ACROCONIDIELLA Lindquist & Alippi
    Hyphomycetes   1 sp.
Acroconidiella tropaeoli (Bond) Lindquist & Alippi causes a disease of nasturtium (CMI 161; Ellis, 1971).

ACTINOPELTE   See *Tubakia*.

AECIDIUM Pers.:Pers.
    Uredinales   600 spp.
Laundon, G. F. 1967. The taxonomy of the imperfect rusts. Trans. Br. Mycol. Soc. 50:349-353. (key to aecial forms)
Sato, T., and Sato, S. 1985. Morphology of aecia of the rust fungi. Trans. Br. Mycol. Soc. 85:223-238. (keys to 14 morphological types of aecia, illustrations)
Aecidium has been used for rusts known only in the aecial form. See Cummins and Hiratsuka (1983) and Wilson and Henderson (1966) for further discussion and illustrations of the five main types of aecia. Sato and Sato (1985) describe 14 aecial types and find that morphological types of aecia correlate with rust genera. *Aecidium physalidis* Burr. causes *Physalis* rust (F. Can. 186). See Uredinales.

AGARICODOCHIUM Liu
    Hyphomycetes   1 sp.
Chi, I. H., and Wei, A. J. 1983. [A study on the sporodochia of pathogenic fungus of soft-rot disease of oil-tea tree.] J. South China Agric. Coll. 4:16-22. (in Chinese with English abstract,

AGARICODOCHIUM - cont.
description and illustration of *A. camelliae*)
Agaricodochium camelliae Liu, Wei & Fan causes soft-rot disease of oil-tea tree in China.

ALBUGO (Pers.) Roussel
    Peronosporales   30 spp.
Biga, M. L. B. 1955. Riesaminazione delle specie del genere *Albugo* in base alla morfologia dei conidi. Sydowia 9:339-358. (in Italian with English summary, key to 30 species by host)
Pound, G. S., and Williams, P. H. 1963. Biological races of *Albugo candida*. Phytopathology 53:1146-1149. (races defined by cross-inoculation, results correlate with conidial size)
Zhang, Z. Y., Wang, Y. X., and Liu, Y. L. 1984. [Taxonomic studies on the family Albuginaceae of China II. A new species of *Albugo* on Acanthaceae and known species of *Albugo* on Cruciferae]. Acta Mycol. Sin. 3:65-71. (in Chinese)
Members of this genus cause white rust diseases including: *Albugo candida* (Pers.) Kuntze, white rust of Brassicaceae (CMI 460); *A. ipomoeae-panduratae* (Schwein.) Swingle, white rust of sweet potato and other Convolvulaceae (CMI 459); and *A. tragopogonis* (DC.) S. F. Gray, white rust of salsify and other Asteraceae (CMI 458). Within *A. candida* several physiological races are recognized.

ALEUROCORTICIUM   See *Aleurodiscus*.

ALEURODISCUS Rabenh. ex J. Schroet.
    Aphyllophorales   15 spp.
Boidin, J., Lanquetin, P., Gilles, G., Candoussau, F., and Hugueney, R. 1985. Contribution a la connaissance des *Aleurodiscoideae* spores amyloides (Basidiomycotina, Corticiaceae). Bull. Trimest. Soc. Mycol. Fr. 101:333-367. (in French with English summary, includes three *Aleurodiscus* spp. and 11 species in related genera, key to genera and species, descriptions and illustrations)
Lemke, P. A. 1964. The genus *Aleurodiscus* (*sensu stricto*) in North America. Can. J. Bot. 42:213-282. (key to amyloid-spored genera of Stereaceae, key to 27 species, descriptions and illustrations)
Lemke, P. A. 1964. The genus *Aleurodiscus* (*sensu lato*) in North America. Can. J. Bot. 42:723-768. (key to 14 *Dendrothele* spp. as *Aleurocorticium*, a segregate of *Aleurodiscus*, descriptions and illustrations, disposition of additional *Aleurodiscus* names)
Lemke, P. A. 1965. *Dendrothele* (1907) vs. *Aleurocorticium* (1963). Persoonia 3:365-367. (rejects *Aleurocorticium*, makes new combinations in *Dendrothele*)
In general, members of this genus are saprophytic. *Aleurodiscus amorphus* (Pers.:Fr.) J. Schroet. can cause a severe infection of suppressed seedlings of *Abies grandis*.

ALTERNARIA Nees
   Hyphomycetes   50 spp.
Groves, J. W., and Skolko, A. J. 1944. Notes on seed-borne fungi II.
   Alternaria. Can. J. Res. Sect. C. 22:217-234. (seven species
   isolated from seeds, descriptions and illustrations)
Joly, P. 1964. Le genre *Alternaria.* Encycl. Mycol. 33:1-250. (in
   French, key to sections and 25 species, descriptions and
   illustrations, discusses 94 additional names)
Neergaard, P. 1945. Danish Species of *Alternaria* and *Stemphylium.*
   Taxonomy, Parasitism, Economical Significance. Einar Munksgaard,
   Copenhagen. 560 pp. (key to subgenera and 17 species,
   descriptions and illustrations, host index)
Rao, V. G. 1969. The genus *Alternaria* - from India. Nova Hedw.
   17:219-258. (includes 57 species with brief descriptions)
Simmons, E. G. 1967. Typification of *Alternaria, Stemphylium,* and
   Ulocladium. Mycologia 59:67-92. (defines genera, descriptions
   and illustrations of type species)
Simmons, E. G. 1981. *Alternaria* themes and variations. Mycotaxon
   13:16-34. (describes and illustrates six species)
Simmons, E. G. 1982. *Alternaria* themes and variations (7-10).
   Mycotaxon 14:17-43. (describes and illustrates four species)
Simmons, E. G. 1982. *Alternaria* themes and variations (11-13).
   Mycotaxon 14:44-57. (describes and illustrates three species)
Simmons, E. G. 1986. *Alternaria* themes and variations (14-16).
   Mycotaxon 25:195-202. (describes and illustrates three species on
   Euphorbia)
Simmons, E. G. 1986. *Alternaria* themes and variations (17-21).
   Mycotaxon 25:203-216. (describes and illustrates four species on
   Helianthus and one on *Chrysanthemum*)
Simmons, E. G. 1986. *Alternaria* themes and variations (22-26).
   Mycotaxon 25:287-308. (*Alternaria* anamorphs of five
   Pleospora-like spp., describes new teleomorph genus *Lewia*)
The genus *Alternaria* is treated in a number of taxonomic and
nomenclatural works. Ellis (1971, 1976) in combination with Simmons'
recent studies provide a useful account of many species. Important
diseases caused by *Alternaria* species include: *Alternaria brassicae*
(Berk.) Sacc., leaf spots of Brassicaceae (CMI 162); *A. brassicicola*
(Schwein.) Wiltshire, leaf spot of cabbage and cauliflower (CMI
163); *A. burnsii* Uppal, Patal & Kamat, blight of cumin (CMI 581); *A.
carthami* Chowdhury, leaf spot of safflower (CMI 241); *A.
chrysanthemi* E. Simmons & Crosier, leaf spot of *Chrysanthemum* (CMI
164); *A. citri* Ellis & Pierce, rots of *Citrus* fruit (CMI 242); *A.
crassa* (Sacc.) Rands, leaf spot of *Datura* (CMI 243); *A. cucumerina*
(Ellis & Everh.) J. A. Elliott, leaf spot of cucumber and melon (CMI
244); *A. helianthi* (Hansf.) Tubaki & Nishihara, leaf spot, head
blight, and other diseases of *Helianthus* spp. (CMI 582); *A. longipes*
(Ellis & Everh.) E. Mason, brown spot of tobacco (CMI 245); *A.*

ALTERNARIA - cont.
macrospora Zimmermann, leaf spot of cotton (CMI 246); *A. padwickii* (Ganguly) M. B. Ellis, stackburn, seedling blight, and leaf spot of rice (CMI 345); *A. passiflorae* Simmonds, brown spot of passion fruit (CMI 247); *A. porri* (Ellis) Cif., purple blotch of onion (CMI 248); *A. radicina* Meier, Drechsler & Eddy, black rot of carrot and other Apiaceae (CMI 346); *A. ricini* (Yoshii) Hansf., diseases of castor (CMI 249); *A. sesami* (S. Kawamura) Mohanty & Behera, diseases of sesame (CMI 250); *A. solani* Sorauer, early blight of potato and tomato (CMI 475); and *A. triticina* Prasada & Prabhu, leaf and sheath blight of wheat, triticale, and barley (CMI 583).

AMEROSPORIUM Speg.
    Coelomycetes   2 spp.
Srivastava, G., Lal, B., and Tandon, M. P. 1981. A new leaf spot of *Rosa alba* L. caused by *Amerosporium circinata* sp. nov. Nat. Acad. Sci. Lett. (India) 4:231-232. (description)
Sutton (1980) currently retains only two species in this genus.

AMYLOSTEREUM Boidin
    Aphyllophorales   5 spp.
Stillwell, M. A. 1966. Woodwasps (Siricidae) in conifers and the associated fungus, *Stereum chailletii*, in eastern Canada. For. Sci. 12:121-128. (description of disease, illustrations)
Talbot, P. H. B. 1977. The *Sirex-Amylostereum-Pinus* association. Ann. Rev. Phytopathol. 15:41-54. (review)
Amylostereum species, in association with woodwasps, cause diseases of conifers. See *Stereum* and Aphyllophorales.

ANGIOSORUS Thirumalachar & O'Brien
    Ustilaginales   1 sp.
O'Brien, M. J., and Thirumalachar, M. J. 1974. The identity of the potato smut. Sydowia 26:199-203. (description and illustrations of *A. solani*)
Angiosorus solani Thirumalachar & O'Brien (syn. *Thecaphora solani* Barrus, not validly published) causes potato smut.

ANGUILLOSPORA   See *Mycopappus*.

ANGUSIA   See *Maravalia*.

ANISOGRAMMA Theissen & Sydow
    Diaporthales   5 spp.
Gottwald, T. R., and Cameron, H. R. 1979. Studies in the morphology and life history of *Anisogramma anomala*. Mycologia 71:1107-1126. (developmental study, description and illustrations)
Anisogramma anomala (Peck) E. Mueller causes filbert blight. See Barr (1978) and Mueller and Arx (1962).

ANNELLOPHORA S. J. Hughes
    Hyphomycetes   7 spp.
Ellis, M. B. 1958. *Clasterosporium* and some allied
    dematiaceae--phragmosporae. I. Mycol. Pap. 70:1-89. (key to
    four *Annellophora* spp., descriptions and illustrations)
Vann, S. R., and Taber, R. A. 1985. *Annellophora* leaf spot of date
    palm in Texas. Plant Dis. 69:903-904. (description and
    illustrations)
Ellis (1971) treats seven species.

ANTHOSTOMELLA Sacc.
    Sphaeriales   50 spp.
Eriksson, O. 1966. On *Anthostomella* Sacc., *Entosordaria* (Sacc.)
    Hoehn. and some related genera (Pyrenomycetes). Sven. Bot.
    Tidskr. 60:315-324. (taxonomic and nomenclatural history)
Francis, S. M. 1975. *Anthostomella* Sacc. (Part I). Mycol. Pap.
    139:1-97. (key to 30 species, illustrations, host list)
Martin, P. 1969. Studies in the Xylariaceae VII. *Anthostomella* and
    Lopadostoma. J. S. Afr. Bot. 35:393-410. (key to 50
    Anthostomella spp., descriptions and illustrations of five
    species)
Francis (1975) gives an extensive treatment; many species are treated
by Arx and Mueller (1954). *Anthostomella* species are weakly
parasitic on the stems and leaves of many plants and are often host
specific. The genus is widely distributed in both temperate and
tropical regions.

ANTHRACOIDEA Bref.
    Ustilaginales   32 spp.
Braun, U., and Hirsch, G. 1978. Uebersicht ueber die europaeischen
    Arten der Gattung *Anthracoidea* Bref. (Ustilaginales). Feddes
    Repert. 89:43-60. (key to 30 European species, descriptions and
    illustrations)
Kukkonen, I. 1963. Taxonomic studies on the genus *Anthracoidea*
    (Ustilaginales). Ann. Bot. Soc. Zool. Bot. Fenn. "Vanamo"
    34:1-122. (key to 24 species, descriptions and illustrations)
Kukkonen, I. 1964. Taxonomic studies on the species of the section
    Echinosporae of *Anthracoidea*. Ann. Bot. Fenn. 1:161-177. (key to
    nine species, descriptions and illustrations)
Nannfeldt, J. A. 1979. *Anthracoidea* (Ustilaginales) on Nordic
    Cyperaceae-Caricoideae, a concluding synopsis. Symb. Bot. Ups.
    22:3:1-41. (key to 34 Nordic species, illustrations, references)
Species in this genus infect members of the Cyperaceae. Mordue and
Ainsworth (1984) provide a key to 11 taxa in the *Anthracoidea
caricis* (Pers.) Bref. group and present descriptions of *A. caricis*,
A. scirpi (Kuehn) Kukkonen, and *A. subinclusa* (Koern.) Bref.

APHANOMYCES de Bary
    Saprolegniales   25 spp.
Dick, M. W. 1973. Saprolegniales. Pages 113-144 in: The Fungi. Vol.
    IVB. G. C. Ainsworth, F. K. Sparrow, and A. S. Sussman, eds.
    Academic Press, New York. (key to genera, list of taxa)
Papavizas, G. C., and Ayers, W. A. 1974. *Aphanomyces* Species and
    their Root Diseases in Pea and Sugarbeet. U.S. Dep. Agric.,
    Tech. Bull. 1485. 158 pp. (review of distribution, ecology,
    pathology, physiology, and economic importance of *A. euteiches*
    and *A. cochlioides*)
Scott, W. W. 1961. A Monograph of the Genus *Aphanomyces*. Va. Agric.
    Exp. Stn. Tech. Bull. 151. 95 pp. (key to subgenera and 25
    species, descriptions and illustrations)
Species of *Aphanomyces* include both saprophytes and parasites of
plants and animals. Five species are considered serious root
pathogens: *Aphanomyces camptostylus* Drechs., parasitic on oats; *A.
cladogamus* Drechs., parasitic on various crop or garden plants; *A.
cochlioides* Drechs., sugar beet black rot; *A. euteiches* Drechs., pea
root rot (CMI 600); and *A. raphani* J. B. Kendrick, parasitic on
radish.

APIOGNOMONIA Hoehn.
    Diaporthales   8 spp.
Monod, M. 1983. Monographie taxonomique des Gnomoniaceae. Sydowia
    Beih. 9:1-315. (in French with English summary, key to 14
    species, descriptions and illustrations)
Barr (1978) provides a key to seven taxa. *Apiognomonia* species that
cause important diseases include *A. erythrostoma* (Pers.) Hoehn.,
cherry leaf scorch, and *A. quercina* (Kleb.) Hoehn., oak anthracnose.

APIOSPORA Sacc.
    Sphaeriales   4 spp.
Hudson, H. J., McKenzie, E. H. C., and Tommerup, I. C. 1976. Conidial
    states of *Apiospora* Sacc. Trans. Br. Mycol. Soc. 66:359-362.
    (compares three species)
The genus is treated by Mueller and Arx (1962).

APIOSPORINA Hoehn.
    Pleosporales   2 spp.
Barr, M. E. 1968. The Venturiaceae in North America. Can. J. Bot.
    46:799-864. (key to two species, descriptions and illustrations)
Apiosporina morbosa (Schwein.) Arx causes black knot of cherry and
other *Prunus* spp. (CMI 224 as *Dibotryon morbosum*, F. Can. 84) and *A.
collinsii* (Schwein.) Hoehn. causes witches broom of *Amelanchier* spp.
(F. Can. 76). The anamorphs belong in *Fusicladium*.

APOSTRASSERIA Nag Raj
  Coelomycetes   4 spp.
Nag Raj, T. R. 1983. Genera coelomycetum. XXI. *Strasseria* and two new anamorph-genera, *Apostrasseria* and *Nothostrasseria*. Can. J. Bot. 61:1-30. (key to four *Apostrasseria* spp., descriptions and illustrations)
These fungi are anamorphs of pathogenic *Phacidium* species. Some were previously placed in *Phacidiopycnis* as anamorphs of *Potebniamyces*. See DiCosmo et al. (1984) under *Phacidium*.

ARCTICOMYCES Savile
  Exobasidiales   1 sp.
Savile, D. B. O. 1959. The botany of Somerset Island, District of Franklin. Can. J. Bot. 37:959-1002. (description and illustrations of *A. warmingii*)
Arcticomyces warmingii (Rostr.) Savile (syn. *Exobasidium warmingii* Rostr.) occurs systemically in Saxifragaceae and has an arctic distribution.

ARMILLARIA (Fr.:Fr.) Staude
  Agaricales   40 spp.
Anderson, J. B. 1986. Biological species of *Armillaria* in North America: Redesignation of groups IV and VIII and enumeration of voucher strains for other groups. Mycologia 78:837-839. (reports that biological species IV is equivalent to V and VIII is equivalent to VI which are designated V and VI, respectively, designates tester strains for all biological species in North America)
Anderson, J. B., Korhonen, K., and Ullrich, R. C. 1980. Relationships between European and North American biological species of *Armillaria mellea*. Exp. Mycol. 4:87-95. (results of pairings of five European and ten North American isolates, some interfertility of biological species from different continents, three interesting interactions discussed)
Anderson, J. B., and Ullrich, R. C. 1979. Biological species of *Armillaria mellea* in North America. Mycologia 71:402-414. (defines ten biological species based on intersterility groups)
Gregory, S. C., and Watling, R. 1985. Occurrence of *Armillaria borealis* in Britain. Trans. Br. Mycol. Soc. 84:47-55. (descriptions and illustrations, potential forest pathogen)
Guillaumin, J. J., and Berthelay, S. 1981. Determination specifique des armillaires par la methode des groupes de compatibilite sexuelle. Specialisation ecologique des especes francaises. Agronomie 1:897-908. (in French with English summary, four species delimited by mating tests, ecological specialization and morphological characters discussed)
Kile, G. A., and Watling, R. 1983. *Armillaria* species from Southeastern Australia. Trans. Br. Mycol. Soc. 81:129-140.

ARMILLARIA - cont.
(descriptions and illustrations of four species)
Korhonen, K. 1978. Interfertility and clonal size in the *Armillaria mellea* complex. Karstenia 18:31-42. (techniques for identification based on mating experiments)
Morrison, D. J., Chu, D., and Johnson, A. L. S. 1985. Species of *Armillaria* in British Columbia. Can. J. Plant Pathol. 7:242-246. (defines six intersterile groups including *A. bulbosa* and *A. ostoyae*)
Raabe, R. D. 1962. Host list of the root rot fungus, *Armillaria mellea*. Hilgardia 33:25-88. (lists several hundred vascular plant hosts and countries)
Rishbeth, J. 1982. Species of *Armillaria* in southern England. Plant Pathol. 31:9-17. (defines five species using mating tests)
Rishbeth, J. 1986. Some characteristics of English *Armillaria* species in culture. Trans. Br. Mycol. Soc. 86:213-218. (discusses cultural characters and their use in identifying four *Armillaria* spp.)
Roll-Hansen, F. 1985. The *Armillaria* species in Europe. A literature review. Eur. J. For. Pathol. 15:22-31. (summarizes nomenclature, distribution, hosts, and characteristics of five species)
Ullrich, R. C., and Anderson, J. B. 1978. Sex and diploidy in *Armillaria mellea*. Exp. Mycol. 2:119-129. (reports bifactorial, multiallelic heterothallism, suggests diploid vegetative stage, reports the occurrence and distribution of biological species)
Wargo, P. M., and Shaw, C. G. III 1985. *Armillaria* root rot: the puzzle is being solved. Plant Dis. 69:826-832. (reviews the disease and its pathogen)
Watling, R., Kile, G. A., and Gregory, N. M. 1982. The genus *Armillaria* - nomenclature, typification, the identity of *Armillaria mellea* and species differentiation. Trans. Br. Mycol. Soc. 78:271-285. (discusses *Armillariella*, descriptions and illustrations)
Serious root rots in trees and other woody plants are caused by *Armillaria* species, including: *Armillaria mellea* (Vahl:Fr.) P. Kumm., the honey fungus (CMI 321 as *Armillariella mellea*); *A. borealis* Marxmueller & Korhonen; and *A. tabescens* (Scop.) Dennis, Orton & Hora. Watling et al. (1982) conclude that *Armillaria mellea* is the correct name for the honey fungus.

ARTHRINIUM Kunze:Fr.
    Hyphomycetes   18 spp.
Ellis, M. B. 1963. Dematiaceous Hyphomycetes. VI. Mycol. Pap. 103:1-46. (key to 18 species, descriptions and illustrations)
Minter, D. W. 1985. A reappraisal of the relationships between *Arthrinium* and other hyphomycetes. Proc. Indian Acad. Sci., Plant Sci. 94:281-308. (reveiws conidial development, hypothesizes nature of basauxic conidiophore development)

ARTHRINIUM - cont.
Ellis (1971, 1976) provides a recent account of this genus. The teleomorph is *Apiospora*.

ARTHROBOTRYS See *Dactylaria*.

ASCOCALYX Naumov
    Helotiales   6 spp.
Groves, J. W. 1968. Two new species of *Ascocalyx*. Can. J. Bot. 46:1273-1278. (key to three species)
Kondo, H., and Kobayashi, T. 1984. A new canker disease of loblolly pine, *Pinus taeda* L., caused by *Ascocalyx pinicola* sp. nov. J. Jpn. For. Soc. 66:60-66. (in Japanese with English summary)
Mueller, E., and Dorworth, C. E. 1983. On the discomycetous genera *Ascocalyx* Naumov and *Gremmeniella* Morelet. Sydowia 36:193-203. (synoptic table, key to six species, anamorph information and key, illustrations)
Schlaepfer-Bernard, E. 1968. Beitrag zur Kenntnis der Discomycetengattungen *Godronia, Ascocalyx, Neogodronia,* und *Encoeliopsis*. Sydowia 22:1-56. (in German with English summary, key to ten related genera, key to three *Ascocalyx* spp., descriptions and illustrations)
Mueller and Dorworth (1983) provide a full account of this genus and discuss the taxonomic and nomenclatural history of *Ascocalyx abietina* (Lagerberg) Schlaepfer-Bernhard (syns. *Godronia abietis* (Naumov) Seaver, *Scleroderris lagerbergii* Gremmen), the cause of *Scleroderris* canker (CMI 369 as *Gremmeniella abietina*). Anamorphs are placed in *Bothrodiscus* and *Brunchorstia*.

ASCOCHYTA Lib.
    Coelomycetes   350 spp.
Boerema, G. H., and Bollen, G. J. 1975. Conidiogenesis and conidial septation as differentiating criteria between *Phoma* and *Ascochyta*. Persoonia 8:111-144. (defines *Ascochyta* with annellidic ontogeny and *Phoma* with phialidic ontogeny)
Boerema, G. H., and Dorenbosch, M. M. J. 1973. The *Phoma* and *Ascochyta* species described by Wollenweber and Hochapfel in their study on fruit-rotting. Stud. Mycol. 3:1-50. (indicates correct identity of 13 species, presents table comparing ten species, descriptions and illustrations)
Cejp, K. 1968. [Contribution to the knowledge of the species of the genus *Ascochyta* Lib. from western Bohemia.] Ceska Mykol. 22:186-188. (in Czechoslovakian with English summary, description of six species)
Melnik, V. A. 1977. [Key to Fungi of the Genus *Ascochyta* Lib.] Leningrad, U.S.S.R. 245 pp. (in Russian, key to 327 species based on host family, descriptions and illustrations, disposes of an additional 414 names)

ASCOCHYTA - cont.
Punithalingam, E. 1979. Graminicolous *Ascochyta* species. Mycol. Pap. 142:1-214. (general key to 79 taxa, keys based on host, descriptions and illustrations)
Punithalingam, E. 1981. Studies on Sphaeropsidales in culture. III. Mycol. Pap. 149:1-42. (table comparing three *Aschochyta* spp. from *Hevea*, descriptions and illustrations)
Punithalingam, E., Gladders, P., and McKeown, B. M. 1985. A new species of *Ascochyta* associated with white leaf blotch of onion in Great Britain. Trans. Br. Mycol. Soc. 85:556-560. (description and illustrations, compares with two other species)
Wollenweber, H. W., and Hochapfel, H. 1936. Beitraege zur Kenntniss parasitaerer und saprophytischer Pilze. I. *Phomopsis, Dendrophoma, Phoma,* and *Ascochyta* und ihre Bezeihung zur Fruchtfaeule. Z. Parasitenkd. 8:561-605. (in German, treats 20 species, descriptions and illustrations)

In spite of extensive treatments by Melnik (1977) and Punithalingam (1979, 1981), both specific and generic limits of *Ascochyta* and related species are unclear. Sutton (1980) is a good source of information. Some species with teleomorphs in *Didymella* and *Mycosphaerella* are included in Sivanesan (1984). Important diseases include: *Ascochyta adzamethica* Schoschiaschvili, net blotch of peanut (CMI 736 under *Didymosphaeria arachidicola*); *A. avenae* (Petr.) Sprague & Johnson, leaf spot or blotch of oats (CMI 731); *A. chrysanthemi* F. Stevens, ray blight of *Chrysanthemum* (CMI 662 under *Didymella ligulicola* as *D. chrysanthemi*); *A. cucumis* Fautr. & Roum., gummy stem blight of Cucurbitaceae (CMI 332, F. Can. 303, under *Didymella bryoniae*); *A. desmazieresii* Cavara, glume and leaf spot of Italian and perennial ryegrasses (CMI 661); *A. fabae* Speg., leaf, stem, and pod spot of broad bean (CMI 461); *A. gossypii* Woron., wet weather blight of cotton (CMI 271); *A. heveae* Petch, leaf scorch or rim blight of rubber leaves (CMI 631); *A. paspali* (Sydow) Punithalingam, leaf blotch or streak of *Paspalum* spp. (CMI 821); *A. pisi* Lib., leaf, stem, and pod spot of pea and other legumes (CMI 334); *A. pinodes* Jones, diseases of pea (CMI 340 under *Mycosphaerella pinodes*); *A. rabiei* (Pass.) Labrousse, blight of chick pea (CMI 337); and *A. sorghi* Sacc., rough leaf spot of sorghum (CMI 632).

ASCOCHYTULINA Petr.
 Coelomycetes   2 spp.
Evans, H. C., and Punithalingam, E. 1985. A new coelomycete from blighted pine needles in Honduras. Trans. Br. Mycol. Soc. 84:568-573. (describes and illustrates *A. pini-acicola*)
Sutton (1980) treats *Ascochytulina deflectens* (P. Karst.) Petr.

ASCOTRICHA Berk.
    Sphaeriales   8 spp.
Arx, J. A. von 1982. The genus *Dicyma*, its synonyms and related fungi. Proc. K. Ned. Akad. Wet., Ser. C. 85:21-28. (key to 11 *Dicyma* spp., some of which are *Ascotricha* anamorphs, discussion, nomenclature, illustrations)
Guarro, J., and Calvo, M. A. 1983. *Dicyma funiculosa* sp. nov. from Spain. Nova Hedw. 37:641-649. (description and illustration, review of other species)
Hawksworth, D. L. 1971. A revision of the genus *Ascotricha* Berk. Mycol. Pap. 126:1-28. (key to eight species, descriptions and illustrations)
Ellis (1971) treats the *Dicyma* anamorph of *Ascotricha chartarum* Berk.

ASHBYA Guilliermond
    Endomycetales   1 sp.
Ashbya gossypii (Ashby & Nowell) Guilliermond causes stigmatomycosis of cotton bolls (CMI 185 as *Nematospora gossypii*).

ASPERGILLUS Mich. ex Link:Fr.
    Hyphomycetes   50 spp.
Al-Musallam, A. 1980. Revision of the Black *Aspergillus* Species. Drukkerij Elinkwink Bv, Utrecht. 92 pp. (dichotomous and synoptic keys, descriptions and illustrations)
Benjamin, C. R. 1955. Ascocarps of *Aspergillus* and *Penicillium*. Mycologia 47:669-687. (keys to five genera, discussion of species)
Christensen, M. 1982. The *Aspergillus ochraceous* group: two new species from western soils and a synoptic key. Mycologia 74:210-225. (synoptic key to 18 species, descriptions)
Christensen, M., and Raper, K. B. 1978. Synoptic key to *Aspergillus nidulans* group species and related *Emericella* species. Trans. Br. Mycol. Soc. 71:177-191. (synoptic key to 30 taxa, brief descriptions, illustrations)
Christensen, M., and States, J. S. 1982. *Aspergillus nidulans* group: *Aspergillus navahoensis*, and a revised synoptic key. Mycologia 74:226-235. (synoptic key to 34 taxa)
Pitt, J. I., and Hocking, A. D. 1985. Interfaces among genera related to *Aspergillus* and *Penicillium*. Mycologia 77:810-824. (discusses generic characteristics, compares similar genera, describes four new species in three genera, illustrations)
Raper, K. B., and Fennell, D. I. 1965. The Genus *Aspergillus*. The Williams and Wilkins Co., Baltimore, MD. 686 pp. (keys to groups and species, descriptions and illustrations)
Samson, R. A. 1979. A compilation of the Aspergilli described since 1965. Stud. Mycol. 18:1-38. (treats 90 taxa, key to teleomorph genera, descriptions of 34 accepted taxa)
Samson, R. A., and Gams, W. 1984. The taxonomic situation in the

ASPERGILLUS - cont.
hyphomycete genera *Penicillium, Aspergillus* and *Fusarium*.
Antonie van Leeuwenhoek 50:815-824. (review)
Samson, R. A., and Pitt, J. I., eds. 1985. Advances in *Penicillium*
and *Aspergillus* Systematics. Plenum Press, New York. 483 pp.
(keys, descriptions and illustrations of groups)
Sarbhoy, A. K. 1985. Cleistothecial states of *Aspergillus* and their
taxonomic position. Mycopathologia 91:133-142. (accepts eight
teleomorph genera, descriptions and illustrations)
Wiley, B. J., and Fennell, D. I. 1973. Ascocarps of *Aspergillus
stromatoides, A. niveus,* and *A. flavipes.* Mycologia 65:752-760.
(descriptions and illustrations, no teleomorph names)
The genus and its teleomorphs *Emericella, Eurotium, Neosartorya,* and others have been well-documented. The species remain difficult to identify. Domsch et al. (1980) provide keys to species in six groups isolated from soil. Some species produce mycotoxins in stored grains and other commodities. Although pathogenic primarily to man and animals, some *Aspergillus* species are associated with plant diseases including: *Aspergillus flavus* Link:Fr., pathogenic on groundnut pods, cotton seedlings, and corn and a major producer of aflatoxin (CMI 91); *A. nidulans* (Eidam) Wint., causing losses in seed germination of bean seedlings (CMI 93); *A. niger* van Tiegh., crown rot of groundnut and several other diseases (CMI 94); and *A. terreus* Thom, rot of apple and fruit (CMI 95).

ASPERISPORIUM Maubl.
Hyphomycetes   4 spp.
Asperisporium caricae (Speg.) Maubl. causes black spot of papaya (CMI 347). Members of this genus are described and illustrated in Ellis (1971, 1976).

ASTEROMELLA   See *Mycosphaerella.*

ATHELIA Pers.
Aphyllophorales   21 spp.
Juelich, W. 1972. Monographie der Athelieae (Cortiaceae,
Basidiomycetes). Willdenowia Beih. 7:1-283. (in German with
English summary, key to genera and species, describes and
illustrates 21 *Athelia* spp.)
Punja, Z. K. 1985. The biology, ecology, and control of *Sclerotium
rolfsii.*   Ann. Rev. Phytopathol. 23:97-127. (includes brief
description)
Punja, Z. K., Grogan, R. G., and Adams, G. C., Jr. 1982. Influence of
nutrition, environment, and the isolate, on basidiocarp
formation, development, and structure in *Athelia (Sclerotium)
rolfsii.* Mycologia 74:917-926. (includes description and
illustrations)
Tu, C. C., and Kimbrough, J. W. 1978. Systematics and phylogeny of

ATHELIA - cont.
fungi in the *Rhizoctonia* complex. Bot. Gaz. 139:454-466.
(compares related genera)
Most species of *Athelia* are saprophytic. *Athelia rolfsii* (Curzi) Tu &
Kimbrough, teleomorph of *Sclerotium rolfsii* Sacc., causes rots and
other diseases of legumes and many other economically important
hosts (CMI 410 as *Corticium rolfsii*). See Aphyllophorales,
especially Juelich (1984) and Juelich and Stalpers (1980).

ATKINSONELLA Diehl
    Clavicipitales   1 sp.
Diehl, W. W. 1950. *Balansia* and the Balansiae in America. U.S. Dep.
Agric., Agric. Monogr. 4. 82 pp. (treats one species and one
variety of *Atkinsonella*)
Rykard, D. M., Luttrell, E. S., and Bacon, C. W. 1984. Conidiogenesis
and conidiomata in the Clavicipitoideae. Mycologia 76:1095-1103.
(description and conidiogenesis of *Ephelis* states of four
Balansia spp. and *Atkinsonella hypoxylon*, discussion of taxonomic
significance of anamorphs)
Atkinsonella hypoxylon (Peck) Diehl causes a choke disease of
grasses.

ATOPOSPORA Petr.
    Pleosporales   2 spp.
Barr, M. E. 1968. The Venturiaceae of North America. Can. J. Bot.
46:799-864. (description and illustration of *A. betulina*)
Mueller, E. (1958)1959. Pilze aus dem Himalaya II. Sydowia
12:160-184. (describes and illustrates *A. taxi* on *Taxus*)
Atopospora betulina (Fr.) Petr. (syn. *Phyllachora betulina* (Fr.)
Fuckel) causes a leaf spot disease of *Betula* in north temperate
areas (F. Can. 88; Sivanesan, 1984).

ATROPELLIS Zeller & Goodding
    Helotiales   4 spp.
Hopkins, J. C. 1961. Studies of the culture of *Atropellis piniphila*.
Can. J. Bot. 39:1521-1529. (descriptions and illustrations)
Hopkins, J. C. 1963. *Atropellis* canker of lodgepole pine: Etiology,
symptoms, and canker development rates. Can. J. Bot.
41:1535-1545. (*A. piniphila*, illustrations of fungus and
symptoms)
Reid, J., and Funk, A. 1966. The genus *Atropellis*, and a new genus of
the Helotiales associated with branch cankers of western hemlock.
Mycologia 58:417-439. (discussion of generic concept, comparison
with similar genera, illustrations, key to four species, most
comprehensive treatment available)
Zeller, S. M., and Goodding, L. N. 1930. Some species of *Atropellis*
and *Scleroderris* on conifers in the Pacific Northwest.
Phytopathology 20:555-567. (original description of genus and two

ATROPELLIS - cont.
species along with *Scleroderris abieticola*, illustrations, description of the disease)
Species in this genus are associated with stem cankers of conifers.

AULOGRAPHINA Arx & E. Mueller
  Dothideales   3 spp.
Wall, E., and Keane, P. J. 1984. Leaf spot of *Eucalyptus* caused by *Aulographina eucalypti*. Trans. Br. Mycol. Soc. 82:257-273. (pathology, description and illustrations)
Mueller and Arx (1962) treat two species.

AUREOBASIDIUM Viala & Boyer
  Hyphomycetes   14 spp.
Cooke, W. B. 1962. A taxonomic study on the "Black yeasts." Mycopathol. Mycol. Appl. 17:1-43. (describes seven species)
Hermanides-Nijhof, E. J. 1977. *Aureobasidium* and allied genera. Stud. Mycol. 15:141-177. (monograph, key to 14 species, descriptions and illustrations)
Pugh, G. J. F., and Buckley, N. G. 1971. *Aureobasidium pullulans*: an endophyte in sycamore and other trees. Trans. Br. Mycol. Soc. 57:227-231. (biology)
Aureobasidium pullulans (de Bary) Arnaud occurs ubiquitously on plant and other surfaces. This species is treated by Ellis (1971).

BALANSIA Speg.
  Clavicipitales   20 spp.
Diehl, W. W. 1950. *Balansia* and the Balansiae in America. U.S. Dep. Agric., Agric. Monogr. 4. 82 pp. (key to 14 species, descriptions and illustrations, especially useful for North and South American species)
Govindu, H. C., and Thirumalachar, M. J. 1961. Studies on some species of *Ephelis* and *Balansia* occurring in India. Mycopathol. Mycol. Appl. 14:189-197. (species based on host, illustrations)
Govindu, H. C., and Thirumalachar, M. J. 1973. *Ephelis* and its ascigerous stage *Balansia* in India. Pages 328-334 in: Taxonomy of Fungi. Part 2. C. V. Subramanian, ed. Amra Press, Madras. (mostly disease descriptions by host and symptoms)
Rykard, D. M., Luttrell, E. S., and Bacon, C. W. 1984. Conidiogenesis and conidiomata in the Clavicipitoideae. Mycologia 76:1095-1103. (description and conidiogenesis of *Ephelis* states of four *Balansia* spp. and *Atkinsonella hypoxylon*, discussion of taxonomic significance of anamorphs)
Ullasa, B. A. 1969. *Balansia claviceps* in artificial culture. Mycologia 61:572-579. (developmental and cultural characteristics, illustrations)
Balansia oryzae-sativae Hashioka causes black choke or false ergot of

BALANSIA - cont.
rice (CMI 640). Anamorphs of *Balansia* are placed in *Ephelis*.

BASIDIOPHORA Roze & Cornu
    Peronosporales   3 spp.
Basidiophora entospora Roze & Cornu causes downy mildew of asters (CMI 681). The CMI description includes a synopsis of all species in the genus and other references. Other species are *B. butleri* (Weston) Thirum. & Whitehead on *Eragrostis aspera* known from Malawa and *B. kellermanii* Swingle ex Sacc. on *Iva xanthifolia* in North America. See Peronosporales.

BATCHELOROMYCES Marasas, Van Wyk & Knox-Davies
    Hyphomycetes   2 spp.
Smit, W. A., Engelbrecht, C., and Knox-Davies, P. S. 1983. Studies on *Batcheloromyces* leaf spot of *Protea cynaroides*. Phytophylactica 15:125-131. (describes and illustrates *B. proteae*)
Wyk, P. S. van, Marasas, W. F. O., and Knox-Davies, P. S. 1985. *Batcheloromyces leucadendri* sp. nov. on *Leucadendron*. South African J. Bot. 5:344-346. (description and illustration)
The two species in this genus cause leaf spots on members of the Proteaceae.

BIFUSELLA Hoehn.
    Rhytismatales   7 spp.
Funk (1985) treats three species occurring on needles of *Pinus*. *Bifusella linearis* (Peck) Hoehn. causes needle blight and needle cast of white pines and tar spot needle cast of pines (CMI 782) and *B. faullii* Darker causes needle cast of *Abies*. See *Lophodermium*.

BIPOLARIS Shoemaker
    Hyphomycetes   45 spp.
Alcorn, J. L. 1981. *Cochliobolus ravenelii* sp. nov. and *C. tripogonis* sp. nov. Mycotaxon 13:339-345. (technique for producing teleomorph)
Alcorn, J. L. 1981. Ascus structure and function in *Cochliobolus* species. Mycotaxon 13:349-360. (includes mechanism of ascospore release)
Alcorn, J. L. 1982. New *Cochliobolus* and *Bipolaris* species. Mycotaxon 15:1-19. (describes and illustrates four *Cochliobolus* spp. and one *Bipolaris* sp.)
Alcorn, J. L. 1982. Ovariicolous *Bipolaris* species on *Sporobolus* and other grasses. Mycotaxon 15:20-48. (comparative table of four species, host list, geographic range)
Alcorn, J. L. 1983. Generic concepts in *Drechslera*, *Bipolaris* and *Exserohilum*. Mycotaxon 17:1-86. (includes list of accepted binomials)
Hardin, H. 1980. *Cochliobolus sativus* (Ito & Kurib.) Drechsl. ex

BIPOLARIS - cont.

Dastur (imperfect stage: *Bipolaris sorokiniana* (Sacc. in Sorok.) Shoem.): a bibliography. Agriculture Canada, Research Branch, Saskatoon. Pages not numbered. (lists nearly 2000 publications dealing with this organism and the diseases it causes, provides a subject index, three supplements have been published with an additional 983 references)

Luttrell, E. S. 1977. Correlations between conidial and ascigerous state characters in *Pyrenophora, Cochliobolus*, and *Setosphaeria*. Rev. Mycol. 41:271-279. (describes and illustrates generic characters)

Shoemaker, R. A. 1959. Nomenclature of *Drechslera* and *Bipolaris*, grass parasites segregated from *Helminthosporium*. Can. J. Bot. 37:879-887. (includes host list)

Sivanesan, A. 1985. New species of *Bipolaris*. Trans. Br. Mycol. Soc. 84:403-421. (reviews history of the genus, describes and illustrates eight new species)

Bipolaris includes species formerly placed in *Helminthosporium*. Alcorn (1983), Luttrell (1977), and Shoemaker (1959) offer good evidence for separating these species into three genera: *Bipolaris, Drechslera*, and *Exserohilum*. Domsch et al. (1980) provide a table comparing the characteristics differentiating these genera. Keys and descriptions are included in Ellis (1971, 1976) under *Drechslera*. Sivanesan (1984) presents keys based on anamorph characters and descriptions under the teleomorph name *Cochliobolus*. Many species in this genus are plant pathogens, including: *Bipolaris hawaiiensis* (M. B. Ellis) Uchida & Aragaki, seed and seedling diseases of graminaceous plants (CMI 728 as *Drechslera hawaiiensis* under *Cochliobolus hawaiiensis*); *B. incurvata* (Ch. Bernard) Alcorn, leaf spot of young coconut (CMI 342 as *Drechslera incurvata*); *B. iridis* (Oudem.) Dickenson, ink disease of iris (CMI 434 as *D. iridis*); *B. maydis* (Nisikado & Miyake) Shoemaker, southern leaf blight of maize (CMI 301 as *D. maydis* under *C. heterostrophus*); *B. nodulosa* (Berk. & M. A. Curtis) Shoemaker, diseases of finger millet (CMI 341 as *D. nodulosa* under *C. nodulosus*); *B. oryzae* (Breda de Haan) Shoemaker, brown spot and seedling blight of rice (CMI 302 as *D. oryzae* under *C. miyabeanus*); *B. sacchari* (E. J. Butler) Shoemaker, eye spot and seedling blight of sugarcane (CMI 305 as *D. sacchari*); *B. setariae* (Sawada) Shoemaker, leaf spot of *Setaria italica* (CMI 473 as *D. setariae* under *C. setariae*); *B. sorghicola* (Lefebvre & Sherwin) Alcorn, target leaf spot of *Sorghum* spp. (CMI 491 as *D. sorghicola*); *B. sorokiniana* (Sacc.) Shoemaker, spot blotch and root rot of cereals (CMI 701 as *D. sorokiniana* under *C. sativus*); *B. stenospila* (Drechs.) Shoemaker, brown stripe of sugarcane (CMI 306 as *D. stenospila*); *B. victoriae* (Meehan & Murphy) Shoemaker, Victoria blight of oats (CMI 703 as *D. victoriae* under *C. victoriae*); and *B. zeicola* (Stout) Shoemaker, leaf spot of corn (CMI 349 as *D. zeicola* under *C. carbonum*).

BISCOGNIAUXIA Kuntze
    Sphaeriales   13 spp.
Eckblad, F.-E., and Granmo, A. 1978. The genus *Nummularia* (Ascomycetes) in Norway. Norw. J. Bot. 25:69-75. (key to two *Biscogniauxia* spp. as *Nummulariella*, descriptions and illustrations)
Jong, S. C., and Benjamin, C. R. 1971. North American species of *Nummularia*. Mycologia 63:862-876. (key to four *Nummularia* spp., descriptions and illustrations)
Petrini, L. E., and Mueller, E. 1986. [Teleomorphs and anamorphs of European species of *Hypoxylon* (Xylariaceae, Sphaeriales) and allied genera.] Mycol. Helv. 1:501-627. (in German with English, French, and Italian summaries, key to five *Biscogniauxia* spp., descriptions of two species, illustrations)
Petrini, L., and Petrini, O. 1985. Xylariaceous fungi as endophytes. Sydowia 38:216-234. (keys to European xylariaceous endophytes based on cultural characters, descriptions of two *Biscogniauxia* anamorphs in culture as *Periconiella*)
Pouzar, Z. 1979. Notes on taxonomy and nomenclature of *Nummularia* (Pyrenomycetes). Ceska Mykol. 33:207-219. (describes *Biscogniauxia*, transfers 13 species from *Nummularia* and the *Hypoxylon nummularium*-group)
Pouzar, Z. 1986. A key and conspectus of Central European species of *Biscogniauxia* and *Obolarina* (Pyrenomycetes). Ceska Mykol. 40:1-10. (key to eight *Biscogniauxia* spp., discusses each species)
Rogers, J. D. 1975. *Nummularia broomeiana*: conidial state and taxonomic aspects. Amer. J. Bot. 62:761-764. (describes *Basidiobotrys* anamorph, illustrations)
Whalley, A. J. S., and Edwards, R. L. 1985. *Nummulariella marginata*: its conidial state, secondary metabolites and taxonomic relationships. Trans. Br. Mycol. Soc. 85:385-390. (discusses generic relationships)
Biscogniauxia was published to replace *Nummularia*, a later homonym. *Biscogniauxia* is similar to *Hypoxylon*, differing primarily in stromal characters. Whalley and Edwards (1985) provide a comprehensive review of the genus. *Biscogniauxia marginata* (Fr.) Pouzar (syns. *Nummulariella marginata* (Fr.) Eckblad & Granmo, *Nummularia discreta* (Schwein.) Tul. & C. Tul.) causes nailhead canker of apple. See *Hypoxylon*.

BLAKESLEA Thaxter
    Mucorales   2 spp.
Kirk, P. M. 1984. A monograph of the Choanephoraceae. Mycol. Pap. 152:1-61. (key to genera, key to two *Blakeslea* spp., descriptions and illustrations)
Mehrotra, B. S., and Baijal, U. 1968. Is *Blakeslea* a valid genus? J.

BLAKESLEA - cont.
Elisha Mitchell Sci. Soc. 84:207. (describes *B. monospora*)
Blakeslea trispora Thaxter occurs as a weak parasite on leaves of tropical plants. See Mucorales.

BLUMERIELLA Arx
    Helotiales   5 spp.
Arx, J. A. von 1961. Ueber *Cylindrosporium padi*. Phytopathol. Z. 42:161-166. (in German, description of *B. jaapii*)
Kaszonyi, S. 1966. Life cycle of *Blumeriella jaapii* (Rehm) v. Arx infecting stone fruits. Acta Phytopathol. Acad. Sci. Hung. 1:93-100. (description and illustrations)
Blumeriella jaapii (Rehm) Arx (syn. *Pseudopeziza jaapii* Rehm) causes a leaf shothole on *Prunus* and *Kerria* in North America and Europe. The anamorph is *Phloeosporella padi* (Lib.) Arx (syn. *Cylindrosporium padi* (Lib.) P. Karst.).

BONDARZEWIA Singer
    Aphyllophorales   3 spp.
Bondarzewia berkeleyi (Fr.) Bond. & Singer (syn. *Polyporus berkeleyi* Fr.) causes a white stringy root and butt rot of living hardwoods. See Aphyllophorales.

BOTHRODISCUS   See *Ascocalyx*.

BOTRYODIPLODIA (Sacc.) Sacc.
    Coelomycetes   17 spp.
Arya, A., Srivastava, R. C., and Lal, B. 1985. *Botryodiplodia* blight of rose in India. Plant Dis. 69:726. (description and illustration of *B. jaczevskii*)
Zambettakis, C. 1954. Recherches sur la systematique des Sphaeropsidales-Phaeodidymae. Bull. Trimest. Soc. Mycol. Fr. 70:219-350. (key to genera, descriptions)
This genus previously included the important tropical plant pathogen known as *Botryodiplodia theobromae* Pat., now placed in *Lasiodiplodia* (Sutton, 1980). See *Diplodia* and *Lasiodiplodia*.

BOTRYOSPHAERIA Ces. & De Not.
    Dothideales   12 spp.
Barr, M. E. 1970. Some amerosporous Ascomycetes on Ericaceae and Empetraceae. Mycologia 62:377-394. (key to 12 species in *Botryosphaeria* and related genera, descriptions and illustrations)
Fulkerson, J. F. 1960. *Botryosphaeria ribis* and its relations to a rot of apples. Phytopathology 50:394-398. (includes morphology and pathogenicity)
Funk, A. 1985. *Botryosphaeria pseudotsugae*; association with a canker of Douglas-fir and observations on its morphology. Can. J. Plant

BOTRYOSPHAERIA - cont.
Pathol. 7:355-358. (describes and illustrates canker and cultural characteristics, lists associated fungi and insects)
Laundon, G. F. 1973. *Botryosphaeria obtusa, B. stevensii*, and *Otthia spiraeae* in New Zealand. Trans. Br. Mycol. Soc. 61:369-374. (table comparing two *Botryosphaeria* spp. associated with black rot cankers of apple)
Pennycook, S. R., and Samuels, G. J. 1985. *Botryosphaeria* and *Fusicoccum* species associated with ripe fruit rot of *Actinidia deliciosa* (kiwifruit) in New Zealand. Mycotaxon 24:445-458. (treats two *Botryosphaeria* spp., three *Fusicoccum* spp., discussion of taxonomy and teleomorph-anamorph connections)
Samuels, G. J., and Singh, B. 1986. *Botryosphaeria xanthocephala*, cause of stem canker in pigeon pea. Trans. Br. Mycol. Soc. 86:295-299. (descriptions and illustrations of *B. xanthocephala* and anamorph *Fusicoccum cajani*)
Shoemaker, R. A. 1964. Conidial states of some *Botryosphaeria* species on *Vitis* and *Quercus*. Can. J. Bot. 42:1297-1301. (key to conidial states of four species, descriptions and illustrations)
Tilak, S. T., and Guikwad, Y. B. 1974. The genus *Botryosphaeria* in India. Botanigue, Nagpur 5:113-118. (lists 12 species and diagnostic characters)

Botryosphaeria species are saprophytic or weakly parasitic on a wide variety of gymnosperms and woody angiosperms, usually maturing in dead tissues. Generic limits between *Botryosphaeria* and *Guignardia* remain controversial. A modern monograph is needed. Sivanesan (1984) treats 11 species and their anamorphs. Barr (1972) considers *Guignardia* a synonym of *Botryosphaeria* and treats 19 species. Diseases caused by *Botryosphaeria* species include: *Botryosphaeria dothidea* (Moug.:Fr.) Ces. & De Not., causing many diseases on many hosts (CMI 395 as *B. ribis*); *B. obtusa* (Schwein.) Shoemaker, canker and dieback of pomaceous fruits and grapevine (CMI 394); *B. rhodorae* (Cooke) Barr, leaf spot of *Rhododendron* (F. Can. 21); and *B. zeae* (Stout) Arx & E. Mueller, grey ear rot of maize (CMI 774). Anamorphs have been placed in *Diplodia, Dothiorella, Fusicoccum, Lasiodiplodia, Macrophoma,* and *Sphaeropsis*.

BOTRYOSPORIUM Corda
  Hyphomycetes 3 spp.
Anderson, T. R., and Welacky, T. W. 1983. Barn mold of burley tobacco caused by *Botryosporium longibrachiatum*. Plant Dis. 67:1158-1159. (pathology)
Hughes, S. J., and Conway, K. E. 1978. *Botryosporium madrasense*. Can. J. Bot. 56:2405-2407. (on various plants, discusses related species)
Mason, E. W. 1928. Annotated account of fungi received at the Imperial Bureau of Mycology List II (Fascicle I). Mycol. Pap. 2:1-43. (discussion of genus)

BOTRYOSPORIUM - cont.
Botryosporium longibrachiatum (Oudem.) Maire causes post-harvest disease of tobacco.

BOTRYOTINIA Whetzel
      Helotiales   18 spp.
Chastagner, G. A. 1983. *Narcissus* fire: prevalence, epidemiology, and control in western Washington. Plant Dis. 67:1384-1386. (describes apothecial production and disease caused by *B. polyblastis*)
Dennis, R. W. G. 1956. A revision of the British Helotiaceae in the herbarium of the Royal Botanic Gardens, Kew, with notes on related European species. Mycol. Pap. 62:1-216. (descriptions of seven species as *Sclerotinia*)
Hennebert, G. L., and Groves, J. W. 1963. Three new species of *Botryotinia* on Ranunculaceae. Can. J. Bot. 41:341-370. (describes and illustrates three species, discusses four additional species)
Botryotinia is a generic segregate of *Sclerotinia*, including species with a *Botrytis* anamorph.

BOTRYTIS Mich. ex Pers.
      Hyphomycetes   50 spp.
Backhouse, D., Willetts, H. J., and Adam, P. 1984. Electrophoretic studies of *Botrytis* species. Trans. Br. Mycol. Soc. 82:625-630. (confirms identity of five species)
Coley-Smith, J. R., Verhoeff, K., and Jarvis, W. R. 1980. The Biology of *Botrytis*. Academic Press, New York. 318 pp. (one chapter on taxonomy)
Hennebert, G. L. 1973. *Botrytis* and *Botrytis*-like genera. Persoonia 7:183-204. (key to related genera and list of accepted species)
Jarvis, W. R. 1977. *Botryotinia* and *Botrytis* species: taxonomy, physiology, and pathogenicity. A guide to the literature. Can. Dep. Agric. Monogr. 15:1-195. (list of accepted species)
Menzinger, W. 1966. Zur Variabilitaet und Taxonomie von Arten und Formen der Gattung *Botrytis* Mich. I. Untersuchungen zur Kulturbedingten Variabilitaet morphologischer Eigenschaften von Formen der Gattung *Botrytis*. Zentralbl. Bakteriol., Parasitenkd., Infektionsk. Hyg. 120:141-178. (in German with English summary, table comparing 12 species)
Morgan, D. J. 1971. Numerical taxonomic studies of the genus *Botrytis* I. The *B. cinerea* complex. Trans. Br. Mycol. Soc. 56:319-325. (uses principal component analysis and cluster analysis, suggests that *B. cinerea* could be grouped into different races)
Morgan, D. J. 1971. Numerical taxonomic studies of the genus *Botrytis*. II. Other *Botrytis* taxa. Trans. Br. Mycol. Soc. 56:327-335. (uses principal component analysis and cluster analysis, key to 12 species)

BOTRYTIS - cont.
Botrytis species are anamorphs of the sclerotiniaceous ascomycete *Botryotinia*. The common grey mold, *Botrytis cinerea* Pers., is frequently parasitic on a variety of hosts (CMI 431 under *Botryotinia fuckeliana* as *Sclerotinia fuckeliana*). Other diseases caused by *Botrytis* species are: *Botrytis aclada* Fresen., neck rot of onions (CMI 433 as *B. allii*, see Hennebert, 1973); *B. fabae* Sardina, chocolate spot of broad beans (CMI 432); *B. paeoniae* Oudem., paeony blight; and *B. tulipae* Lind, tulip fire.

BRACHYBASIDIUM Gaeumann
    Brachybasidiales   1 sp.
Brachybasidium pinangae Gaeumann is parasitic on *Pinanga* (Arecaceae). See Aphyllophorales, especially McNabb and Talbot (1973).

BRASILIOMYCES Viegas
    Erysiphales   4 spp.
Hanlin, R. T., and Tortolero, O. 1984. An unusual tropical powdery mildew. Mycologia 76:439-442. (key to three *Brasiliomyces* spp.)
Hodges, C. S., Jr. 1985. Hawaiian forest fungi VI. A new species of *Brasiliomyces* on *Sapindus oahuensis*. Mycologia 77:977-981. (description and illustration)
Zheng, R.-y. 1984. The genus *Brasiliomyces* (Erysiphaceae). Mycotaxon 19:281-289. (key to three species)
Members of this genus of powdery mildews are primarily tropical.

BREMIA Regel
    Peronosporales   2 spp.
Bai, H. C., Cheng, X. Y., and Meng, Y. R. 1985. [*Bremia betae*, a new species of *Bremia*.] Acta Mycol. Sin. 4:141-143. (not seen)
Crute, I. R., and Dixon, G. R. 1981. Downy mildew diseases caused by the genus *Bremia* Regel. Pages 423-460 in: The Downy Mildews. D. M. Spencer, ed. Academic Press, New York. (table comparing described species, two accepted)
Savulescu, O. 1962. A systematic study of the genera *Bremia* Regel and *Bremiella* Wilson. Rev. Roum. Biol. 7:43-62. (not seen)
Skidmore, D. I., and Ingram, D. S. 1985. Conidial morphology and the specialization of *Bremia lactucae* Regel (Peronosporaceae) on hosts in the family Compositae. Bot. J. Linn. Soc. 91:503-522. (finds no consistent morphological differences between isolates from different hosts, based on cross-infection studies recommends formae speciales designations)
Bremia lactucae Regel causes lettuce downy mildew (CMI 682). See Peronosporales.

BREMIELLA G. W. Wils.
    Peronosporales   2 spp.
Constantinescu, O. 1979. Revision of *Bremiella* (Peronosporales).

BREMIELLA - cont.
Trans. Br. Mycol. Soc. 72:510-515. (describes and illustrates two species)
See Peronosporales.

BRIOSIA Cavara
    Hyphomycetes   2 spp.
Chant, S. R., and Gbaja, I. S. 1984. Scanning electron microscopy of colonization of *Rhododendron* by *Pycnostysanus azaleae*. Trans. Br. Mycol. Soc. 83:233-238. (illustrations of *B. azaleae* as *P. azaleae*)
Sigler, L., and Carmichael, J. W. 1976. Taxonomy of *Malbranchea* and some other Hyphomycetes with arthroconidia. Mycotaxon 4:349-488. (description, discussion of two *Briosia* spp.)
Sigler and Carmichael (1976) include *Briosia azaleae* (Peck) Dearn. (syn. *Pycnostysanus azaleae* (Peck) E. Mason), the cause of bud and twig blight of azaleas and rhododendrons. Ellis (1976) provides a description of both species, treating *B. azaleae* under *Pycnostysanus*.

BRUNCHORSTIA   See *Ascocalyx*.

BUTLERELFIA Weresub & Illman
    Aphyllophorales   1 sp.
Weresub, L. K., and Illman, W. I. 1980. *Corticium centrifugum* reisolated from fisheye rot of stored apples. Can. J. Bot. 58:137-146. (describes *B. eustacei* to replace the name *C. centrifugum*, illustrations)
Butlerelfia eustacei Weresub & Illman (syn. *Corticium centrifugum* (Lev.) Bres.) causes fisheye rot of stored apples.

BYSSOCHLAMYS   See *Paecilomyces*.

CALICIOPSIS Peck
    Coryneliales   10 spp.
Funk, A. 1963. Studies in the genus *Caliciopsis*. Can. J. Bot. 41:503-543. (descriptions, illustrations, and biology of three species causing cankers on conifers)
Several species on conifers are described and illustrated by Funk (1981).

CALONECTRIA De Not.
    Hypocreales   7 spp.
Alfieri, S. A., Jr., Linderman, R. G., Morrison, R. H., and Sobers, E. K. 1972. Comparative pathogenicity of *Calonectria theae* and *Cylindrocladium scoparium* to leaves and roots of azalea. Phytopathology 62:647-650. (comparative table and illustrations, describes differences in the shape of the apical vesicle)

CALONECTRIA - cont.
Boesewinkel, H. J. 1982. *Cylindrocladiella*, a new genus to accommodate *Cylindrocladium parvum* and other small-spored species of *Cylindrocladium*. Can. J. Bot. 60:2288-2294. (key to five *Cylindrocladiella* spp., descriptions and illustrations, transfers one *Calonectria* teleomorph to *Nectria*)
El-Gholl, N. E., Kimbrough, J. W., Barnard, E. L., Alfieri, S. A., Jr., and Schoulties, C. L. 1986. *Calonectria spathulata* sp. nov. Mycotaxon 26:151-164. (descriptions and illustrations of teleomorph and *Cylindrocladium* anamorph causing leaf spot on *Eucalyptus*)
El-Gholl, N. E., Schoulties, C. L., and Alfieri, S. A., Jr. 1983. Homothallism in *Calonectria theae*. Mycologia 75:162-163. (technique for production of sexual state)
Rossman, A. Y. 1979a. *Calonectria* and its type species, *C. daldiniana*, a later synonym of *C. pyrochroa*. Mycotaxon 8:321-328. (descriptions and illustrations)
Rossman, A. Y. 1979b. A preliminary account of the taxa described in *Calonectria*. Mycotaxon 8:485-558. (an alphabetical list of all taxa and their taxonomic disposition based on an examination of type specimens)
Rossman, A. Y. 1983. The phragmosporous species of *Nectria* and related genera. Mycol. Pap. 150:1-164. (key to 53 species including five *Calonectria* spp., descriptions and illustrations)
Sobers, E. K. 1972. Morphology and pathogenicity of *Calonectria floridana*, *Calonectria kyotensis* and *Calonectria uniseptata*. Phytopathology 62:485-487. (provides a comparative synopsis of morphological characteristics, considers these species synonymous)
This genus has been used as a repository for *Nectria*-like species with multiseptate ascospores. Rossman (1979a, 1983) defines the genus to include only those species with *Cylindrocladium* anamorphs, many of which are pathogenic. Boesewinkel (1982) places *Calonectria* spp. with small, one-septate ascospores in *Nectria* and places the anamorphs in *Cylindrocladiella*. Several diseases are caused by *Calonectria* species including: *Calonectria colhounii* Peerally, foliage disease of tea (CMI 430); *C. crotalariae* (Loos) Bell & Sobers, *Cylindrocladium* black rot of peanuts (CMI 429); *C. hederae* C. Booth & Murray, disease of *Hedera helix* (CMI 426); *C. kyotensis* Terashita, root rot of peach trees (CMI 421); *C. quinqueseptata* Figueiredo & Namekata, disease of clove seedlings (CMI 423); and *C. theae* Loos, *Cercosporella* disease of tea bushes (CMI 424).

CALOTHYRIOPSIS Hoehn.
Pleosporales   3 spp.
Subhedar, A. W., and Rao, V. G. 1980. An undescribed species of *Calothyriopsis* on apple. Reinwardtia 9:421-424. (describes and illustrates *C. mali* causing fruit blemish)

CALOTHYRIOPSIS - cont.
Mueller and Arx (1962) treat two species.

CATENOPHORA Luttrell
    Coelomycetes   2 spp.
Luttrell, E. S. 1940. An undescribed fungus on Japanese cherry. Mycologia 32:530-536. (descriptions and illustrations of morphology and development in *C. pruni*)
Nag Raj, T. R. 1977. Miscellaneous microfungi. II. Can. J. Bot. 55:757-765. (description and illustrations of *C. yuccae*)
Sutton (1980) treats *Catenophora pruni* Luttrell and discusses *C. yuccae* Nag Raj, which causes lesions on *Yucca* in the southeastern U.S.

CENANGIUM Fr.:Fr.
    Helotiales   25 spp.
Korf, R. P., and Kohn, L. M. 1976. Notes on *Phibalis*, type genus of the Encoelioideae (Discomycetes). Mem. N. Y. Bot. Gard. 28:109-118. (considers *Cenangium* a synonym of *Phibalis*, descriptions)
Torkelson, A.-E., and Eckblad, F.-E. 1977. Encoelioideae (Ascomycetes) of Norway. Norw. J. Bot. 24:133-149. (key to two species, descriptions and illustrations)
Vloten, H. van, and Gremmen, J. 1953. Studies in the Discomycete genera *Crumenula* De Not. and *Cenangium* Fr. Acta Bot. Neerl. 2:226-241. (describes *Cenangium ferruginosum* and *C. acicolum*)
Cenangium ferruginosum Fr.:Fr. (syn. *C. abietis* (Pers.) Duby) causes dieback of pines and is included in Funk (1981). See *Encoelia*.

CENTROSPORA   See *Mycocentrospora*.

CEPHALEUROS Kunze
    Trentepohliaceae   13 spp.
Chapman, R. L., and Henk, M. C. 1985. Observations on the habit, morphology and ultrastructure of *Cephaleuros parasiticus* (Chlorophyta) and a comparison with *C. virescens*. J. Phycol. 21:513-522. (descriptions and illustrations)
Holcomb, G. E. 1985. New hosts of the parasitic lichen *Strigula*. Plant Dis. 69:1100. (lists 16 new hosts)
Holcomb, G. E. 1986. Hosts of the parasitic alga *Cephaleuros virescens* in Louisiana and new host records for the continental United States. Plant Dis. 70:1080-1083. (lists 218 hosts, 167 are new U. S. records)
Marlatt, R. B., and Alfieri, S. A., Jr. 1881. Hosts of a parasitic alga, *Cephaleuros* Kunze, in Florida. Plant Dis. 65:520-522. (lists 157 hosts)
Printz, H. 1940. Vorarbeiten zu einer Monographie der Trentepohliaceen. Nytt Mag. Naturvidenskapene 80:137-210. (in

CEPHALEUROS - cont.
German, key to 13 *Cephaleuros* spp., descriptions and illustrations)
Santesson, R. 1952. Foliicolous lichens I. A revision of the taxonomy of the obligately foliicolous, lichenized fungi. Symb. Bot. Ups. 12:1-590. (discusses nomenclature and relationship of *Cephaleuros* and *Strigula* spp.)
Wolf, F. A. 1930. A parasitic alga, *Cephaleuros virescens* Kunze, on *Citrus* and certain other plants. J. Elisha Mitchell Sci. Soc. 45:187-205. (describes alga and disease, illustrations)
Although *Cephaleuros* is an alga, the genus is included here because *C. virescens* Kunze causes a leaf spot disease of numerous hosts and is occasionally mistaken for a fungus. The foliicolous, parasitic lichens *Strigula elegans* (Fee) Muell. Arg. and *S. complanata* (Fee) Mont. contain the phycobiont *C. virescens*. According to Santesson (1952), the alga may be associated with a fungus to varying degrees, thus the distinction between the lichen *Strigula* and the alga *Cephaleuros* is unclear.

CEPHALOSPORIUM   See *Acremonium* and *Hymenella*.

CERACEOPSORA Kakishima, T. Sato & S. Sato
    Uredinales   1 sp.
Kakishima, M., Sato, T., and Sato, S. 1984. *Ceraceopsora*, a new genus of Uredinales from Japan. Mycologia 76:969-974. (description and illustration)
This rust causes a disease on *Anemone flaccida* and *Eleagnus* spp. in Japan.

CERACEOSORUS   See *Dicellomyces*.

CERATOBASIDIUM D. P. Rogers
    Tulasnellales   5 spp.
Jackson, H. S. 1949. Studies of Canadian Thelephoraceae IV. *Corticium anceps* in North America. Can. J. Res. Sect. C. 27:241-252. (description and illustrations of *Ceratobasidium anceps* and the disease it causes, host list)
Murray, D. I. L., and Burpee, L. L. 1984. *Ceratobasidium cereale* sp. nov., the teleomorph of *Rhizoctonia cerealis*. Trans. Br. Mycol. Soc. 82:170-172. (descriptions and illustrations)
Talbot, P. H. B. 1965. Studies of "*Pellicularia*" and associated genera of hymenomycetes. Persoonia 3:371-406. (keys to related genera and five *Ceratobasidium* spp.)
Talbot, P. H. B. 1970. Taxonomy and nomenclature of the perfect state. Pages 20-31 in: *Rhizoctonia solani*: Biology and Pathology. J. R. Parmeter, Jr., ed. University of California Press, Berkeley. (discusses taxonomic relationship to *Thanatephorus*)

CERATOBASIDIUM - cont.
Warcup, J. H., and Talbot, P. H. B. 1967. Perfect states of Rhizoctonias associated with orchids. New Phytol. 66:631-641. (descriptions and illustrations of three *Ceratobasidium* spp.)
Warcup, J. H., and Talbot, P. H. B. 1971. Perfect states of Rhizoctonias associated with orchids II. New Phytol. 70:35-40. (description and illustration of one *Ceratobasidium* sp.)
Warcup, J. H., and Talbot, P. H. B. 1980. Perfect states of Rhizoctonias associated with orchids III. New Phytol. 86:267-272. (descriptions and illustrations of three *Ceratobasidium* spp.)
Many *Ceratobasidium* species have *Rhizoctonia* anamorphs. *Ceratobasidium anceps* (Bres. & Sydow) H. Jacks. (syn. *Corticium anceps* (Bres. & Sydow) Gregor) causes leaf lesions on a variety of hosts including ferns, monocots, and dicots. The anamorph is *Sclerotium deciduum* J. J. Davis. See *Rhizoctonia, Sclerotium, Thanatephorus,* and Aphyllophorales.

CERATOCYSTIS Ellis & Halst.
    Ophiostomatales   60 spp.
Butin, H., and Aquilar, A. M. 1984. Blue-stain fungi on *Nothofagus* from Chile-including two new species of *Ceratocystis* Ellis & Halst. Phytopathol. Z. 109:80-89. (describes and illustrates four species)
Davidson, R. W. 1971. New species of *Ceratocystis*. Mycologia 63:5-15. (describes five new species and one new variety from sapwood of forest trees)
Griffin, H. D. 1968. The genus *Ceratocystis* in Ontario. Can. J. Bot. 46:689-718. (key to 60 species, descriptions and illustrations)
Hoog, G. S. de 1974. The genera *Blastobotrys, Sporothrix, Calcarisporium,* and *Calcarisporiella* gen. nov. Stud. Mycol. 7:1-84. (list of 20 accepted *Ceratocystis* spp., descriptions and illustrations of over 20 *Sporothrix* and *Ophiostoma* spp., key to 28 species in various genera, references)
Hoog, G. S. de, and Scheffer, R. J. 1984. *Ceratocystis* versus *Ophiostoma*: a reappraisal. Mycologia 76:292-299. (summary of evidence for recognizing *Ophiostoma*, 14 *Ceratocystis* spp. without *Chalara* anamorphs transferred to *Ophiostoma*)
Hunt, J. 1956. Taxonomy of the genus *Ceratocystis*. Lloydia 19:1-58. (key and descriptions of 39 species, illustrations)
Olchowecki, A., and Reid, J. 1974. Taxonomy of the genus *Ceratocystis* in Manitoba. Can. J. Bot. 52:1675-1711. (key to 70 species, descriptions and illustrations)
Samuels, G. J., and Mueller, E. 1978. Life-history studies of Brazilian Ascomycetes 5. Two new species of *Ophiostoma* and their *Sporothrix* anamorphs. Sydowia 31:169-179. (descriptions and illustrations of two species, discussion of distinction between *Ceratocystis* and *Ophiostoma*)

CERATOCYSTIS - cont.
Upadhyay, H. P. 1981. A Monograph of *Ceratocystis* and *Ceratocystiopsis*. University of Georgia Press, Athens. 176 pp. (keys to 75 species, descriptions and illustrations, generic concepts controversial)
Wingfield, M. J., and Marasas, W. F. O. 1980. *Ceratocystis ips* associated with *Orthotomicus erosus* (Coleoptera: Scolytidae) on *Pinus* spp. in the Cape Province of South Africa. Phytophylactica 12:65-69. (compares *C. ips* with *C. montia*)
Wright, E. F., and Cain, R. F. 1961. New species of the genus *Ceratocystis*. Can. J. Bot. 39:1215-1230. (describes and illustrates four species on conifers)
The genus has been monographed by Upadhyay (1981) who recognizes several segregate genera. His generic concepts are controversial. De Hoog and Scheffer (1984) give a summary of the evidence for the recognition of *Ophiostoma* for some species formerly placed in *Ceratocystis*. The cause of dutch elm disease is now placed in *Ophiostoma* as *O. ulmi* (Buisman) Nannf. (CMI 361 as *Ceratocystis ulmi*). Synanamorphs of *Ceratocystis* sensu de Hoog belong to *Chalara, Chalaropsis*, and *Thielaviopsis*. Important diseases are caused by: *Ceratocystis fagacearum* (T. W. Bretz) W. Hunt, the agent of oak wilt; *C. fimbriata* Ellis & Halst., on various hosts (CMI 141); *C. moniliformis* (Hedgc.) C. Moreau, on various woody plants (CMI 142); and *C. paradoxa* (Dade) C. Moreau, diseases on various hosts (CMI 143).

CERCOSEPTORIA Petr.
  Hyphomycetes   10 spp.
Deighton, F. C. 1976. Studies on *Cercospora* and allied genera. VI. *Pseudocercospora* Speg., *Pantospora* Cif. and *Cercoseptoria* Petr. Mycol. Pap. 140:1-168. (defines *Cercoseptoria* including ten species, describes and illustrates four species)
Deighton, F. C. 1983. Studies on *Cercospora* and allied genera. VIII. Further notes on *Cercoseptoria* and some new species and redispositions. Mycol. Pap. 151:1-13. (includes *C. theae* and similar species on *Camellia*)
Evans, H. C. 1984. The genus *Mycosphaerella* and its anamorphs *Cercoseptoria, Dothistroma* and *Lecanosticta* on pines. Mycol. Pap. 153:1-102. (includes *C. pini-densiflorae* and the teleomorph *Mycosphaerella gibsonii*, extensive treatment, descriptions and illustrations)
Cercoseptoria, a segregate of *Cercospora*, is similar to *Pseudocercospora*. *Cercoseptoria handelii* (Bubak) Deighton causes leaf spot of *Rhododendron* spp. (F. Can. 22 as *Cercospora handelii*). *Cercoseptoria pini-densiflorae* (Hori & Nambu) Deighton causes brown needle blight or *Cercospora* blight of pines (CMI 329 as *Cercospora pini-densiflorae*). See *Cercospora*.

CERCOSPORA Fresen.
Hyphomycetes   1270 spp.
Boedijn, K. B. 1961. The genus *Cercospora* in Indonesia. Nova Hedw.
    3:411-437. (lists 90 species by host family with comments)
Brown, L. G., and Morgan-Jones, G. 1976. Notes on Hyphomycetes. XI.
    Additions to the genera *Cercosporidium, Passalora* and
    Phaeoisariopsis. Mycotaxon 4:299-306. (redescribes, illustrates,
    and reclassifies four *Cercospora* spp. from *Cassia* and *Psoralea*)
Brown, L. G., and Morgan-Jones, G. 1977. Studies on Hyphomycetes. XX.
    "*Cercospora*-complex" fungi of *Cassia* and *Psoralea*. Mycotaxon
    6:261-276. (redescribes and illustrates seven *Cercospora* spp.,
    six of which are reclassified in segregate genera)
Chupp, G. 1953. A Monograph of the Fungus Genus *Cercospora*. Cornell
    University Press, Ithaca, NY. 667 pp. (the most comprehensive
    work on this genus, based primarily on host, generic concepts
    outdated, keys, descriptions and illustrations)
Deighton, F. C. 1959. Studies on *Cercospora* and allied genera. I.
    Cercospora species with coloured spores on *Phyllanthus*
    (Euphorbiaceae). Mycol. Pap. 71:1-23. (key to six species on
    Phyllanthus, descriptions and illustrations)
Deighton, F. C. 1967. Studies on *Cercospora* and allied genera. II.
    Passalora, Cercosporidium and some species of *Fusicladium* on
    Euphorbia. Mycol. Pap. 112:1-80. (treats 17 *Cercosporidium* spp.,
    key to five *Cercosporidium* spp. on Apiaceae)
Deighton, F. C. 1969. Studies on *Cercospora* and allied genera. IV.
    Cercosporella Sacc. *Pseudocercosporella* gen. nov. and
    Pseudocercosporidium gen. nov. Mycol. Pap. 133:1-62. (treats
    nearly 30 taxa mostly in *Cercosporella* and *Pseudocercosporella*,
    describes and illustrates two *Cercospora* spp.)
Deighton, F. C. 1971. Studies on *Cercospora* and allied genera. III.
    Centrospora. Mycol. Pap. 124:1-13. (treats four *Mycocentrospora*
    spp. as *Centrospora*)
Deighton, F. C. 1974. Studies on *Cercospora* and allied genera. V.
    Mycovellosiella Rangel, and a new species of *Ramulariopsis*.
    Mycol. Pap. 137:1-73. (describes and illustrates 33
    Mycovellosiella spp., many transferred from *Cercospora*)
Deighton, F. C. 1976. Studies on *Cercospora* and allied genera. VI.
    Pseudocercospora Speg., *Pantospora* Cif. and *Cercoseptoria* Petr.
    Mycol. Pap. 140:1-168. (treats species in *Pseudocercospora*,
    Cercoseptoria, and *Pantospora*, many transferred from *Cercospora*)
Deighton, F. C. 1979. Studies on *Cercospora* and allied genera. VII.
    New species and redispositions. Mycol. Pap. 144:1-56. (treats
    species in *Pseudocercospora, Mycovellosiella, Phaeoramularia,
    Paracercospora*, and *Stenella*, many transferred from *Cercospora*)
Deighton, F. C. 1983. Studies on *Cercospora* and allied genera. VIII.
    Further notes on *Cercoseptoria* and some new species and
    redispositions. Mycol. Pap. 151:1-13. (treats species in
    Cercoseptoria and *Stigmina*, distinguishes four taxa on *Camellia*,

CERCOSPORA - cont.
many transferred from *Cercospora*)
Hino, T., and Tokeshi, H. 1978. Some pathogens of cercosporiosis collected in Brazil. Trop. Agric. Res. Cen., Min. Agric., Jpn., Tech. Bull. 11:1-131. (lists 302 species according to host, describes and illustrates 81 species from Brazil)
Katsuki, S. 1965. Cercosporae of Japan. Trans. Mycol. Soc. Jpn., extra issue 1:1-100. (describes 226 species according to host family)
Katsuki, S., and Kobayashi, T. 1982. *Cercospora* of Japan and allied genera, (Supplement 5). Trans. Mycol. Soc. Jpn. 23:41-49. (the most recent in a series, descriptions and illustrations)
Ondrej, M. 1984. Funde von parasitischen imperfekten Pilzen *Cercospora* Fres. aus der Tschechoslowakei (Teil III). Ceska Mykol. 38:230-234. (in Czechoslovakian, lists 40 species and recent references)
Pollack, F. G. 1987. An Annotated Compilation of *Cercospora* Names. Mycol. Mem. 12:1-187. (in press, includes over 3000 names with bibliographic information, annotations, host index)
Sivanesan, A. 1985. Teleomorphs of *Cercospora sesami* and *Cercoseptoria sesami*. Trans. Br. Mycol. Soc. 85:397-404. (key to six *Cercospora* or *Cercospora*-like species on *Sesamum*, descriptions and illustrations)
Sontirat, P., Phitakpraiwan, P., Choobamroong, W., and Giatgong, P. 1980. Plant Pathogenic Cercosporae in Thailand. Mycology Branch, Plant Pathology and Microbiology Division, Dept. of Agriculture, Ministry of Agriculture and Cooperatives, Bangkok, Thailand. 51 pp. (describes and illustrates 22 species)
Thaung, M. M. 1984. Some fungi of *Cercospora* complex from Burma. Mycotaxon 19:425-452. (a list of 200 taxa based on host, synoptic table to genera segregated from *Cercospora* by Deighton)
Vasudeva, R. S. 1963. Indian Cercosporae. Indian Council of Agric. Res., New Delhi. 245 pp. (key based on host, descriptions of 260 species)
Yen, J.-M., Kar, A. K., and Das, B. K. 1982. Studies on hyphomycetes from West Bengal, India. I. *Cercospora* and allied genera of West Bengal, 1. Mycotaxon 16:35-57. (descriptions and illustrations of 15 species including 10 new species and 1 new combination)
Yen, J.-M., and Lim, G. 1980. *Cercospora* and allied genera of Singapore and the Malay Peninsula. Gard. Bull., Singapore 33:151-263. (descriptions and illustrations of 98 species, host list)
Over 2,000 species have been described in *Cercospora*. No modern comprehensive account of the genus exists. Chupp (1953) treated most species described up to 1953 using a broad generic concept. Since then many species have been placed in segregate genera based mainly on conidium morphology and mode of conidiogeny. Most of these are

CERCOSPORA - cont.
included in Ellis' (1976) keys to genera. Deighton (1976) discusses generic concepts of these segregate genera. Deighton has treated species in this complex in the Mycological Papers cited above and under the segregate genera. They are useful but are often difficult to use due to lack of keys and the lack of a comprehensive compilation of the species treated. Ellis (1971, 1976) is the most useful resource for identifying common species in the *Cercospora* complex. Many species in this complex have a *Mycosphaerella* teleomorph. Sivanesan (1984) provides a key to ten *Mycosphaerella* spp. with *Cercospora* anamorphs. *Cercospora* species cause a number of leaf spot diseases including: *Cercospora arachidicola* Hori, early leaf spot of peanut (CMI 411 under *Mycosphaerella arachidis*); *C. avicularis* Wint., leaf spot of *Polygonum* spp. (F. Can. 62); *C. beticola* Sacc., leaf spot of beet (CMI 721); *C. brassicicola* Henn., leaf spot of *Brassica* spp. (CMI 722); *C. canescens* Ellis & G. Martin, leaf spot of *Phaseolus* spp. (CMI 462); *C. capsici* Heald & Wolf, leaf spot of *Capsicum* spp. (CMI 723); *C. carthami* Sundararaman & T. G. Ramakrishnan, leaf spot of safflower (CMI 626); *C. coffeicola* Berk. & Cooke, brown eye spot of coffee (CMI 415); *C. elaeidis* Steyaert, freckle of oil palm (CMI 464); *C. kikuchii* (Matsumato & Tomoyasu) M. W. Gardner, purple seed stain of soybean and other legumes (CMI 466); *C. longipes* Butler, brown spot of sugarcane (CMI 418); *C. nicotianae* Ellis & Everh., frog eye of tobacco (CMI 416); *C. oryzae* Miyake, narrow brown leaf spot of rice (CMI 420); *C. sequoiae* Ellis & Everh., *Cercospora* blight of cypress (CMI 366); *C. sesami* Zimmermann, leaf and pod spot of sesame (CMI 627); *C. sorghi* Ellis & Everh., grey leaf spot of *Sorghum* spp. (CMI 419); and *C. zeae-maydis* Tehon & Daniels, gray leaf spot of maize.

CERCOSPORELLA Sacc.
    Hyphomycetes   80 spp.
Deighton, F. C. 1973. Studies on *Cercospora* and allied genera. IV. *Cercosporella* Sacc., *Pseudocercosporella* gen. nov. and *Pseudocercosporidium* gen. nov. Mycol. Pap. 133:1-62. (treats 15 species in *Cercosporella*, describes *Pseudocercosporella* and treats ten species, describes *Pseudocercosporidium* with one species, illustrations)
This genus is a segregate of *Cercospora*. Sivanesan (1984) includes a key to five *Mycosphaerella* spp. with *Cercosporella* anamorphs. *Cercosporella pastinaceae* P. Karst. causes parsnip leaf spot, and *C. virgaureae* (Thuem.) Allesch. occurs on Asteraceae, particularly *Aster, Solidago*, and *Erigeron* (F. Can. 61). See *Cercospora*.

CERCOSPORIDIUM Earle
    Hyphomycetes   17 spp.
Deighton, F. C. 1967. Studies on *Cercospora* and allied genera II. *Passalora, Cercosporidium* and some species of *Fusicladium* on

CERCOSPORIDIUM - cont.
Euphorbia. Mycol. Pap. 112:1-80. (descriptions and illustrations of 17 species, key to five *Cercosporidium* spp. on Apiaceae)
Liu, X.-j., and Guo, Y.-l. 1982. [Studies on some species of the genus *Cercosporidium* in China.] Acta Mycol. Sin. 1:88-102. (in Chinese with English summary, describes and illustrates 18 species)
This generic segregate of *Cercospora* is included in Ellis (1971, 1976). *Cercosporidium personatum* (Berk. & M. A. Curtis) Deighton causes late leaf spot of peanut (CMI 412 under *Mycosphaerella berkeleyi*). See *Cercospora.*

CEREBELLA Ces.
    Hyphomycetes   1 sp.
Langdon, R. F. N. 1955. The genus *Cerebella.* Mycol. Pap. 61:1-18. (description and illustration, most described species placed in synonymy with *C. andropogonis*)
Schol-Schwarz, M. B. 1959. The genus *Epicoccum* Link. Trans. Br. Mycol. Soc. 42:149-173. (includes *C. andropogonis* in *Epicoccum*, descriptions and illustrations)
Cerebella andropogonis Ces. in Rabenh. occurs on spikelets of grasses infected with ergot, *Claviceps purpurea* (Fr.) Tul., and is often mistaken for a member of the Ustilaginales. Ellis (1971) provides a description and illustration.

CEROTELIUM Arth.
    Uredinales   20 spp.
Cerotelium fici (E. J. Butler) Arth. causes fig rust (CMI 281). See Uredinales.

CEUTHOSPORA Grev.
    Coelomycetes   100 spp.
Sutton (1980) and DiCosmo et al. (1984 under *Phacidium*) are the best references for the identification of these species, many of which are anamorphs of *Phacidium. Ceuthospora lunata* Shear associated with *Strasseria oxycocci* Shear causes black rot of cranberries, a fresh market storage problem. See *Strasseria.*

CHACONIA Juel
    Uredinales   8 spp.
Eboh, D. O. 1985. A re-evaluation of *Ypsilospora.* Trans. Br. Mycol. Soc. 85:39-46. (*Ypsilospora* is considered a synonym of *Chaconia*, describes and illustrates two species)
Ono, Y., and Hennen, J. F. 1983. Taxonomy of the Chaconiaceous genera (Uredinales). Trans. Mycol. Soc. Jpn. 24:369-402. (keys to ten *Goplana* spp., six *Chrysocelis* spp., six *Chaconia* spp., and seven *Olivea* spp. all based on host, descriptions)
Eboh (1985) synonymizes *Ypsilospora* with *Chaconia.* See Cummins and

CHACONIA - cont.
Hiratsuka (1983).

CHAETOCHALARA See *Chalara*.

CHAETOMELLA Fuckel
    Coelomycetes   7 spp.
Sarbhoy, A. K. 1984. The problematic genus *Chaetomella*, its correct taxonomic position. Pages 502-505 in: Taxonomy of Fungi, Part 2. C. V. Subramanian, ed. Amra Press, Madras. (key to seven species, brief discussion)
Stolk, A. C. 1963. The genus *Chaetomella* Fuckel. Trans. Br. Mycol. Soc. 46:409-425. (describes and illustrates three species, gives taxonomic disposition of 28 names in *Chaetomella*)
Sutton, B. C., and Sarbhoy, A. K. 1976. Revision of *Chaetomella*, and comments upon *Vermiculariopsis* and *Thyriochaetum*. Trans. Br. Mycol. Soc. 66:297-303. (key to five taxa, descriptions and illustrations)
Sutton (1980) provides brief descriptions of five taxa, an illustration of *Chaetomella raphigera* Swift, and numerous references.

CHAETOMIUM Kunze:Fr.
    Sphaeriales   200 spp.
Ames, L. M. 1961. A monograph of the Chaetomiaceae. U. S. Army Res. Dev., Ser. 2:1-125. (comprehensive monograph with keys, descriptions, and illustrations, outdated)
Arx, J. A. von 1984. A reevaluation of *Chaetomium* and the Chaetomiaceae. Persoonia 12:169-179. (key to 32 species in *Chaetomium* and related genera)
Arx, J. A. von, Guarro, J., and Figueras, M. J. 1986. The Ascomycete Genus *Chaetomium*. J. Cramer, Berlin. 162 pp. (not seen)
Cannon, P. F. 1986. A revision of *Achaetomium, Achaetomiella* and *Subramaniula*, and some similar species of *Chaetomium*. Trans. Br. Mycol. Soc. 87:45-76. (keys to species in the above-mentioned genera including nearly 30 *Chaetomium* spp. with inconspicuous or no ascomatal hairs, descriptions and illustrations)
Dreyfuss, M. (1975)1976. Taxonomische Untersuchungen innerhalb der Gattung *Chaetomium* Kunze. Sydowia 28:50-133. (in German with English summary, synoptic key to 16 species of the *C. globosum* group)
Hawksworth, D. L., and Wells, H. 1973. Ornamentation on the terminal hairs in *Chaetomium* Kunze ex Fr. and some allied genera. Mycol. Pap. 134:1-24. (evaluates 92 species)
Millner, P. D. 1975. Ascomycetes of Pakistan: *Chaetomium*. Biologia (Lahore) 21:39-73. (descriptions and illustrations)
Seth, H. K. (1970) 1972. A monograph of the genus *Chaetomium*. Nova Hedw. Beih. 37:1-133. (key to 110 species, descriptions and

CHAETOMIUM - cont.
illustrations)
Udagawa, S. 1960. A taxonomic study of the Japanese species of *Chaetomium*. J. Gen. Appl. Microbiol. 6:223-251. (key to 24 species, descriptions and illustrations)
Udagawa, S., and Cain, R. F. 1969. Some new or noteworthy species of the genus *Chaetomium*. Can. J. Bot. 47:1939-1951. (describes and illustrates six species)
Species in this genus are cosmopolitan in distribution and generally cellulolytic. Domsch et al. (1980) provide a key to soil-inhabiting species with descriptions, illustrations, and numerous references.

CHAETOSEPTORIA Tehon
    Coelomycetes   2 spp.
Stevenson, J. A. 1946. Fungi novi denominati-II. Mycologia 38:524-533. (describes *C. wellmanii*)
Sutton, B. C. 1964. Coelomycetes III. *Annellolacinia* gen. nov., *Aristastoma, Phaeocytostroma, Seimatosporium*, etc. Mycol. Pap. 97:1-42. (description of *C. vignae*)
Chaetoseptoria wellmanii J. A. Stevenson causes brown leaf spot of *Phaseolus* and *Vigna* (CMI 822), and *C. vignae* Tehon causes leaf spot of *Vigna sinensis* (Sutton, 1964).

CHALARA (Corda) Rabenh.
    Hyphomycetes   70 spp.
Holubova-Jechova, V. 1984. Lignicolous hyphomycetes from Czechoslovakia 7. *Chalara, Exochalara, Fusichalara*, and *Dictyochaeta*. Folia Geobot. Phytotaxon. 19:387-438. (keys to related genera and 20 *Chalara* spp., descriptions and illustrations)
Kiffer, E., and Delon, R. 1983. *Chalara elegans* (: *Thielaviopsis basicola*) and allied species. II - Validation of two taxa. Mycotaxon 18:165-174. (describes and illustrates three species)
Kirk, P. M., and Spooner, B. M. 1984. An account of the fungi of Arran, Gigha and Kintyre. Kew Bull. 38:503-597. (reduces *Chaetochalara* to synonymy with *Chalara*, describes two *Chalara* spp.)
Nag Raj, T. R., and Kendrick, B. 1975. A Monograph of *Chalara* and Allied Genera. Wilfred Laurier University Press, Waterloo, Ontario. 200 pp. (comprehensive, keys, descriptions, and illustrations)
Nag Raj and Kendrick (1975) provide an excellent monograph but it is becoming outdated as new species are described. Teleomorphs of *Chalara* belong to *Ceratocystis*. *Chalara elegans* Nag Raj & Kendrick, the synanamorph of *Thielaviopsis basicola* (Berk. & Broome) Ferraris, causes black root rot on a number of crops (CMI 170 under *Thielaviopsis basicola*).

CHEILARIA Lib.
    Coelomycetes   1 sp.
Cheilaria agrostis Lib., also known erroneously as *Septogloeum oxysporum* Bommer, M. Rousseau & Sacc., causes blotch and char spot of Poaceae in temperate areas (CMI 488; Sutton, 1980).

CHOANEPHORA Curr.
    Mucorales   2 spp.
Kirk, P. M. 1984. A monograph of the Choanephoraceae. Mycol. Pap. 152:1-61. (key to genera, key to two *Choanephora* spp., descriptions and illustrations)
Choanephora cucurbitarum (Berk. & Ravenel) Thaxter causes diseases of various vegetables.

CHONDROPLEA   See *Cryptodiaporthe*, *Discosporium*, and *Dothichiza*.

CHONDROSTEREUM Pouzar
    Aphyllophorales   1 sp.
Chondrostereum purpureum (Pers.:Fr.) Pouzar (syn. *Stereum purpureum* Pers.:Fr.) causes silver leaf disease of *Prunus* spp., apple, and other fruit trees. See Aphyllophorales.

CHRYSOCELIS Lagerh. & Dietel
    Uredinales   6 spp.
The genus occurs on a variety of hosts. See *Chaconia*.

CHRYSOMYXA Unger
    Uredinales   20 spp.
Savile, D. B. O. 1950. North American species of *Chrysomyxa*. Can. J. Res. Sect. C. 28:318-330. (keys to 11 species based on aecia and based on uredinia or telia, descriptions)
This genus includes rusts with coniferous and ericaceous alternate hosts. *Chrysomyxa abietis* (Wallr.) Unger causes spruce needle rust (CMI 576), and *C. weirii* H. Jacks. causes needle rust of spruce (CMI 221). Ziller (1974) provides a key and treats ten species.

CIBORIA Fuckel
    Helotiales   15 spp.
Schumacher, T. 1978. A guide to the amenticolous species of the genus *Ciboria* in Norway. Norw. J. Bot. 25:145-155. (key to nine taxa, descriptions and illustrations)
The stromatic tissues of these fungi form on catkins of hardwood trees.

CIBORINIA Whetzel
    Helotiales   9 spp.
Batra, L. R. 1960. The species of *Ciborinia* pathogenic to *Salix*, *Magnolia* and *Quercus*. Am. J. Bot. 47:819-827. (key to 13

CIBORINIA - cont.
species, descriptions and illustrations)
Batra, L. R., and Korf, R. P. 1959. The species of *Ciborinia* pathogenic to herbaceous angiosperms. Am. J. Bot. 46:441-450. (descriptions and illustrations of four species, information on biology of *C. erythronii*, amends generic concept)
Groves, J. W., and Bowerman, C. A. 1955. The species of *Ciborinia* on *Populus*. Can. J. Bot. 33:577-590. (treats four species, descriptions and illustrations, cultural characters)
Kohn, L. M., and Nagasawa, E. 1984. A taxonomic reassessment of *Sclerotinia camelliae* Hara (=*Ciborinia camelliae* Kohn), with observations on flower blight of camellia in Japan. Trans. Mycol. Soc. Jpn. 25:149-161. (descriptions and illustrations of the fungus causing sclerotial flower blight of camellia)
See *Sclerotinia*.

CLADOSPORIUM Link:Fr.
Hyphomycetes   50 spp.
Gottwald, T. R. 1982. Taxonomy of the pecan scab fungus *Cladosporium caryigenum*. Mycologia 74:382-390. (description and illustration)
Iwatsu, T. 1984. A new species of *Cladosporium* from Japan. Mycotaxon 20:521-533. (compares four medically important species)
Kirk, P. M., and Crompton, J. G. 1984. Pathology and taxonomy of *Cladosporium* leaf blotch of onion (*Allium cepa*) and leek (*A. porrum*). Plant Pathol. 33:317-324. (describes and illustrates *C. allii* on leeks and *C. allii-cepae* on onions)
Traquair, J. A., Meloche, R. B., and Jarvis, W. R. 1984. Hyperparasitism of *Puccinia violae* by *Cladosporium uredinicola*. Can. J. Bot. 62:181-184. (hyperparasitic species, illustrations)
Vries, G. A. de 1952. Contribution to the Knowledge of the Genus *Cladosporium* Link ex Fr. Uitgeverij and Drukkerij, Hollandia, Baarn. 121 pp. (a classic work, now outdated, key to 23 species, descriptions, illustrations, cultural characters)
Ellis (1971, 1976) provides the best modern treatment of this genus including keys to numerous species, both pathogenic and saprophytic, with descriptions and illustrations. Teleomorphs of *Cladosporium* species belong to *Mycosphaerella* and *Venturia* (Sivanesan, 1984). See Domsch et al. (1980) for a key to and descriptions of the ubiquitous, saprophytic species. Diseases caused by *Cladosporium* species include: *Cladosporium allii* (M. B. Ellis & G. W. Martin) P. M. Kirk & Crompton, leaf blotch of *Allium* spp. (CMI 841); *C. allii-cepae* (Ranojevic) M. B. Ellis, leaf blotch of *Allium cepa* (CMI 679, 842); *C. carpophilum* Thuem., peach freckle (CMI 402 under *Venturia carpophila*); *C. cucumerinum* Ellis & Arth., scab of cucurbits (CMI 348); and *C. iridis* (Fautr. & Roum.) de Vries, leaf spot or blotch of *Iris* (CMI 435 under *Mycosphaerella macrospora*).

CLATHROSPORA Rabenh.
Pleosporales   18 spp.
Eriksson, O. 1967. On graminicolous pyrenomycetes from Fennoscandia 1. Dictyosporous species. Ark. Bot. 6:339-380. (key to six *Clathrospora* spp., descriptions)
Harr, J. 1970. Einfluss ausserer Factoren auf die Entwicklung einiger arten der Gattung *Clathrospora* Rab. Nova Hedw. 20:865-901. (describes cultural characteristics of three species)
Wehmeyer, L. E. 1961. A World Monograph of the Genus *Pleospora* and its segregates. University of Michigan, Ann Arbor. 451 pp. (key to eight *Clathrospora* spp., descriptions and illustrations)
This genus includes *Pleospora*-like species that produce flattened ascospores with numerous vertical septa. See *Pleospora*.

CLAVICEPS Tul.
Clavicipitales   35 spp.
Bove, F. J. 1970. The Story of Ergot. S. Karger, New York. 297 pp. (all aspects)
Brady, L. R. 1962. Phylogenetic distribution of parasitism by *Claviceps* species. Lloydia 25:1-36. (host range)
Chahal, S. S., Rao, V. P., and Thakur, R. P. 1985. Variation in morphology and pathogenicity in *Claviceps fusiformis*, the causal agent of pearl millet ergot. Trans. Br. Mycol. Soc. 84:325-332. (discusses broad variability)
Langdon, R. F. N. 1954. The origin and differentiation of *Claviceps* species. Univ. Queensl. Pap., Dep. Bot. 3:61-68. (lists 26 species and their hosts)
Rykard, D. M., Luttrell, E. S., and Bacon, C. W. 1984. Conidiogenesis and conidiomata in the Clavicipitoideae. Mycologia 76:1095-1103. (describes anamorphs of *Balansia, Claviceps*, and other genera, discusses their taxonomic significance)
Several important diseases are caused by *Claviceps* species and their *Sphacelia* anamorphs. No comprehensive monograph of this genus exists. *Claviceps purpurea* (Fr.) Tul. causes ergot of many grasses.

CLIMACOCYSTIS Kotl. & Pouzar
Aphyllophorales   1 sp.
Climacocystis borealis (Fr.:Fr.) Kotl. & Pouzar (syn. *Polyporus borealis* Fr.:Fr.) causes a butt and root rot of living conifers. See Aphyllophorales.

CLYPEOPORTHE Hoehn.
Diaporthales   2 spp.
See Barr (1978). *Clypeoporthe iliau* (Lyon) Barr causes iliau disease of sugarcane (CMI 705).

COCCOMYCES De Not.
  Rhytismatales   50 spp.
Johnston, P. R. 1986. Rhytismataceae in New Zealand 1. Some
  foliicolous species of *Coccomyces* de Notaris and *Propolis* (Fries)
  Corda. N. Z. J. Bot. 24:89-124. (dichotomous and synoptic keys
  to ten *Coccomyces* spp., key to four *Propolis* spp., descriptions
  and illustrations, anamorph information)
Sherwood, M. A. 1980. Taxonomic studies in the Phacidiales: The
  genus *Coccomyces* (Rhytismataceae). Occasional Pap. Farlow
  Herbarium 15:1-120. (comprehensive with key, descriptions and
  illustrations of 50 species, host index)
This genus has been monographed by Sherwood (1980). Leaf spot of
mango is caused by *Coccomyces vilis* Sydow & Butler (CMI 792).

COCHLIOBOLUS Drechs.
  Pleosporales   32 spp.
Alcorn, J. L. 1982. New *Cochliobolus* and *Bipolaris* species.
  Mycotaxon 15:1-19. (describes four new *Cochliobolus* spp. and
  their anamorphs)
Alcorn, J. L. 1983. On the genera *Cochliobolus* and
  Pseudocochliobolus. Mycotaxon 16:353-379. (lists 32 *Cochliobolus*
  spp. and their anamorphs, considers *Pseudocochliobolus* a synonym
  of *Cochliobolus*, references)
Ammon, H. U. 1963. Ueber einige Arten aus den Gattungen *Pyrenophora*
  Fries und *Cochliobolus* Drechsler mit *Helminthosporium* als
  Nebenfruchtform. Phytopathol. Z. 47:244-300. (keys)
Chidambaram, P., Mathur, S. B., and Neergaard, P. 1973.
  Identification of seed-borne *Drechslera* species. Friesia
  10:165-207. (descriptions and illustrations of 25 *Drechslera*
  spp., most now in *Bipolaris*)
Hardin, H. 1980. *Cochliobolus sativus* (Ito & Kurib.) Drechsl. ex
  Dastur (imperfect stage: *Bipolaris sorokiniana* (Sacc. in Sorok.)
  Shoem.): a bibliography. Agriculture Canada, Research Branch,
  Saskatoon. Pages not numbered. (lists nearly 2000 publications
  dealing with this organism and the diseases it causes, provides a
  subject index, three supplements have been published with an
  additional 983 references)
Luttrell, E. S. 1977. Correlations between conidial and ascigerous
  state characters in *Pyrenophora, Cochliobolus*, and *Setosphaeria*.
  Rev. Mycol. 41:271-279. (reviews basis for generic segregation)
Luttrell, E. S., and Rogerson, C. T. 1959. Homothallism in an
  undescribed species of *Cochliobolus* and in *Cochliobolus kusanoi*.
  Mycologia 51:195-202. (key to 11 species, describes and
  illustrates one species)
Sivanesan, A. 1985. The teleomorph of *Curvularia tuberculata*. Trans.
  Br. Mycol. Soc. 84:548-551. (key to six *Cochliobolus* spp. with
  Curvularia anamorphs, description and illustrations of
  Cochliobolus tuberculatus)

COCHLIOBOLUS - cont.
Anamorphs of *Cochliobolus* belong to *Bipolaris* and *Curvularia*. Ellis (1971, 1976) provides keys, descriptions, and illustrations for anamorphs under *Drechslera* and *Curvularia*. Sivanesan (1984) provides keys to 22 *Cochliobolus* spp. but refers to *Bipolaris* spp. as *Drechslera*. Species of *Cochliobolus* cause serious diseases including: *Cochliobolus carbonum* Nelson, leaf spot of corn (CMI 349); *C. cymbopogonis* J. A. Hall & Sivanesan, diseases of Andropogoneae (CMI 726); *C. geniculatus* Nelson, seed and seedling blight of many plants (CMI 727); *C. hawaiiensis* Alcorn, seed and seedling diseases (CMI 728); *C. heterostrophus* (Drechs.) Drechs., southern leaf blight of corn (CMI 301); *C. lunatus* Nelson & Haasis, leaf spot and seedling blight of many angiosperms (CMI 474); *C. miyabeanus* (Ito & Kuribayashi) Drechs. ex Dastur, brown spot and seedling blight of rice (CMI 302); *C. nodulosus* Luttrell, diseases of finger millet (CMI 341); *C. sativus* (Ito & Kuribayashi) Drechs. ex Dastur, spot blotch of cereals in temperate regions (CMI 701); *C. setariae* (Ito & Kuribayashi) Drechs. ex Dastur, leaf spot of *Setaria italica* (CMI 473); *C. spicifer* Nelson, diseases on many hosts, including spring dead spot of Bermuda grass (CMI 702); and *C. victoriae* Nelson, Victoria blight of oats (CMI 703).

COLEOSPORIUM Lev.
    Uredinales   80 spp.
Kaneko, S. 1981. The species of *Coleosporium*, the causes of pine needle rusts, in the Japanese Archipelago. Rep. Tottori Mycol. Inst. 19:1-159. (key to 28 species, descriptions, illustrations, host index)
This genus includes rust species that cause diseases on conifers and have dicotyledonous, rarely monocotyledonous, alternate hosts. Ziller (1974) presents keys by host, telial stage, and aecial stage to the four species found in western Canada. *Coleosporium campanulae* Lev. ex Kickx occurs on *Campanula* spp. and *Pinus* spp. (F. Can. 218), and *C. ipomoeae* (Schwein.) Burrill causes orange rust of sweet potato and a needle rust of pine (CMI 282). See Uredinales, especially Cummins and Hiratsuka (1983).

COLEROA (Fr.) Rabenh.
    Pleosporales   8 spp.
Corlett, M., and Barr, M. E. 1986. *Hormotheca* for species of *Coleroa* with hemispherical ascomata. Mycotaxon 25:255-257. (discussion of generic characters, transfers three *Coleroa* spp. to *Hormotheca*)
Mueller and Arx (1962) provide an account of these fungi with descriptions and illustrations of eight species. Corlett and Barr (1986) place some *Coleroa* species in *Hormotheca*. Species include *Coleroa chaetomium* (Kunze:Fr.) Rabenh., on living leaves of some *Rubus* spp. (F. Can. 20).

COLLETOTRICHUM Corda
Coelomycetes   22 spp.
Arx, J. A. von 1957. Die Arten der Gattung *Colletotrichum* Cda. Phytopathol. Z. 29:413-468. (in German with English summary, key to 11 species, descriptions and illustrations, lists several hundred synonyms of *C. gloeosporioides*)
Arx, J. A. von 1970. A revision of the fungi classified as *Gloeosporium*. Bibl. Mycol. 24:1-203. (key to 29 related genera, lists 734 names in *Gloeosporium*, indicates their taxonomic disposition, many as *Colletotrichum* spp., descriptions and illustrations)
Baxter, A. P., and Westhuizen, G. C. A. van der 1984. A synoptic key to South African isolates of *Colletotrichum*. South African J. Bot. 3:265-266. (includes 11 species)
Baxter, A. P., Westhuizen, G. C. A. van der, and Eicker, A. 1983. Morphology and taxonomy of South African isolates of *Colletotrichum*. South African J. Bot. 2:259-289. (key to 11 species, descriptions and illustrations of species both in culture and on host tissue)
Baxter, A. P., Westhuizen, G. C. A. van der, and Eicker, A. 1985. A review of literature on the taxonomy, morphology and biology of the fungal genus *Colletotrichum*. Phytophylactica 17:15-18. (review, numerous references)
Manandhar, J. B., Hartman, G. L., and Sinclair, J. B. 1986. *Colletotrichum destructivum*, the anamorph of *Glomerella glycines*. Phytopathology 76:282-285. (description and illustrations, pathology)
Pennycook, S. R. 1983. *Colletotrichum gloeosporioides* (Penzig) Penzig et Saccardo. Mycotaxon 16:507-508. (nomenclature)
Colletotrichum is a well-documented genus. The use of cultural characteristics has helped to clarify species concepts. Sutton (1980) provides a useful key with brief descriptions and illustrations for 22 species, most of which are pathogenic. *Colletotrichum gloeosporioides* (Penz.) Penz. & Sacc., the anamorph of *Glomerella cingulata* (Stoneman) Spauld. & Schrenk, commonly causes anthracnose and other diseases of a wide variety of hosts (CMI 315 under *Glomerella cingulata*). Other species include: *Colletotrichum acutatum* Simmonds ex Simmonds, black spot of strawberry (CMI 630); *C. capsici* (Sydow) E. J. Butler & Bisby, diseases of various hosts (CMI 317); *C. coccodes* (Wallr.) S. J. Hughes, diseases of various hosts, especially Cucurbitaceae, Fabaceae, and Solanaceae (CMI 131, F. Can. 51); *C. falcatum* Went., diseases of *Saccharum* (CMI 133 under *Glomerella tucumanensis*); *C. graminicola* (Ces.) Wilson, diseases of *Sorghum* and other Poaceae (CMI 132); *C. lindemuthianum* (Sacc. & Magnus) Briosi & Cavara, anthracnose of beans and other legumes (CMI 316); and *C. musae* (Berk. & M. A. Curtis) Arx, diseases of banana (CMI 222).

CONIELLA Hoehn.
    Coelomycetes   8 spp.
Sutton (1980) provides a key to eight species. *Coniella diplodiella* (Speg.) Petr. & Sydow causes white rot of *Vitis vinifera* (CMI 82), and *C. pulchella* Hoehn. occurs on *Pisum sativum* and *Vicia faba* (F. Can. 65).

CONIOTHYRIUM Corda
    Coelomycetes   25 spp.
Biga, M. L. B., Ciferri, R., and Bestagno, G. (1958)1959. Ordinamento artificiale delle specie del genere *Coniothyrium* Corda. Sydowia 12:258-320. (in Italian with English summary, lists over 125 taxa by host and conidium size)
Sutton (1980) reviews the systematics of the genus stating that many species previously placed in *Coniothyrium* now belong in *Microsphaeropsis*. Domsch et al. (1980) treat four species and cite numerous references on their biology. See the teleomorph genera *Leptosphaeria* and *Paraphaeosphaeria*. *Coniothyrium fuckelii* Sacc. causes cane blight of raspberry and other *Rubus* spp. (CMI 663 under *Leptosphaeria coniothyrium*), *C. minitans* Campbell is a hyperparasite on sclerotia of many phytopathogenic fungi (CMI 732), and *C. wernsdorffiae* Laub. causes brand canker of rose.

CORDANA G. Preuss
    Hyphomycetes   8 spp.
Hoog, G. S. de, Oorschot, C. A. N. van, and Hijwegen, T. 1983. Taxonomy of the *Dactylaria* complex. II. *Dissoconium* gen. nov. and *Cordana* Preuss. Proc. K. Ned. Akad. Wet., Ser. C. 86:197-206. (key to eight *Cordana* spp., brief account of each)
Cordana musae (Zimmermann) Hoehn. causes leaf blotch or spot of banana (CMI 350). See the related genus *Dactylaria*.

CORNICULARIELLA   See *Foveostroma*.

CORTICIUM Pers.
    Aphyllophorales   25 spp.
Most pathogenic *Corticium* species have been placed in segregate genera such as *Butlerelfia, Ceratobasidium, Erythricium*, and *Scytinostroma*. *Corticium* in the narrow sense includes species previously placed in *Laeticorticium*. Corticioid species causing pink thread diseases of turf grasses are now placed in *Laetisaria* and *Limonomyces*. See also *Sclerotium, Thanatephorus*, and Aphyllophorales.

CORYNELIA Achar.:Fr.
    Coryneliales   7 spp.
Benny, G. L., Samuelson, D. A., and Kimbrough, J. W. 1985. Studies on

CORYNELIA - cont.
the Coryneliales. I. *Fitzpatrickella*, a monotypic genus on the
fruits of *Drimys*. Bot. Gaz. 146:232-237. (description and
illustrations of *F. operculata*, key to genera of Coryneliales)
Benny, G. L., Samuelson, D. A., and Kimbrough, J. W. 1985. Studies on
the Coryneliales. II. Taxa parasitic on Podocarpaceae: *Corynelia*.
Bot. Gaz. 146:238-251. (monograph, accepts seven *Corynelia* spp.,
descriptions and illustrations)
Benny, G. L., Samuelson, D. A., and Kimbrough, J. W. 1985. Studies on
the Coryneliales. III. Taxa parasitic on Podocarpaceae:
Lagenulopsis and *Tripospora*. Bot. Gaz. 146:431-436. (key to
three *Tripospora* spp., one *Lagenulopsis* sp., descriptions and
illustrations)
Benny, G. L., Samuelson, D. A., and Kimbrough, J. W. 1985. Studies on
the Coryneliales. IV. *Caliciopsis, Coryneliopsis*, and
Coryneliospora. Bot. Gaz. 146:437-448. (key to two
Coryneliospora spp., two *Coryneliopsis* spp., and 25 *Caliciopsis*
spp., descriptions and illustrations of selected taxa)
Corynelia species and many other members of the Coryneliales are
parasitic on Podocarpaceae.

CORYNESPORA Guessow
   Hyphomycetes   20 spp.
Swart, H. J. 1985. Australian leaf-inhabiting fungi. XVIII.
   Corynespora acaciae sp. nov. Trans. Br. Mycol. Soc. 84:175-177.
   (describes and illustrates one new species on *Acacia*)
Ellis (1971, 1976) provides the most comprehensive account of
Corynespora species. *Corynespora cassiicola* (Berk. & M. A. Curtis)
Wei is a cosmopolitan species causing target spot on a wide range of
economically important hosts (CMI 303).

CORYNEUM Nees
   Coelomycetes   20 spp.
Muthumary, J., and Sutton, B. C. 1986. *Coryneum quercinum* sp. nov. on
   Quercus alba from India. Trans. Br. Mycol. Soc. 86:512-515. (key
   to seven *Coryneum* spp. on *Quercus*, description and illustration
   of new species)
Sutton, B. C. 1975. Coelomycetes. V. *Coryneum*. Mycol. Pap.
   138:1-224. (keys to 20 taxa, descriptions and illustrations,
   treats 182 species described as *Coryneum*, indicates their
   taxonomic disposition)
Sutton, B. C. 1986. Improvizations on conidial themes. Trans. Br.
   Mycol. Soc. 86:1-38. (transfers *Murogenella* species to *Coryneum*)
Sutton, B. C., and Rizwi, M. A. 1980. Two problematical Coelomycetes
   with distoseptate conidia. Nova Hedw. 32:341-346. (*Coryneum
   indicum* sp. nov., description, illustration, and comparison with
   similar taxa, unusual in that it produces paraphyses)
Sutton (1980) treats 20 species. Most *Coryneum* species are host

CORYNEUM - cont.
specific. *Pseudovalsa* is the reported teleomorph.

CRINIPELLIS Pat.
    Agaricales   30 spp.
Pegler, D. N. 1978. *Crinipellis perniciosa* (Agaricales). Kew Bull. 32:731-736. (key to three species, descriptions and illustrations)
Pegler, D. N. 1983. Agaric flora of the Lesser Antilles. Kew Bull. Add. Ser. 9:1-668. (key to six *Crinipellis* spp., descriptions and illustrations)
Singer, R. 1976. Marasmieae (Basidiomycetes-Tricholomataceae). Flora Neotropica, Monograph 17. New York Botanical Garden, Bronx. 347 pp. (key to 41 species, descriptions and illustrations)
Crinipellis perniciosa (Stahel) Singer (syn. *Marasmius perniciosus* Stahel) causes witches broom of cacao (CMI 223).

CRISTULARIELLA Hoehn.
    Hyphomycetes   2 spp.
Cline, M. N., Crane, J. L., and Cline, S. D. 1983. The teleomorph of *Cristulariella moricola*. Mycologia 75:988-994. (describes the discomycetous teleomorph *Grovesinia*)
Redhead, S. A. 1975. The genus *Cristulariella*. Can. J. Bot. 53:700-707. (describes and illustrates two species)
Cristulariella moricola (Hino) Redhead (syn. *C. pyramidalis* Waterm. & Marshall), the teleomorph of which belongs in *Grovesinia*, causes a zonate leaf spot on numerous woody and annual plants.

CRONARTIUM Fr.
    Uredinales   20 spp.
Burdsall, H. H., Jr., and Snow, G. A. 1977. Taxonomy of *Cronartium quercuum* and *C. fusiforme*. Mycologia 69:503-508. (considers these species synonyms, recognizes four formae speciales within *C. quercuum* based on host specificity)
Cummins, G. B. 1984. Two new rust fungi (Uredinales). Mycotaxon 20:617-618. (describes new species on *Pinus ponderosa*)
Hiratsuka, Y., and Powell, J. M. 1976. Pine stem rusts of Canada. Can. For. Serv., For. Tech Rep. 4:1-83. (keys to five *Cronartium* spp. based on host, aecial state, and macroscopic features, descriptions and illustrations)
Peterson, R. S. 1973. Studies of *Cronartium* (Uredinales). Rep. Tottori Mycol. Inst. 10:203-223. (descriptions of 19 species)
Cronartium species are widespread and occur alternatively on *Pinus* and dicotyledonous plants. As with other genera of rust fungi, species are usually considered on a geographical or host basis. Ziller (1974) provides a key according to host, telial, or aecial stages for the four species occurring in western Canada. Species include: *Cronartium coleosporioides* Arth., stalactiform blister rust

CRONARTIUM - cont.
of pines (CMI 577); *C. comandrae* Peck, Comandra blister rust of
pines (CMI 578); *C. comptoniae* Arth., sweet fern blister rust of
pines (CMI 579); *C. flaccidum* (Albertini & Schwein.) Wint., scotch
pine blister rust of pines (CMI 580); *C. quercuum* (Berk.) Miyabe ex
Shirai f. sp. *fusiforme*, fusiform rust of loblolly and slash pine;
and *C. ribicola* J. C. Fisch., white pine blister rust (CMI 283).

CRUMENULA   See *Godronia* and *Cenangium*.

CRYPHONECTRIA (Sacc.) Sacc.
  Diaporthales   6 spp.
Hodges, C. S., Jr., Alfenas, A. C., and Ferreira, F. A. 1986. The
 conspecificity of *Cryphonectria cubensis* and *Endothia eugeniae*.
 Mycologia 78:343-350. (description and illustrations, isozyme
 analyses)
Micales, J. A., and Stipes, R. J. 1986. The differentiation of
 Endothia and *Cryphonectria* species by exposure to selected
 fungitoxicants. Mycotaxon 26:99-117. (discusses rationale for
 generic segregation)
Walker, J., Old, K. M., and Murray, D. I. L. 1985. *Endothia gyrosa* on
 Eucalyptus in Australia with notes on some other species of
 Endothia and *Cryphonectria*. Mycotaxon 23:353-370. (discusses the
 generic distinction, describes *E. gyrosa* and other *Endothia*
 species on *Eucalyptus*)
Barr (1978) places species of *Endothia* with one-septate ascospores in
Cryphonectria and presents justification for this separation based
on stromal structure and position of perithecia within the stroma.
These species are generally more pathogenic than *Endothia* species in
the restricted sense (Micales and Stipes, 1986). Anamorphs are
placed in *Endothiella*. *Cryphonectria parasitica* (Murrill) Barr is
the cause of chestnut blight (CMI 704) and *C. cubensis* (Bruner) C.
S. Hodges causes acute dieback of clove and *Eucalyptus* canker (CMI
449 as *Endothia eugeniae*).

CRYPTOCLINE Petr.
  Coelomycetes   15 spp.
Morgan-Jones, G. 1973. Genera coelomycetarum. VII. *Cryptocline*
 Petrak. Can. J. Bot. 51:309-325. (describes and illustrates 14
 Cryptocline spp.)
Petrini, O. 1984. *Cryptocline arctostaphyli* sp. nov., ein endophyt
 von *Arctostaphylos uva-ursi* und anderen Ericaceae. Sydowia
 37:238-241. (in German, description and illustration)
Sutton (1980) discusses the problem of delimiting genera and lists
species still placed in *Cryptocline*.

CRYPTODIAPORTHE Petr.
  Diaporthales   20 spp.
Butin, H. 1958. Ueber die auf *Salix* und *Populus* vorkommenden Arten der Gattung *Cryptodiaporthe* Petrak. Phytopathol. Z. 32:399-415. (in German with English summary, key to four species, descriptions and illustrations)
Barr (1978) treats 13 species and discusses others. *Dothichiza* canker of poplar is caused by *Cryptodiaporthe populea* (Sacc.) Butin (CMI 364). The anamorph is *Discosporium populeum* (Sacc.) Sutton (syns. *Dothichiza populea* Sacc. & Briard, *Chondroplea populea* (Sacc.) Kleb.). See *Diaporthe*, *Diplodina*, and *Dothichiza*.

CRYPTOSPHAERIA Grev.
  Diatrypales   15 spp.
Glawe, D. A., and Rogers, J. D. 1984. Diatrypaceae in the Pacific Northwest. Mycotaxon 20:401-460. (key to two *Cryptosphaeria* spp., taxonomy, biology, host and fungus index)
Hinds, T. E. 1981. *Cryptosphaeria* canker and *Libertella* decay of aspen. Phytopathology 71:1137-1145. (descriptions of the disease and its biology)
Literature on the Diatrypaceae is scattered and generally inadequate for identifying species, except on a regional basis. *Cryptosphaeria populina* (Pers.:Fr.) Sacc. causes a canker of *Populus* (Funk, 1981). See *Libertella*.

CRYPTOSPORELLA Sacc.
  Diaporthales   15 spp.
Nag Raj, T. R., and DiCosmo, F. 1981. A monograph of *Harknessia* and *Mastigosporella*. Bibl. Mycol. 80:1-62. (descriptions and illustrations of four *Cryptosporella* spp., presumptively teleomorphs of *Harknessia* spp.)
Barr (1978) reviews the genus and describes and illustrates *Cryptosporella hypodermia* (Fr.) Sacc. on *Ulmus* sp. *Cryptosporella anomala* (Peck) Sacc. causes *Corylus* blight, *C. umbrina* (Jenk.) Jenk. & Wehm. causes brown canker of rose, and *C. viticola* (Reddick) Shear causes dead-arm disease of *Vitis* spp. Anamorphs are placed in *Harknessia*.

CRYPTOSPORIOPSIS Bubak & Kab.
  Coelomycetes   10 spp.
Senula, A. 1985. [Studies on the morphology and physiology of *Cryptosporiopsis malicortis* (Cordl.) Nannf. and *Phlyctaena vagabunda* Desm.] Arch. Phytopathol. Pflanzenschutz 21:273-286. (in German with English summary, describes characters used to differentiate the two species)
Taylor, G. S. 1983. *Cryptosporiopsis* canker of *Acer rubrum*: Some relationships among host, pathogen, and vector. Plant Dis. 67:984-986. (associated with tree crickets)

CRYPTOSPORIOPSIS - cont.
Sutton (1980) includes ten species with brief descriptions and illustrations. These fungi are anamorphs of *Neofabraea, Ocellularia,* and *Pezicula. Cryptosporiopsis malicorticis* (Cordley) Nannf. (syn. *Gloeosporium perennans* Zeller & Childs) causes a perennial canker of apple.

CRYPTOSTICTIS See *Seimatosporium.*

CRYPTOSTROMA P. Gregory & S. Waller
    Hyphomycetes   1 sp.
Cryptostroma corticale (Ellis & Everh.) P. Gregory & S. Waller causes sooty bark disease of sycamore, *Acer* spp., and other hardwoods (CMI 539).

CUCURBITARIA S. F. Gray
    Pleosporales   40 spp.
Mirza, F. 1968. Taxonomic investigations on the ascomycetous genus *Cucurbitaria* S. F. Gray. Nova Hedw. 16:161-213. (keys to 23 species, descriptions and illustrations)
Sivanesan (1984) provides a key to 19 species. *Cucurbitaria staphula* Dearn. ex Arnold & Russell occurs as a secondary organism on galls and rough-bark of *Populus* spp. caused by *Diplodia tumefaciens* (F. Can. 17). Anamorphs of *Cucurbitaria* are placed in *Camarosporium, Dichomera, Diplodia,* and *Pyrenochaeta.*

CUMMINSIELLA Arth.
    Uredinales   8 spp.
Baxter, J. W. 1957. The genus *Cumminsiella.* Mycologia 49:864-873. (key to six species, descriptions and illustrations)
McCain, J. N., and Hennen, J. F. 1982. Is the taxonomy of *Berberis* and *Mahonia* (Berberidaceae) supported by their rust pathogens, *Cumminsiella santa* sp. nov. and other *Cumminsiella* species (Uredinales)? Syst. Bot. 7:48-59. (synoptic and dichotomous keys to eight species)
Species in this rust genus occur on *Mahonia* and *Berberis* in the Berberidaceae. *Cumminsiella mirabilissima* (Peck) Nannf., Lundell & Nannf. causes rust of *Mahonia* (CMI 261, F. Can. 288).

CUNNINGHAMELLA Matr.
    Mucorales   7 spp.
Lunn, J. A., and Shipton, W. A. 1983. Reevaluation of taxonomic criteria in *Cunninghamella* Trans. Br. Mycol. Soc. 81:303-312. (key to three species, amended descriptions, summary of characteristics)
Samson, R. A. 1969. Revision of the genus *Cunninghamella* (Fungi, Mucorales). Proc. K. Ned. Akad. Wet., Ser. C. 72:322-335. (key to seven species, descriptions and illustrations)

CUNNINGHAMELLA - cont.
Weitzman, I. 1984. The case for *Cunninghamella elegans, C. bertholletiae* and *C. echinulata* as separate species. Trans. Br. Mycol. Soc. 83:527-529. (argues for distinct taxa based on a number of characters)
Some species of *Cunninghamella* cause postharvest diseases of economically-important fruits and vegetables. *Cunninghamella echinulata* (Thaxter) Thaxter occurs on *Cucurbita* spp. (CMI 103).

CURVULARIA Boedijn
      Hyphomycetes   29 spp.
Benoit, M. A., and Mathur, S. B. 1970. Identification of species of *Curvularia* on rice seed. Proc. Int. Seed Test. Assoc. 35:99-119. (synoptic table with diagnostic characters of 12 species, illustrations)
Corbetta, G. 1965. Rassegna delle specie del genera *Curvularia*. Riso 14:1-23. (in Italian, synoptic table to 24 species, descriptions)
Ellis, M. B. 1966. Dematiaceous Hyphomycetes VIII. *Curvularia, Brachysporium*, etc. Mycol. Pap. 106:1-57. (key to 31 *Curvularia* spp., descriptions and illustrations)
Sivanesan, A. 1985. The teleomorph of *Curvularia tuberculata*. Trans. Br. Mycol. Soc. 84:548-551. (key to six *Cochliobolus* spp. using *Curvularia* anamorph characters)
Somal, B. S. 1976. A key to the species of *Curvularia*. Indian J. Mycol. & Plant Pathol. 6:59-64. (key to 15 species, brief discussion of species in India)
Tsuda, M., Nagakubo, T., Taga, M., and Ueyama, A. 1985. Sexuality for the teleomorph formation and conidial variability in *Curvularia lunata*. Trans. Mycol. Soc. Jpn. 26:27-39. (compares related species, references)
Tsuda, M., and Ueyama, A. 1985. Two new *Pseudocochliobolus* and a new species of *Curvularia*. Trans. Mycol. Soc. Jpn. 26:321-330. (compares four *Curvularia* taxa)
Curvularia is related to *Bipolaris, Drechslera,* and *Exserohilum*, all of which occur mostly as tropical and subtropical facultative plant pathogens. The genus is included in Ellis (1971, 1976). Teleomorphic states are placed in *Cochliobolus* and *Pseudocochliobolus*. Diseases caused by *Curvularia* species include: *Curvularia cymbopogonis* (Dodge) Groves & Skolko, seed and seedling blights, leaf spots of citronella and lemon grass (CMI 726 under *Cochliobolus cymbopogonis*); *C. geniculata* (Tracy & Earle) Boedijn, seed and seedling blights on many hosts (CMI 727 under *Cochliobolus geniculatus*); *C. lunata* (Wakker) Boedijn, leaf spots and seedling blight of monocots (CMI 474 under *Cochliobolus lunatus*); *C. spicifera* (Bainier) Boedijn, various diseases including spring dead spot of Bermuda grass (CMI 702 under *Cochliobolus spicifer*); and *C. trifolii* (Kauffm.) Boedijn f. sp. *gladioli* Parmelee & Luttrell, diseases of cultivated gladiolas (CMI 307).

CYCLANEUSMA DiCosmo, Peredo & Minter
Rhytismatales   2 spp.
Cannon, P. F., and Minter, D. W. 1986. The Rhytismataceae of the Indian subcontinent. Mycol. Pap. 155:1-123. (key to genera of Rhytismataceae, description and illustration of *C. minus*)
DiCosmo, F., Nag Raj, T. R., and Kendrick, W. B. 1984. A revision of the Phacidiaceae and related anamorphs. Mycotaxon 21:1-234. (discusses disposition of species previously placed in *Naemacyclus*, many now in *Cyclaneusma*)
DiCosmo, F., Peredo, H., and Minter, D. W. 1983. *Cyclaneusma* gen. nov., *Naemacyclus* and *Lasiostictis*, a nomenclatural problem resolved. Eur. J. For. Pathol. 13:206-212. (describes new genus for *Naemacyclus minor* and *N. niveus*, descriptions and illustrations)
DiCosmo et al. (1983, 1984) include *Cyclaneusma minus* (Butin) DiCosmo, Peredo & Minter (CMI 659 as *Naemacyclus minor*) and *C. niveus* (Pers.:Fr.) DiCosmo, Peredo & Minter (CMI 660 as *Naemacyclus niveus*), both causing needle casts of pine. Funk (1985) treats both species as *Naemacyclus*.

CYLINDROCARPON Wollenw.
Hyphomycetes   27 spp.
Booth, C. 1966. The genus *Cylindrocarpon*. Mycol. Pap. 104:1-56. (key, descriptions and illustrations)
Booth, C., and Evans, H. C. 1984. A new species of *Cylindrocarpon* associated with pine galls in Honduras. Trans. Br. Mycol. Soc. 82:745-748. (describes and illustrates *C. carneum*)
Booth, C., and Stover, R. H. 1974. *Cylindrocarpon musae* sp. nov., commonly associated with burrowing nematode (*Radopholus similis*) lesions on bananas. Trans. Br. Mycol. Soc. 63:503-507. (description and illustrations)
Gerlach, W., and Ershad, D. 1970. Beitrag zur Kenntnis der *Fusarium*-und *Cylindrocarpon*-arten in Iran. Nova Hedw. 20:725-784. (in German with English summary, descriptions and illustrations of four *Cylindrocarpon* spp.)
MacDonald, J. D., and Butler, E. E. 1981. *Cylindrocarpon* root rot of tulip poplar. Plant Dis. 65:154-157. (describes and illustrates *C. liriodendri*)
Samuels, G. J. 1978. Some species of *Nectria* having *Cylindrocarpon* imperfect states. N. Z. J. Bot. 16:73-82. (descriptions and illustrations of species with *Nectria* teleomorph)
Cylindrocarpon is similar to *Fusarium*, differentiated by the cylindrical conidia having rounded ends. Like *Fusarium*, these fungi are anamorphs of *Nectria* spp. Diseases caused by *Cylindrocarpon* species include: *Cylindrocarpon carneum* C. Booth & H. Evans, galls on *Pinus maximinoi* in Honduras (Booth and Evans, 1984); *C. cylindroides* Wollenw., canker of balsam fir and other Pinaceae (CMI

CYLINDROCARPON - cont.
623 under *Nectria macrospora*); *C. cylindroides* Wollenw. var. *tenue* Wollenw., wound parasite and dieback of Pinaceae (CMI 624); *C. destructans* (Zinssm.) Scholten, various diseases (CMI 148 under *Nectria radicicola*); *C. faginatum* C. Booth, beech bark disease (CMI 533 under *Nectria coccinea* var. *faginata*); and *C. heteronema* (Berk. & Broome) Wollenw., canker disease of apple trees (CMI 147 under *Nectria galligena* as *Cylindrocarpon mali*).

CYLINDROCLADIELLA Boesewinkel
    Hyphomycetes   5 spp.
Boesewinkel, H. J. 1982. *Cylindrocladiella*, a new genus to accommodate *Cylindrocladium parvum* and other small-spored species of *Cylindrocladium*. Can. J. Bot. 60:2288-2294. (key to five species, descriptions and illustrations)
This genus was established for species that are like *Cylindrocladium* but have one-septate conidia. The *Calonectria*-like teleomorphs with one-septate ascospores are placed in *Nectria*. *Cylindrocladiella camelliae* (Venkataramani & Ram) Boesewinkel causes root rot of tea (CMI 428 as *Cylindrocladium camelliae*).

CYLINDROCLADIUM Morg.
    Hyphomycetes   10 spp.
Boedijn, K. B., and Reitsma, J. 1950. Notes on the genus *Cylindrocladium*. Reinwardtia 1:51-60. (keys, descriptions and illustrations)
Mohanan, C., and Sharma, J. K. 1985. *Cylindrocladium* causing seedling diseases of *Eucalyptus* in Kerala, India. Trans. Br. Mycol. Soc. 84:538-539. (discusses two species)
Numerous diseases caused by members of this genus have been reported in the last ten years. The teleomorphs are placed in *Calonectria*. Species with one-septate conidia are placed in *Cylindrocladiella*. Species include: *Cylindrocladium braziliensis* (Batista & Cif.) Peerally, diseases of *Eucalyptus* spp. (CMI 427); *C. clavatum* C. S. Hodges & May, associated with root diseases of various hosts (CMI 422); *C. colhounii* Peerally, foliage diseases of tea (CMI 430 under *Calonectria colhounii*); *C. crotalariae* (Loos) Bell & Sobers, cause of collar rot of *Crotalaria* and *Cylindrocladium* black rot of peanuts (CMI 429 under *Calonectria crotalariae*); *C. floridanum* Sobers & Seymour, root rot of peach trees and diseases of other hosts (CMI 421 under *Calonectria kyotensis*); *C. hederae* (Arnaud) Peerally, disease of *Hedera helix* (CMI 426 under *Calonectria hederae*); *C. ilicicola* (Hawley) Boedijn & Reitsma, leaf spot of *Ilex aquifolia* (CMI 425); *C. quinqueseptatum* Boedijn & Reitsma, disease of clove seedlings (CMI 423 under *Calonectria quinqueseptata*); *C. scoparium* Morg., many diseases, especially of young *Eucalyptus* and pine (CMI 362); and *C. theae* (Petch) C. V. Subramanian, *Cercosporella* disease of tea bushes (CMI 424 under *Calonectria theae*).

CYLINDROSPORIUM Grev.
    Coelomycetes   1 sp.
Sutton (1980) includes *Cylindrosporium concentricum* Grev., the anamorph of *Pyrenopeziza brassicae* Sutton & Rawlinson, cause of light leaf spot of *Brassica* spp. (CMI 536). See *Blumeriella*.

CYMADOTHEA Wolf
    Dothideales   1 sp.
Barr (1972) recognizes this genus as *Mycosphaerella* section *Cymadothea* (Wolf) Arx. *Cymadothea trifolii* (Pers.:Fr.) Wolf causes black blotch or sooty blotch of clover (CMI 393). The anamorph is *Polythrincium trifolii* Kunze.

CYSTOSTEREUM Pouzar
    Aphyllophorales   6 spp.
Davidson, R. W., Campbell, W. A., and Lorenz, R. C. 1941. Association of *Stereum murrayi* with heart rot and cankers of living hardwoods. Phytopathology 31:82-87. (description of disease and fungus in culture)
Cystostereum murraii (Berk. & M. A. Curtis) Pouzar (syn. *Stereum murraii* (Berk. & M. A. Curtis) Burt) causes heart rot and cankers on living trees. See Aphyllophorales.

CYSTOTHECA Berk. & M. A. Curtis
    Erysiphales   3 spp.
Katumoto, K. 1973. Notes on the genera *Lanomyces* Gaeum. and *Cystotheca* Berk. & Curt. Rep. Tottori Mycol. Inst. 10:437-446. (key to three *Cystotheca* spp., synoptic table of characters, illustrations)
This genus is similar to *Sphaerotheca*.

CYTOPLEA Bizz. & Sacc.
    Coelomycetes   3 spp.
Kobayashi, T. 1965. *Neopycnodothis* (Sphaeroidaceae, Fungi Imperfecti), a new genus of bamboo inhabitant. Ann. Phytopathol. Soc. Jpn. 30:153-155. (descriptions and illustrations of the genus, now a synonym of *Cytoplea*)
Morgan-Jones, G. 1974. Icones Generum Coelomycetum VII. Univ. Waterloo, Biol. Ser. 14:1-42. (describes and illustrates *Cytoplea phyllostachydis* as *Neopycnodothis*)
Sutton (1980) describes and illustrates three species including *Cytoplea phyllostachydis* (T. Kobayashi) Sutton (syn. *Neopycnodothis phyllostachydis* T. Kobayashi) on bamboo.

CYTOSPORA Ehrenb.:Fr.
    Coelomycetes   100 spp.
Gaiova, V. P., and Merezhko, T. O. 1984. Cultural and morphological

CYTOSPORA - cont.
characteristics of *Cytospora* Ehr. ex Fr. species. Ukr. Bot. Zh. 41:55-59. (describes cultural characteristics of three species)
Grosclaude, C. 1979. Les chancres a *Cytospora* sur pecher dans la moyenne Vallee du Rhone. Rev. Zool. Agric. Pathol. Veg. (in French with English summary, compares *C. cincta* with *C. leucostoma* on peaches)
Gvritishvili, M. N. 1982. Fungi of the genus *Cytospora* in the USSR. Tbilici: Izdatelstve Sabchota Sakarstvelo. 214 pp. (in Russian)
Kastirr, U., and Ficke, W. 1984. *Cytospora personata* Fr.-ein bedeutender Rindenbranderreger am Apfel in der DDR. Arch. Phytopathol. Pflanzerschutz 20:383-399. (in German with English summary, compares four *Cytospora* spp. on apple)
Scharpf, R. F. 1983. Temperature-influenced growth and pathogenicity of *Cytospora abietis* on white fir. Plant Dis. 67:137-139. (description of canker)
Spielman, L. J. 1985. A monograph of *Valsa* on hardwoods in North America. Can. J. Bot. 63:1355-1378. (keys to genera and species, descriptions and illustrations of seven *Cytospora* spp. as anamorphs of *Valsa* spp., lists disposition of excluded *Cytospora* names)
Urban, Z. 1958. Revise ceskolovenskych zastpcu rodu *Valsa, Leucostoma*, a *Valsella*. Rozpr. Cesk. Akad. Ved., Rada Mat. Prir. Ved. 68:1-101. (in Czechoslovakian, keys to teleomorphs, detailed descriptions of both teleomorphs and anamorphs, host lists)
Members of this genus are anamorphs of *Valsa*. At present no workable system exists for species identification. The only comprehensive key is by Gvritishvili (1982), an excellent work but only available in Russian. Grove (1935) is outdated but useful. Spielman (1985) and Urban (1958) deal primarily with the *Valsa* teleomorph but include descriptions of the *Cytospora* anamorph. *Cytospora sacchari* E. Butler causes sheath rot of sugarcane (CMI 777).

DACTULIOPHORA Leakey
Agonomycetes   4 spp.
Datnoff, L. E., Levy, C., Naik, D. M., and Sinclair, J. B. 1986. *Dactuliophora glycines*, a sclerotial state of *Pyrenochaeta glycines*. Trans. Br. Mycol. Soc. 87:297-301. (describes and illustrates fungus and disease, presents evidence for the connection between *D. glycines* and *P. glycines*)
Leakey, C. L. A. 1964. *Dactuliophora*, a new genus of mycelia sterilia from tropical Africa. Trans. Br. Mycol. Soc. 47:341-350. (describes and illustrates four species)
These species cause leaf spots of Poaceae and Fabaceae, primarily in Africa. *Dactuliophora glycines* Leakey is the sclerotial state of *Pyrenochaeta glycines* R. B. Stewart.

DACTYLARIA Sacc.
Hyphomycetes   35 spp.
Bhatt, G. C., and Kendrick, W. B. 1968. The generic concepts of *Diplorhinotrichum* and *Dactylaria*, and a new species of *Dactylaria* from soil. Can. J. Bot. 46:1253-1257. (*Diplorhinotrichum* considered a synonym of *Dactylaria*, describes one new species, illustrations)
Chowdhry, P. N. 1982. Notes on Indian Hyphomycetes-IV. *Dactylaria arundica* sp. nov. Indian Phytopathol. 35:532-534. (compares one-septate species)
Hoog, G. S. de 1985. Taxonomy of the *Dactylaria* complex. IV. *Dactylaria, Neta, Subulispora* and *Scolecobasidium*.   Stud. Mycol. 26:1-60. (keys, descriptions and illustrations)
Hoog, G. S. de, Hennebert, G. L., and Hijwegen, T. 1983. Taxonomy of the *Dactylaria* complex, III. A pleomorphic species of *Isthmolongispora*. Proc. K. Ned. Akad. Wet., Ser. C. 86:343-346. (description and illustration)
Hoog, G. S. de, and Oorschot, C. A. N. van 1983. Taxonomy of the *Dactylaria* complex. I. Notes on the genus *Dichotomophthora*. Proc. K. Ned. Akad. Wet., Ser. C. 86:55-61. (treats two species, revision of taxonomy, descriptions and illustrations)
Hoog, G. S. de, and Oorschot, C. A. N. van 1985. Taxonomy of the *Dactylaria* complex, VI. Key to the genera and checklist of epithets. Stud. Mycol. 26:97-122. (key to genera, annotated checklist of epithets, illustrations)
Hoog, G. S. de, Oorschot, C. A. N. van, and Hijwegen, T. 1983. Taxonomy of the *Dactylaria* complex. II. *Dissoconium* gen. nov. and *Cordana* Preuss. Proc. K. Ned. Akad. Wet., Ser. C. 86:197-206. (key, descriptions and illustrations)
Onofri, S., and Zucconi, L. 1984. Rare or interesting hyphomycetes from tropical forest litter V. *Dactylaria fusiformis*. Notes on the generic concept of *Dactylaria*. Mycotaxon 19:523-528. (description and illustration)
Oorschot, C. A. N. van 1985. Taxonomy of the *Dactylaria* complex, V. A review of *Arthrobotrys* and allied genera. Stud. Mycol. 26:61-96. (keys to 38 taxa in this complex, descriptions and illustrations)
Schenck, S., Kendrick, W. B., and Pramer, D. 1977. A new nematode-trapping hyphomycete and a reevaluation of *Dactylaria* and *Arthrobotrys*. Can. J. Bot. 55:977-985. (transfers many *Dactylaria* spp. to *Arthrobotrys*)
This genus includes many lignicolous, weakly pathogenic species. Some nematode-trapping fungi were described in *Dactylaria* but have been transferred to *Arthrobotrys*. The complex of genera has been treated in a series of papers by de Hoog and others as cited above. Ellis (1976) includes a key to five species.

DALDINIA Ces. & De Not.
  Sphaeriales   13 spp.
Child, M. 1932. The genus *Daldinia*. Ann. Missouri Bot. Gard. 19:429-496. (key to 13 species, descriptions and illustrations, outdated)
Eckblad, F.-E. 1969. The genera *Daldinia, Ustulina* and *Xylaria* in Norway. Nytt Magn. Bot. 16:139-145. (distribution of *D. concentrica* in Norway, discussion)
Martin, P. 1969. Studies in the Xylariaceae: VI. *Daldinia, Nummulariola* and their allies. J. S. Afr. Bot. 35:267-320. (key to 13 *Daldinia* spp., descriptions and illustrations)
Perez-Silva, E. 1973. El genero *Daldinia* (Pyrenomycetes) en Mexico. Bol. Soc. Mex. Micol. 7:51-58. (in Spanish, describes cultures of *D. occidentalis*)
Petrini, L. E., and Mueller, E. 1986. Teleomorphs and anamorphs of European species of *Hypoxylon* (Xylariaceae, Sphaeriales) and allied genera. Mycol. Helv. 1:501-627. (in German with English, French, and Italian summaries, key to five *Daldinia* spp., descriptions and illustrations)
Petrini, L., and Petrini, O. 1985. Xylariaceous fungi as endophytes. Sydowia 38:216-234. (keys to European xylariaceous endophytes based on cultural characters, describes culture of *D. occidentalis*)
Thind, K. S., and Dargan, J. S. 1978. Xylariaceae of India - IV. The genus *Daldinia*. Kavaka 6:15-24. (key to six taxa, descriptions and illustrations)
Daldinia spp. are saprophytic on deciduous trees.

DARKERA Whitney, J. Reid & Pirozynski
  Rhytismatales   2 spp.
DiCosmo, F., Nag Raj, T. R., and Kendrick, W. B. 1984. A revision of the Phacidiaceae and related anamorphs. Mycotaxon 21:1-234. (key to two *Darkera* spp., descriptions and illustrations)
Whitney, H. S., Reid, J., and Pirozynski, K. A. 1975. Some new fungi associated with needle blight of conifers. Can. J. Bot. 53:3051-3063. (descriptions and illustrations of *D. abietis*, *D. parca*, and their *Tiarosporella* anamorphs)
Species in this genus cause needle blight of conifers and have *Tiarosporella* anamorphs.

DAVISOMYCELLA Darker
  Rhytismatales   7 spp.
Darker, G. D. 1967. A revision of the genera of the Hypodermataceae. Can. J. Bot. 45:1399-1444. (key to genera of Hypodermataceae, lists six *Davisomycella* spp.)
Darker, G. D. 1967. A new *Davisomycella* species on *Pinus banksiana*. Can. J. Bot. 45:1445-1449. (describes and illustrates *D. fragilis*)

DAVISOMYCELLA - cont.
This genus contains fungi on needles of *Pinus* spp., many of which were formerly placed in *Hypodermella*. *Davisomycella ampla* (J. J. Davis) Darker causes periodic acute defoliation of *P. banksiana* in Canada (CMI 561). See *Lophodermium*.

DEIGHTONIELLA S. J. Hughes
    Hyphomycetes   10 spp.
Constantinescu, O. 1983. Deightoniella on *Phragmites*. Proc. K. Ned. Akad. Wet., Ser. C. 86:137-141. (compares two species on *Phragmites*)
Ellis, M. B. 1957. Some species of *Deightoniella*. Mycol. Pap. 66:1-12. (key to five species, descriptions and illustrations)
Ondrej, M. 1984. The genus *Deightoniella* Hughes in Czechoslovakia. Ceska Mykol. 38:39-45. (in Czechoslovakian with English summary, describes three species)
Pollack, F. G., and Matthews, F. D. 1976. *Deightoniella argemonensis*, a new fungus on Mexican pricklepoppy, associated with *Cercosporidium guanicense*. Mycologia 68:1093-1097. (description and illustration)
Species of *Deightoniella* occur on various plants especially in the tropics. Ellis (1971, 1976) provides a key to eight species and describes a new species. *Deightoniella argemonensis* Pollack & Matthews causes a leaf spot of *Argemone mexicana*, Mexican pricklepoppy; *D. arundinacea* (Corda) S. J. Hughes occurs on *Phragmites communis*; *D. papuana* D. Shaw causes veneer blotch of *Saccharum* spp.; and *D. torulosa* (Sydow) M. B. Ellis causes black tip disease of bananas and other diseases of *Musa textilis* (CMI 165).

DENDROPHOMA Sacc.
    Coelomycetes   5 spp.
Sutton, B. C. 1965. Typification of *Dendrophoma* and a reassessment of *D. obscurans*. Trans. Br. Mycol. Soc. 48:611-616. (description and illustration of *Dinemasporium [Dendrophoma]cytosporoides* and *Phomopsis [Dendrophoma] obscurans*)
Dendrophoma is a synonym of *Dinemasporium*, however, most species await revisionary work (Sutton, 1965). Grove (1935) includes a key to five British species. Sivanesan (1984) uses *Dendrophoma* for anamorphs of some *Didymosphaeria* and *Keissleriella* spp. See also *Phomopsis*.

DENDROSEPTORIA Bausa Alcalde
    Coelomycetes   2 spp.
Punithalingam, E. 1981. New microfungi from cereals and grasses. II. Nova Hedw. 34:67-95. (key to two *Dendroseptoria* spp., descriptions and illustrations)
Dendroseptoria oryzopsidis Punithalingam occurs in lesions on *Oryzopsis miliacea* with other fungi suggesting that it is weakly

DENDROSEPTORIA - cont.
pathogenic (Punithalingam, 1981). Sutton (1980) treats *D. arrhenatheri* Bausa Alcalde.

DENDRYPHION Wallr.
    Hyphomycetes   4 spp.
Ellis (1971, 1976) treats four species including *Dendryphion penicillatum* (Corda) Fr., the anamorph of *Pleospora papaveracea* (De Not.) Sacc., the cause of leaf blight of opium poppy (CMI 730).

DERMATODOTHIS Racib.
    Pleosporales   6 spp.
Katumoto, K. 1983. Notes on some plant-inhabiting Ascomycotina from western Japan (3). Trans. Mycol. Soc. Jpn. 24:259-269.
    (describes three *Dermatodothis* spp.)
Mueller, E. (1975)1976. Die Ascomycetengattung *Dermatodothis* Raciborski. Sydowia 28:148-154. (in German with English summary, key to four species, descriptions and illustrations)
Species of *Dermatodothis* usually occur on living leaves. This genus is included in Arx and Mueller (1975).

DERMEA Fr.
    Helotiales   22 spp.
Funk, A. 1976. The genus *Dermea* and related conidial states on Douglas fir. Can. J. Bot. 54:2852-2856. (describes and illustrates two *Dermea* spp. and four *Micropera* spp.)
Groves, J. W. 1946. North American species of *Dermea*. Mycologia 38:351-431. (key to 16 species, descriptions and illustrations)
Species of *Dermea* occur on hardwoods and as weak pathogens of conifers. Funk (1981) treats four species. Many have anamorphs in *Gelatinosporium* and in *Foveostroma* described as *Micropera* spp.

DEUTEROPHOMA   See *Phoma*.

DIACHORA J. Mueller
    Polystigmatales   4 spp.
Mueller, E. 1986. On the genus *Diachora* J. Mueller (Ascomycetes). Trans. Bot. Soc. Edinburgh, 150th Anniv. Suppl., Pp. 69-75. (key to four species, descriptions and illustrations, host index)
Diachora species are parasitic on members of the Fabaceae and have *Diachorella* anamorphs.

DIAPORTHE Nitschke
    Diaporthales   75 spp.
Gilman, J. C., Tiffany, L. H., and Lewis, R. M. 1959. Iowa Ascomycetes III. Diaporthaceae: Diaportheae. Iowa State Coll. J. Sci. 33:325-393. (key to 12 genera, treats 27 *Diaporthe* spp., descriptions and illustrations)

DIAPORTHE - cont.

Kobayashi, T. 1970. Taxonomic studies of Japanese Diaporthaceae with special reference to their life-histories. Bull. Gov. For. Exp. Stn. (Jpn.) 226:1-242. (key to genera of Diaporthaceae, keys to species, key to 19 *Diaporthe* spp., descriptions and illustrations)

Wehmeyer, L. E. 1933. The genus *Diaporthe* Nitschke and its segregates. Univ. Mich. Stud. Sci. Ser. 9:1-349. (outdated but useful for identification, key to related genera, key to 70 species, descriptions and illustrations)

A modern revision of this large, important genus is needed. The literature is scattered, outdated, or incomplete and deals primarily with the *Phomopsis* anamorph. Wehmeyer (1933) is still the most useful reference in combination with Kobayashi (1970) for the identification of these fungi. Diseases include: *Diaporthe alleghaniensis* R. Arnold, on *Betula* spp. (F. Can. 70); *D. capsici* Punithalingam, dieback and fruit rot of *Capsicum* spp. (CMI 733); *D. citri* Wolf, melanose of *Citrus* spp. (CMI 396); *D. manihotis* Punithalingam, leaf spot of cassava (CMI 734); *D. phaseolorum* (Cooke & Ellis) Sacc., diseases of soybean, dry rot of sweet potato, and pod blight of lima bean (CMI 336); and *D. woodii* Punithalingam, *Phomopsis* stem blight of lupine and lupinosis of sheep (CMI 476).

DIAPORTHOPSIS Fabre
    Diaporthales   8 spp.

Bonar, L. 1966. A new *Diaporthopsis* causing brooming in *Baccharis*. Am. J. Bot. 53:181-184. (describes and illustrates *D. sclerophila*)

Mueller, E., and Ahmad, S. 1955. New or noteworthy Ascomycetes from Pakistan I. Sydowia 9:233-245. (in German, description and illustration of *D. spiraeae*)

Roane, M. K., and Fosberg, F. R. 1983. A new pyrenomycete associated with *Metrosideros collina* subspecies *polymorpha* (Myrtaceae). Mycologia 75:163-166. (describes and illustrates *D. metrosideri*, now placed in *Endothia* by Barr, 1983)

Wehmeyer, L. E. 1933. The genus *Diaporthe* Nitschke and its segregates. Univ. Mich. Stud. Sci. Ser. 9:1-349. (outdated but useful for identification, key to six *Diaporthopsis* spp., descriptions and illustrations)

Diaporthopsis is a segregate of *Diaporthe*. Four species are treated by Arx and Mueller (1954).

DIATRACTIUM Sydow
    Diaporthales   2 spp.

Baker, R. E. D., and Dale, W. T. 1951. Fungi of Trinidad and Tobago. Mycol. Pap. 33:1-123. (lists two *Diatractium* spp., illustration)

Diatractium cordiana (Ellis & Kelsey) Sydow is parasitic on *Cordia* leaves and *D. ingae* (Allesch.) Sydow occurs on leaves of *Inga*. Both

DIATRACTIUM - cont.
species are treated in Mueller and Arx (1962) and Barr (1978).

DIATRYPE Fr.
  Diatrypales   50 spp.
Glawe, D. A., and Rogers, J. D. 1982. Observations on the anamorphs of six species of *Diatrype* and *Diatrypella*. Can. J. Bot. 60:245-251. (descriptions and illustrations)
Glawe, D. A., and Rogers, J. D. 1984. Diatrypaceae in the Pacific Northwest. Mycotaxon 20:401-460. (key to genera, key to ten *Diatrype* and similar species, taxonomy, biology, host/fungus index)
Patil, M. S., and Patil, S. D. 1983. Studies in Pyrenomycetes of Maharashtra-II. Genus *Diatrype*. Indian J. Mycol. & Plant Pathol. 13:134-178. (treats 22 species with brief discussion and host information)
Rogers, J. D., and Glawe, D. A. 1983. *Diatrype whitmanensis* sp. nov. and the anamorphs of *Diatrype bullata* and *Eutypella sorbi*. Mycotaxon 18:73-80. (description and illustrations)
Tiffany, L. H., and Gilman, J. C. 1965. Iowa Ascomycetes IV. Diatrypaceae. Iowa State Coll. J. Sci. 40:121-161. (key to four genera, key to two *Diatrype* spp, descriptions and illustrations)
Species of *Diatrype* and Diatrypaceae occur on dead or declining woody angiosperms; some are pathogenic. Glawe and Rogers (1984) discuss taxa found in the Pacific Northwest, including their anamorphs. *Diatrype albopruinosa* (Schwein.) Cooke occurs on various hardwoods (F. Can. 72), and *D. virescens* (Schwein.) Ravenel is found on the dead branches and twigs of *Fagus* (F. Can. 73). See *Eutypa* and Ascomycotina.

DIATRYPELLA (Ces. & De Not.) Sacc.
  Diatrypales   30 spp.
Croxall, H. E. 1950. Studies on British Pyrenomycetes III. The British species of the genus *Diatrypella* Cesati & De Notaris. Trans. Br. Mycol. Soc. 33:45-72. (describes two *Diatrypella* spp. with anamorphs, illustrations)
Glawe, D. A. 1986. Taxonomic notes on *Diatrypella discoidea, Diatrypella decorata*, and *Diatrypella placenta*. Mycotaxon 25:19-25. (descriptions, taxonomy, nomenclature)
Glawe, D. A., and Rogers, J. D. 1982. Observations on the anamorphs of six species of *Diatrype* and *Diatrypella*. Can. J. Bot. 60:245-251. (descriptions and illustrations)
Glawe, D. A., and Rogers, J. D. 1984. Diatrypaceae in the Pacific Northwest. Mycotaxon 20:401-460. (key to seven *Diatrypella* spp., taxonomy, biology, host/fungus index)
This genus is similar to *Diatrype* but the asci contain more than eight ascospores. The anamorphs are discussed by Glawe and Rogers (1982, 1984).

DIBOTRYON   See *Apiosporina*.

DICELLOMYCES Olive
   Dacrymycetales   2 spp.
Cunningham, J. L., Bakshi, B. K., Lentz, P. L., and Gilliam, M. S. 1976. Two new genera of leaf-parasitic fungi (Basidiomycetidae: Brachybasidiaceae). Mycologia 68:640-654. (description and illustrations of *Ceraceosorus bombacis*, key to four genera of Brachybasidiaceae)
Olive, L. S. 1945. A new *Dacrymyces*-like parasite of *Arundinaria*. Mycologia 37:543-552. (description and illustrations of *Dicellomyces gloeosporus*)
Reid, D. A. 1976. *Dicellomyces scirpi* (Basidiomycetes) - new to Britain. Trans. Br. Mycol. Soc. 66:536-538. (description and illustrations)
Dicellomyces gloeosporus Olive is parasitic on *Arundinaria*, and *D. scirpi* Raitviir is parasitic on *Scirpus*. *Ceraceosorus bombacis* (Bakshi) Bakshi (syn. *Dicellomyces bombacis* Bakshi) causes a leaf spot of *Bombax*.

DICHOMITUS Reid
   Aphyllophorales   2 spp.
Dichomitus squalens (P. Karst.) Reid (syn. *Polyporus anceps* Peck) causes a white pocket rot in living and dead conifers. See Aphyllophorales.

DICHOTOMOPHTHORA Mehrlich & Fitzp. ex M. B. Ellis
   Hyphomycetes   2 spp.
Hoog, G. S. de, and Oorschot, C. A. N. van 1983. Taxonomy of the *Dactylaria* complex. I. Notes on the genus *Dichotomophthora*. Proc. K. Ned. Akad. Wet., Ser. C. 86:55-61. (treats two species, descriptions and illustrations)
Rao, P. N. 1966. A new species of *Dichotomophthora* on *Portulaca oleracea* from Hyderabad-India. Mycopathol. Mycol. Appl. 28:137-140. (includes two species, both invalid fide Ellis, 1971)
Ellis (1971) gives a description and illustration of *Dichotomophthora portulacae* which causes a leaf spot of purslane. See de Hoog and van Oorschot (1983) for the most recent treatment of the genus.

DICYMA   See *Ascotricha*.

DIDYMELLA Sacc. ex Sacc.
   Pleosporales   75 spp.
Corbaz, R. 1957. Recherches sur le genre *Didymella* Sacc. Phytopathol. Z. 28:375-414. (in French, key to 19 species, descriptions and illustrations)
Corlett, M. 1981. A taxonomic survey of some species of *Didymella* and

DIDYMELLA - cont.
Didymella-like species. Can. J. Bot. 59:2016-2042. (key to 15 species, descriptions, illustrations, host index)
Holm, L. 1953. Taxonomical notes on ascomycetes. III. The herbicolous Swedish species of the genus *Didymella* Sacc. Sven. Bot. Tidskr. 47:520-525. (key to ten species)
Mueller, E. 1952. Pilzliche Erreger der Getreideblattduerre. Phytopathol. Z. 19:403-416. (in German with English summary, *D. caudata* and *D. exitialis*)
Walker, J., and Baker, K. F. 1983. The correct binomial for the chrysanthemum ray blight pathogen in relation to its geographical distribution. Trans. Br. Mycol. Soc. 80:31-38. (reports *Didymella ligulicola* the correct name for ray blight pathogen, compares with *D. chrysanthemi*, descriptions and illustrations)
Didymella species occur as saprophytes, parasites, or hyperparasites on other fungi. The anamorphs are placed in *Ascochyta*, *Phloeospora*, *Phoma*, and *Stagonospora*. Corlett (1981) treats 15 species and Sivanesan (1984) includes 29 species and their anamorphs. See the similar genus *Mycosphaerella*. Diseases caused by *Didymella* species include: *Didymella applanata* (Niessl) Sacc., spur blight of *Rubus* spp. (CMI 735, F. Can. 49); *D. bryoniae* (Auersw.) Rehm (syns. *Mycosphaerella citrullina* Gross., *M. melonis* (Pass.) Chiu & J. C. Walker), gummy stem blight of Cucurbitaceae (CMI 332, F. Can. 303); *D. delphinii* Earle, on *Arnica latifolia* (F. Can. 131); *D. exitialis* (Morini) E. Mueller, leaf scorch of barley and wheat (CMI 633); *D. ligulicola* (K. Baker, Dimock & L. H. Davis) Arx, ray blight of chrysanthemum (CMI 622 as *D. chrysanthemi*); and *D. lycopersici* Kleb., stem and fruit rot of tomato (CMI 272).

DIDYMOSPHAERIA Fuckel
    Pleosporales   20 spp.
Scheinpflug, H. 1958. Untersuchungen ueber die Gattung *Didymosphaeria* und einige verwandte Gattungen. Ber. Schweiz Bot. Ges. 68:325-385. (in German with English summary, key to 18 species, descriptions and illustrations)
Didymosphaeria species are saprophytic, parasitic, or hyperparasitic on other fungi. Sivanesan (1984) includes a key to five species and Mueller and Arx (1962) treat eight species. Species causing disease include: *Didymosphaeria arachidicola* (Chockrjakov) Alcorn, Punithalingam & McCarthy, web-blotch of peanut (CMI 736); *D. brunneola* Niessl, severe blight of asparagus fern; *D. donacina* (Niessl) Sacc., leaf spot of cluster yam; *D. oregonensis* Gooding, canker of *Alnus* sp.; *D. populina* Vuill., leaf lesions of *Populus* sp.; and *D. taiwanensis* Yen & Chi, leaf blast of sugarcane.

DIDYMOSPORINA Hoehn.
    Coelomycetes   2 spp.
Sutton (1980) includes *Didymosporina aceris* (Lib.) Hoehn. (syn.

DIDYMOSPORINA - cont.
Marsonnina truncatula (Sacc.) P. Magn.), the cause of leaf blight of *Acer* spp., and mentions *D. africana* H. Sydow described from leaves of *Rhus viminalis* in Africa.

DILOPHIA   See *Lidophia.*

DILOPHOSPORA Desmaz.
    Coelomycetes   1 sp.
Walker, J. 1980. *Gaeumannomyces, Linocarpon, Ophiobolus,* and several other genera of scolecospored Ascomycetes and *Phialophora* conidial states, with a note on hyphopodia. Mycotaxon 11:1-129. (discussion of possible genetic connection of *D. alopecuri* with *Lidophia graminis*)
Walker, J., and Sutton, B. C. 1974. *Dilophia* Sacc. and *Dilophospora* Desm. Trans. Br. Mycol. Soc. 62:231-241. (describes and illustrates *D. alopecuri*, discusses other species)
Sutton (1980) includes *Dilophospora alopecuri* (Fr.:Fr.) Fr., the cause of twist of grasses (CMI 490), the teleomorph of which is probably *Lidophia graminis* (Sacc.) J. Walker & Sutton.

DIMERIELLA Speg.
    Pleosporales   1 sp.
Farr, M. L. 1979. The didymosporous dimeriaceous fungi described on Asteraceae. Mycologia 71:243-271. (discusses status of genus, description and illustration of *Didymella hirtula*)
Punithalingam, E. 1984. New microfungi from Malaysia and Papua New Guinea. Nova Hedw. 39:57-74. (reviews taxonomy of *Dimeriella, Lasiostemma, Wentiomyces* and related genera)
The generic limits of *Dimeriella* are obscure; many species have been transferred to other genera (Farr, 1979; Punithalingam, 1984). *Dimeriella sacchari* (Breda da Haan) Hansf. ex Abbott causes red leaf spot of sugarcane (CMI 775).

DINEMASPORIUM Lev.
    Coelomycetes   3 spp.
Morgan-Jones, G. 1971. A new species of *Dinemasporium* from Ontario. Can. J. Bot. 49:1363-1365. (compares *D. canadense* with similar species)
Nag Raj, T. R., and Kendrick, B. 1986. On *Dinemasporium adeanum* Petrak. Mycotaxon 25:15-18. (description and illustration, considered a synonym of *Vermiculariopsiella immersa*)
Sutton, B. C. 1969. *Minimidochium setosum* n. gen., n. sp. and *Dinemasporium aberrans* n. sp. from West Africa. Can. J. Bot. 47:2095-2100. (descriptions and illustrations)
Webster, J. 1955. Graminicolous Pyrenomycetes. V. Conidial states of *Leptosphaeria michotii, L. microscopica, Pleospora vagans* and the perfect state of *Dinemasporium graminum.* Trans. Br. Mycol.

DINEMASPORIUM - cont.
Soc. 38:347-365. (treats *D. graminum*, now a synonym of *D. strigosum* [Sutton, 1980], describes and illustrates the teleomorph *Phomatospora dinemasporium*)
Sutton (1980) treats three species. See *Dendrophoma*.

DIPLOCARPON Wolf
    Helotiales   6 spp.
No comprehensive treatment of this genus exists. *Diplocarpon earliana* (Ellis & Everh.) Wolf causes strawberry leaf scorch (CMI 486), *D. mespili* (Sorauer) Sutton causes leaf blight of Rosaceae (CMI 481 as *D. maculatum*, syn. *Fabraea maculata* Atk.), and *D. rosae* Wolf causes black spot of roses (CMI 485). See the anamorphs *Entomosporium* and *Marssonina*.

DIPLODIA Fr.
    Coelomycetes   24 spp.
Laundon, G. 1984. *Diplodia pittospororum* and *Diplodia pittospori*. Trans. Br. Mycol. Soc. 83:164-166. (tranfers *D. pittospororum* to *Microsphaeropsis*, considers *D. pittospori* a synonym of *Diplodia mutila*, descriptions and illustrations)
Wollenweber, H. W., and Hochapfel, H. 1943. Beitraege zur Kenntnis parasitaerer und saprophytischer Pilze. V. 1. *Diplodia* und ihre Beziehung zur Fruchtfaeule. Z. Parasitenkd. 12:165-250. (in German, descriptions and illustrations of eight species)
Zambettakis, C. 1954. Recherches sur la systematique des Sphaeropsidales-Phaeodidymae. Bull. Trimest. Soc. Mycol. Fr. 70:219-350. (in French, nomenclature generally not accepted, lacks adequate keys and descriptions)
Many species have been placed indiscriminately in *Diplodia*. The group is badly in need of revision. Sutton (1980) treats only the type species. Although the names are generally outdated, Grove (1937) includes many *Diplodia* species arranged alphabetically by host. Sivanesan (1984) includes *Diplodia* spp. with known teleomorphs which include *Botryosphaeria, Cucurbitaria*, and *Otthia*. See also *Sphaeropsis* and *Stenocarpella*.

DIPLODINA Westendorp
    Coelomycetes   3 spp.
Sutton (1980) treats three *Diplodina* spp. with *Cryptodiaporthe* teleomorphs. Many species previously placed in *Diplodina* have been transferred to other genera such as *Ascochyta* and *Phoma*.

DIPLORHINOTRICHUM   See *Dactylaria*.

DISCELLA   See *Sirococcus*.

DISCOCHORA   See *Guignardia*.

DISCOGLOEUM Petr.
    Coelomycetes   3 spp.
Morgan-Jones, G. 1971. An addition to the genus *Discogloeum* Petrak. Can. J. Bot. 49:1461-1462. (transfers *Exosporium concentricum* to *Discogloeum*, description and illustration, discussion of two other species)
Sutton (1980) includes *Discogloeum veronicae* (Lib.) Petr. the cause of leaf lesions on *Veronica tournefortii*. He indicates that this genus is similar to *Cylindrosporium* differing in the more irregularly-shaped conidia and prominent phialidic apparatus of *Discogloeum*.

DISCOSIA Lib.
    Coelomycetes   20 spp.
Chandra-Reddy, K. R. 1984. Taxonomic study of the genus *Discosia* Libert. Pages 493-501 in: Taxonomy of Fungi. Part 2. C. V. Subramanian, ed. Amra Press, Madras. (lists about 13 names in *Discosia* and their taxonomic disposition)
Subramanian, C. V., and Chandra-Reddy, K. R. 1974. The genus *Discosia* I. Taxonomy. Kavaka 2:57-89. (type studies, descriptions)
Most species of *Discosia* are saprophytic although *Discosia strobilina* Lib. (syn. *D. theae* Cavara) may be pathogenic on conifers (Sutton, 1980).

DISCOSPORIUM Hoehn.
    Coelomycetes   3 spp.
Sutton (1980) treats three species, including *Discosporium populeum* (Sacc.) Sutton, the cause of *Dothichiza* canker of poplar (CMI 364 as *Chondroplea populea* under *Cryptodiaporthe populea*). Teleomorphs belong in *Cryptodiaporthe*.

DISCOSTROMA Clements
    Sphaeriales   8 spp.
Brockmann, I. 1975. Untersuchungen ueber die Gattung *Discostroma* Clements (Ascomycetes). Sydowia 28:275-338. (in French with English summary, key to eight species, descriptions and illustrations)
These fungi are usually found as the *Seimatosporium* anamorph.

DISCULA Sacc.
    Coelomycetes   12 spp.
Arx, J. A. von 1970. A revision of the fungi classified as *Gloeosporium*. Bibl. Mycol. 24:1-203. (key to related genera, includes 12 *Discula* spp., descriptions and illustrations)
Petrak, F. 1962. Ueber die Gattung *Discula* Sacc. Sydowia 15:221-223. (in German, amends genus, describes *D. melanotricha*)
Petrak, F. (1970)1971. Kritische Bemerkungen zur Nomenklatur der Gattung *Discula* Sacc. Sydowia 24:270-273. (in German,

DISCULA - cont.
nomenclature)
Salogga, D. S., and Ammirati, J. F. 1983. *Discula* species associated with anthracnose of dogwood in the Pacific Northwest. Plant Dis. 67:1290. (new disease)
Species of *Discula* cause leaf spots and wilt of twigs. *Discula umbrinella* (Berk. & Broome) Sutton occurs on *Platanus*, *Quercus*, and other broad-leaved trees (Sutton, 1980). The teleomorphs belong to *Apiognomonia* and *Gnomonia* (Barr, 1978).

DOTHICHIZA Lib. ex Roum.
   Coelomycetes 15 spp.
Alfieri, S. A., Jr. 1982. *Dothichiza* leaf spot of blueberries. Fla. Dep. Agric., Div. Plant Ind., Plant Pathol. Circular 242:1-2. (describes and illustrates *D. caroliniana*)
Petrak, F. 1957. Ueber die Gattungen *Dothichiza* Lib. und *Chrondroplea* Kleb. Sydowia 10:201-235. (in German, describes *D. sorbi*)
Taylor, J., and Clayton, C. N. 1959. Comparative studies on *Gloeosporium* stem and leaf fleck and *Dothichiza* leaf spot of high bush blueberry. Phytopathology 49:65-67. (illustrates *D. caroliniana*)
Sutton (1977, 1980) discusses the nomenclatural problems in this genus and includes several species in other genera. Barr (1972) and Sivanesan (1984) list *Dothichiza* anamorphs for several *Dothiora* spp. See *Cryptodiaporthe*.

DOTHIDEA Fr.:Fr.
   Dothideales 20 spp.
Loeffler, W. 1957. Untersuchungen ueber die Ascomyceten-Gattung *Dothidea* Fr. Phytopathol. Z. 30:349-386. (in German, key to eight species, biology, nomenclature)
Dothidea has been redefined by Barr (1972) and Mueller and Arx (1962), who treat eight species. Sivanesan (1984) includes many names in other genera.

DOTHIORA Fr.
   Dothideales 14 spp.
Froidevaux, L. 1972. Contribution a l'etude des Dothioracees (Ascomycetes). Nova Hedw. 23:679-734. (in French, key to 14 *Dothiora* spp., descriptions and illustrations)
Barr (1972) treats 11 species with their *Dothichiza* anamorphs; Sivanesan (1984) provides a key to nine species. *Dothiora ribesia* (Fr.:Fr.) Barr causes black pustule of currant (CMI 707).

DOTHIORELLA Sacc.
   Coelomycetes 50 spp.
Maas, J. L., and Uecker, F. A. 1984. *Botryosphaeria dothidea* cane canker of thornless blackberry. Plant Dis. 68:720-726.

DOTHIORELLA - cont.
(description and illustrations of anamorph of *B. dothidea*, pathology)
Shahin, E. A., and Claflin, L. E. 1980. Twig blight of Douglas fir: a new disease caused by *Dothiorella dothidea*. Plant Dis. 64:47-50. (pathology, description and illustrations)
The taxonomic status of *Dothiorella* is in question but the genus has been used for anamorphs of *Botryosphaeria* spp. See also *Fusicoccum*.

DOTHISTROMA Hulbary
    Coelomycetes   1 sp.
Evans, H. C. 1984. The genus *Mycosphaerella* and its anamorphs *Cercoseptoria, Dothistroma* and *Lecanosticta* on pines. Mycol. Pap. 153:1-102. (extensive treatment of *D. septospora* and teleomorph *Mycosphaerella pini*, descriptions and illustrations)
Gibson, I. A. S. 1972. *Dothistroma* blight of *Pinus radiata*. Ann. Rev. Phytopathol. 10:51-72. (reviews the causal fungus and the disease)
Ivory, M. H. 1967. A new variety of *Dothistroma pini* in Kenya. Trans. Br. Mycol. Soc. 50:289-297. (description of a third variety, var. *keniensis*)
Roux, C. 1984. The morphology of *Dothistroma septospora* on *Pinus canariensis* from South Africa. South African J. Bot. 3:397-401. (describes conidial variability in culture and on host)
Dothistroma septospora (Doroguine) Morelet var. *septospora* causes *Dothistroma* blight or red-band needle blight of pine (CMI 368 under the teleomorph *Mycosphaerella pini* as *Scirrhia pini*) and is treated by Sutton (1980), Sivanesan (1984), and Funk (1985 as *Dothistroma pini*) in addition to those references listed above.

DRECHSLERA Ito
    Hyphomycetes   20 spp.
Alcorn, J. L. 1983. Generic concepts in *Drechslera, Bipolaris* and *Exserohilum*. Mycotaxon 17:1-86. (includes list of accepted binomials)
Chidambaram, P., Mathur, S. B., and Neergaard, P. 1973. Identification of seed-borne *Drechslera* species. Friesia 10:165-207. (key to 25 species, descriptions and illustrations)
Luttrell, E. S. 1977. Correlations between conidial and ascigerous state characters in *Pyrenophora, Cochliobolus*, and *Setosphaeria*. Rev. Mycol. 41:271-279. (reviews characteristics of the genus *Drechslera*, compares with related genera and teleomorph)
Shoemaker, R. A. 1959. Nomenclature of *Drechslera* and *Bipolaris*, grass parasites segregated from *Helminthosporium*. Can. J. Bot. 37:879-887. (includes host list)
Shoemaker, R. A. 1962. *Drechslera* Ito. Can. J. Bot. 40:809-836. (key to 15 species on host and in culture, descriptions and illustrations)

DRECHSLERA - cont.
Drechslera, Bipolaris, and *Exserohilum* constitute a group of related genera of grass pathogens that have been separated from *Helminthosporium sensu lato* on the basis of their teleomorphs, biology, and mode of conidium production. Alcorn (1983), Luttrell (1977), and Shoemaker (1959) offer convincing evidence for the acceptance of these genera. Domsch et al. (1980) provide a table of their distinguishing characteristics. *Drechslera* species have teleomorphs in *Pyrenophora*. In addition to the references listed above, Sivanesan (1984) under the teleomorph *Cochliobolus* and Ellis (1971, 1976) provide keys for species identification. Some important species and the diseases they cause are: *Drechslera avenae* (Eidam) Scharif, leaf stripe and seedling blight of oats (CMI 389 under *Pyrenophora avenae*); *D. dictyoides* (Drechs.) Shoemaker, net blotch and leaf spot of fescues (CMI 493 under *P. dictyoides*); *D. graminea* (Rabenh.) Shoemaker, leaf stripe of barley (CMI 388 under *P. graminea*); *D. heveae* (Petch) M. B. Ellis, bird's eye spot of rubber (CMI 343); *D. siccans* (Drechs.) Shoemaker, brown blight of *Festuca, Lolium,* and other Poaceae (CMI 492); *D. teres* (Sacc.) Shoemaker, net blotch of barley (CMI 390 under *P. teres*); and *D. tritici-repentis* (Died.) Shoemaker, yellow leaf spot of cereals and grasses (CMI 494 under *P. tritici-repentis*).

DREPANOPEZIZA (Kleb.) Hoehn.
    Helotiales   9 spp.
Gremmen, J. 1965. Three poplar-inhabiting *Drepanopeziza* species and their life-history. Nova Hedw. 9:170-176. (key to three species of *Drepanopeziza* and their *Marssonina* anamorphs, descriptions)
Rimpau, R. H. 1961. Untersuchungen ueber die Gattung *Drepanopeziza* (Kleb.) Hoehnel. Phytopathol. Z. 43:257-306. (in German with English summary, key to eight species, descriptions and illustrations)
Spiers, A. G. 1983. Studies of *Marssonina* and *Drepanopeziza* species pathogenic to poplars. Soil Conserv. Centre, Aokautere, New Zealand. Publ. 4. 41 pp. (recognizes three *Drepanopeziza-Marssonina* holomorphs, key to three *Marssonina* spp., discussion, pathology)
Drepanopeziza species cause anthracnose diseases on a variety of hosts. Rimpau (1961) provides a comprehensive account of the genus. The anamorphs have been placed in *Gloeosporidiella, Marssonina,* and *Monostichella*. *Drepanopeziza ribis* (Kleb.) Hoehn. causes anthracnose of currants and gooseberries (CMI 638).

DUOSPORIUM Thind & Rawla
    Hyphomycetes   1 sp.
Thind, K. S., and Rawla, G. S. 1961. A new fungus on *Cyperus iria.* Am. J. Bot. 48:859-862. (description of genus and species, illustrations)

DUOSPORIUM - cont.
Tsuda, M., and Ueyama, A. 1982. *Duosporium yamadanum*, a pathogen of *Cyperus* spp. Mycotaxon 14:145-148. (morphological data and taxonomic discussion)
Duosporium yamadanum (Matsuura) Tsuda & Ueyama (syn. *D. cyperi* Thind & Rawla) causes a leaf stripe of *Cyperus* spp. and is apparently related to *Drechslera*. Ellis (1971) treats the species as *Duosporium cyperi*.

ECHINODONTIUM Ellis & Everh.
    Aphyllophorales   4 spp.
Gross, H. L. 1964. The Echinodontiaceae. Mycopathol. Mycol. Appl. 24:1-26. (key to six *Echinodontium* spp., two are now placed in *Laurilia*, descriptions and illustrations)
Echinodontium tinctorium (Ellis & Everh.) Ellis & Everh., commonly known as the Indian paint fungus, causes a serious heart rot of firs in the western United States. See Aphyllophorales.

ELSINOE Racib.
    Dothideales   40 spp.
Arx, J. A. von 1963. Die Gattungen der Myriangiales. Persoonia 2:421-475. (key to genera, treats 16 *Elsinoe* spp., brief descriptions)
Elsinoe species cause scab and anthracnose of various tropical plants. Species are differentiated mainly by host; most are morphologically indistinguishable. No comprehensive treatment exists for *Elsinoe* or its anamorph *Sphaceloma*. Most species are described in the publications of A. E. Jenkins, often with A. A. Bitancourt. Sivanesan (1984) treats nine species. Diseases caused by *Elsinoe* species include: *Elsinoe ampelina* Shear, grape anthracnose (CMI 439; Sutton, 1980); *E. australis* Bitancourt & Jenk., sour orange scab (CMI 440); *E. canavaliae* Racib., scab of sword and jack beans (CMI 313); *E. fawcettii* Bitancourt & Jenk., sour orange scab (CMI 438); *E. phaseoli* Jenk., scab of lima beans (CMI 314); and *E. veneta* (Burkholder) Jenk., cane spot or anthracnose of raspberry (CMI 484).

ELYTRODERMA Darker
    Rhytismatales   2 spp.
Diamandis, S., and Minter, D. W. 1979. *Elytroderma torres-juanii* sp. nov. from Greece. Trans. Br. Mycol. Soc. 72:169-172. (description and illustration)
Elytroderma is a segregate of *Hypoderma*. *Elytroderma deformans* (J. R. Weir) Darker causes a witches broom of pine (CMI 655) and *E. torres-juanii* Diamandis & Minter (syn. *E. hispanicum* (Torres Juan) Darker) causes needle blight of pines (CMI 654). See *Lophodermium*.

EMBELLISIA E. Simmons
    Hyphomycetes   14 spp.
Simmons, E. G. 1971. *Helminthosporium allii* as type of a new genus.
    Mycologia 63:380-386. (describes and illustrates *Embellisia* with
    two species)
Simmons, E. G. 1983. An aggregation of *Embellisia* species. Mycotaxon
    17:216-241. (key to 13 species, descriptions)
A comprehensive treatment of *Embellisia* is presented by Simmons (1983). Ellis (1976) includes a key to three common species with brief descriptions and illustrations. *Embellisia allii* (Campanile) E. Simmons occurs on bulb scales of garlic, and *E. hyacinthi* de Hoog & E. Mueller causes spotting on bulb scales of *Freesia, Hyacinthus,* and *Scilla.*

EMERICELLA Berk. & Broome
    Eurotiales   15 spp.
Christensen, M., and Raper, K. B. 1978. Synoptic key to *Aspergillus nidulans* group species and related *Emericella* species. Trans.
    Br. Mycol. Soc. 71:177-191. (synoptic key to 30 taxa, brief
    descriptions and illustrations)
Emericella is the teleomorph of some *Aspergillus* species. *Emericella nidulans* (Eidam) Vuill. is the cause of aspergillosis, a disease of humans (CMI 93 under *Aspergillus nidulans*).

ENCOELIA (Fr.) P. Karst.
    Helotiales   7 spp.
Spooner, B. M., and Trigaux, G. 1985. A new *Encoelia* (Helotiales) from *Prunus spinosa* in France. Trans. Br. Mycol. Soc.
    85:547-552. (description and illustration)
Encoelia pruinosa (Ellis & Everh.) Torkelson & Eckblad (syn. *Cenangium singulare* (Rehm) Davidson & Cash) causes sooty bark canker of trembling aspen (Funk, 1981).

ENDOCALYX Berk. & Broome
    Hyphomycetes   4 spp.
Montemarini, C. 1962. Revisione della fammiglia Graphiolaceae (Deuteromycetes). Atti Ist. Bot. Univ. Pavia 20:253-275. (in Italian with English summary, key to three *Endocalyx* spp., descriptions and illustrations)
Okada, G., and Tubaki, K. 1984. A new species and a new variety of *Endocalyx* (Deuteromycotina) from Japan. Mycologia 76:300-313. (key to five taxa, descriptions and illustrations)
Sharma, M. P., and Prasha, I. B. 1982. *Endocalyx melanoxanthus* (Berk. & Br.) Petch: an addition to the Indian Stilbaceae. Res. Bull. Panjab Univ., Sci. 33(III-IV):149-151. (key to two *Endocalyx* spp., descriptions and illustrations)
Species of *Endocalyx* are commonly found as weak pathogens on members of Palmaceae, rarely on Vitaceae and Liliaceae. Ellis (1971)

ENDOCALYX - cont.
includes two species.

ENDOCRONARTIUM Y. Hiratsuka
    Uredinales   2 spp.
Hiratsuka, Y. 1969. *Endocronartium*, a new genus for autoecious pine stem rusts. Can. J. Bot. 47:1493-1495. (description of *Endocronartium* for two *Peridermium* spp., discussion) *Endocronartium harknessii* (J. P. Moore) Y. Hiratsuka causes western gall rust on *Pinus* spp. in North America (Ziller, 1974), and *E. pini* (Pers.:Pers.) Y. Hiratsuka occurs in Europe. Cummins and Hiratsuka (1983) describe the genus with comments on the taxonomy and biology. See *Peridermium*.

ENDOTHIA Fr.
    Diaporthales   10 spp.
Barr, M. E. 1983. On *Diaporthopsis metrosideri*. Mycologia 75:930-931. (discussion of generic characters, transfers this species to *Endothia*)
Kobayashi, T., and Ito, K. 1956. Notes on the genus *Endothia* in Japan. I. Species of *Endothia* collected in Japan. Bull. Gov. For. Exp. Stn. (Jpn.) 92:81-98. (describes and illustrates seven species, key to ten species in Japanese)
Micales, J. A., and Stipes, R. J. 1986. The differentiation of *Endothia* and *Cryphonectria* species by exposure to selected fungitoxicants. Mycotaxon 26:99-117. (discusses rationale for generic segregation)
Roane, M. K., Griffin, G. J., and Elkins, J. R. 1986. Chestnut Blight, other *Endothia* Diseases, and the Genus *Endothia*. Monograph Series. American Phytopathological Society, St. Paul, MN. 53 pp. (key to genera, 11 species treated as *Endothia*, descriptions and illustrations)
Stipes, R. J., Emert, G. H., and Brown, R. D., Jr. 1982. Differentiation of *Endothia gyrosa* and *Endothia parasitica* by disc electrophoresis of intramycelial enzymes and proteins. Mycologia 74:138-141. (results corroborate morphological species concepts)
Walker, J., Old, K. M., and Murray, D. I. L. 1985. *Endothia gyrosa* on *Eucalyptus* in Australia with notes on some other species of *Endothia* and *Cryphonectria*. Mycotaxon 23:353-370. (discussion of generic concepts, illustrations)
Barr (1978) segregates some species of *Endothia* into *Cryphonectria* including *Cryphonectria parasitica* (Murrill) Barr (syn. *Endothia parasitica* (Murrill) H. W. & P. J. Anderson), the cause of chestnut blight (CMI 704). Justification for this separation is discussed by Walker et al. (1985) and Micales and Stipes (1986). The remaining species of *Endothia* are mainly opportunistic pathogens (Micales and Stipes, 1986). Mueller and Arx (1962) treat ten species. *Endothia*

ENDOTHIA - cont.
gyrosa (Schwein.:Fr.) Fr. causes canker of oak and other hardwoods (CMI 449). Anamorphs are placed in *Endothiella*.

ENDOTHIELLA Sacc.
    Coelomycetes   5 spp.
Funk, A. 1984. *Endothiella aggregata* n. sp. (Phialostromatineae) on western conifers. Can. J. Bot. 62:154-155. (description and illustration)
Species of *Endothiella* are anamorphs of *Cryphonectria* and *Endothia*; descriptions can generally be found with those of the teleomorph. Sutton (1980) includes the *Endothiella* state of *Cryphonectria cubensis* (Bruner) C. S. Hodges (as *Endothia eugeniae* (Nutman & Roberts) J. Reid & C. Booth) on clove and *Eucalyptus*.

ENTODESMIUM Riess
    Pleosporales   6 spp.
Holm, L. 1957. Etudes taxonomiques sur les Pleosporacees. Symb. Bot. Ups. 14:1-188. (in French, key to six *Entodesmium* spp., descriptions)
Shoemaker, R. A. 1984. Canadian and some extralimital *Nodulosphaeria* and *Entodesmium* species. Can. J. Bot. 62:2730-2753. (key to six *Entodesmium* spp., descriptions and illustrations)
Entodesmium species occur on members of the Fabaceae. See the related genus *Leptosphaeria*.

ENTOMOSPORIUM Lev.
    Coelomycetes   1 sp.
Horie, H., and Kobayashi, T. 1980. *Entomosporium* leaf spot of Pomoideae (Rosaceae) in Japan. III. Additional basis for identification of the fungus, and distribution of the disease. Eur. J. For. Pathol. 10:225-235. (compares morphological variability of conidia on various hosts, host list, geographic distribution)
Stowell, E. A., and Backus, M. P. 1966. Morphology and cytology of *Diplocarpon maculatum* on *Crataegus*. I. The *Entomosporium* stage. Mycologia 58:949-960. (host-parasite relationship, developmental morphology of anamorph)
Stowell, E. A., and Backus, M. P. 1967. Morphology and cytology of *Diplocarpon maculatum* on *Crataegus*. II. Initiation and development of the apothecium. Mycologia 59:623-636. (developmental morphology of teleomorph)
Zwet, T. van der, and Stroo, H. F. 1985. Effects of cultural conditions on sporulation, germination, and pathogenicity of *Entomosporium maculatum*. Phytopathology 75:94-97. (physiological studies)
Entomosporium mespili (DC. ex Duby) Sacc. (syn. *E. maculatum* Lev.) (Sutton, 1980) causes *Entomosporium* leaf blight of Rosaceae (CMI 481

ENTOMOSPORIUM - cont.
under *Diplocarpon mespili* as *D. maculatum*).

ENTYLOMA de Bary
    Ustilaginales   100 spp.
Savile, D. B. O. 1947. A study of the species of *Entyloma* on North American composites. Can. J. Res. Sect. C. 25:105-120. (discusses species according to host, comparative table of morphological characters)
Species of *Entyloma* are parasitic, forming sori mainly in leaves. A modern treatment of this genus is needed. Mordue and Ainsworth (1984) provide a key to 16 British species. Species include: *Entyloma arnicale* Ellis & Everh., on *Arnica* spp. (F. Can. 112); *E. calendulae* (Oudem.) de Bary, leaf spot of *Calendula* (CMI 801); *E. calendulae* f. *dahliae* (Sydow) Viegas, leaf spot of *Dahlia* (CMI 802); *E. fuscum* J. Schroet., leaf spot of *Papaver* spp. (CMI 803); and *E. oryzae* Sydow & P. Sydow, leaf smut of rice (CMI 296). See Ustilaginales.

EPHELIS   See *Balansia* and *Myriogenospora*.

EPICHLOE (Fr.) Tul. & C. Tul.
    Clavicipitales   8 spp.
Doguet, G. 1960. Morphologie, organogenie et evolution nucleaire de l'*Epichloe typhina*. La place des Clavicipitaceae dans la classification. Bull. Trimest. Soc. Mycol. Fr. 76:171-203. (in French, developmental study)
Morgan-Jones, G., and Gams, W. 1982. Notes on Hyphomycetes. XLI. An endophyte of *Festuca arundinacea* and the anamorph of *Epichloe typhina*, new taxa in one of two new sections of *Acremonium*. Mycotaxon 15:311-318. (description of *Acremonium coenophialum*, the endophyte frequently misidentified as *E. typhina*, description of *Acremonium typhinum*, anamorph of *E. typhina*)
Rykard, D. M., Luttrell, E. S., and Bacon, C. W. 1984. Conidiogenesis and conidiomata in the Clavicipitoideae. Mycologia 76:1095-1103. (describes anamorphs of *Balansia, Epichloe*, and other genera, discusses taxonomic significance)
No monograph of *Epichloe* exists. *Epichloe typhina* (Pers.:Fr.) Tul. causes choke disease of grasses (CMI 639). The anamorph is *Acremonium typhinum* Morgan-Jones & Gams (syn. *Sphacelia typhina* Sacc.). See *Acremonium*.

EPICOCCUM Link:Fr.
    Hyphomycetes   2 spp.
Cannon, P. F. 1986. International Commission on the Taxonomy of Fungi (ICTF): name changes in fungi of microbiological, industrial and medical importance. Part 1. Microbiol. Sci. 3:168-171. (presents explanation for use of *E. nigrum*, rather than *E. purpurascens*)

EPICOCCUM - cont.
Punithalingam, E., Tulloch, M., and Leach, C. M. 1972. *Phoma epicoccina* sp. nov. on *Dactylis glomerata*. Trans. Br. Mycol. Soc. 59:341-345. (description and illustration of *Epicoccum* state of *Phoma glomerata*)
Schol-Schwarz, M. B. 1959. The genus *Epicoccum* Link. Trans. Br. Mycol. Soc. 42:149-173. (reduces most described *Epicoccum* names to synonymy with *E. nigrum*, description and illustration)
Epicoccum nigrum Link (syn. *E. purpurascens* Ehrenb. ex Schlecht.) is now the correct name for this ubiquitous saprophyte or weak parasite on many hosts (CMI 680; Ellis, 1971). See *Cerebella*.

EREMOTHECIUM Borzi
    Endomycetales   2 spp.
Batra, L. R. 1973. Nematosporaceae (Hemiascomycetidae): Taxonomy, Pathogenicity, Distribution, and Vector Relations. U.S. Dep. Agric., Tech. Bull. 1469. 71 pp. (descriptions, illustrations, biology)
Boedijn, K. B. 1960. Stigmatomycosis in Indonesia. Mycopathologia 13:243-246. (describes *E. cymbalariae*, illustration)
Watkins, G. M. 1981. Compendium of Cotton Diseases. American Phytopathological Society, St. Paul, MN. 87 pp. (discusses *E. cymbalariae* as the cause of internal boll infection of cotton in the United States)
Eremothecium ashbyi Guill. causes cotton boll rot, a pathogen transmitted by insects (CMI 181).

ERYSIPHE R. Hedw. ex DC.:Fr.
    Erysiphales   10 spp.
Braun, U. 1981. Taxonomic studies in the genus *Erysiphe* I. Generic delimitation and position in the system of the Erysiphaceae. Nova Hedw. 34:679-719. (key to 19 genera, discussion and references, synopsis of the genus *Erysiphe*, discusses generic placement of *E. graminis*, *E. trifolii*, and others)
Braun, U. 1981. Miscellaneous notes on the Erysiphaceae (II). Feddes Repert. 92:499-513. (describes and illustrates new *Erysiphe* and *Microsphaera* taxa)
Braun, U. 1982. Descriptions of new species and combinations in *Microsphaera* and *Erysiphe*. Mycotaxon 14:369-374. (describes and discusses two *Erysiphe* spp.)
Braun, U. 1983. Descriptions of new species and combinations in *Microsphaera* and *Erysiphe*. (III). Mycotaxon 16:417-424. (describes and illustrates *E. cichoracearum* var. *poonaensis*)
Braun, U. 1984. Descriptions of new species and combinations in *Microsphaera* and *Erysiphe*. (V). Mycotaxon 19:375-383. (describes and illustrates *E. geraniacearum*, *E. greeneana*, *E. poeltii*, and *E. mayorii* var. *cicerbitae*)
Braun, U. 1984. Descriptions of new species and combinations in

ERYSIPHE - cont.
Microsphaera and *Erysiphe*. (VI). Mycotaxon 20:491-498.
(describes and illustrates six new *Erysiphe* taxa with notes on four other taxa)
Braun, U. 1985. The *Erysiphe-Microsphaera* complex on Fabaceae. Zbl. Mikrobiol. 140:393-417. (key to 29 species in the complex, descriptions and illustrations)
Junell, L. 1967. A revision of *Erysiphe communis* (Wallr.) Fr. sensu Blumer. Sven. Bot. Tidskr. 61:209-230. (recognizes seven species in the *E. communis* complex, key, descriptions)
Parmelee, J. A. 1977. The fungi of Ontario. II. Erysiphaceae (mildews). Can. J. Bot. 55:1940-1983. (keys to six genera and 28 species, descriptions and illustrations, review of taxonomy of *E. polygoni*, host list, useful for northeastern North America)
Stavely, J. R., and Hanson, E. W. 1966. Pathogenicity and morphology of isolates of *Erysiphe polygoni*. Phytopathology 56:309-318. (discussion and table comparing *E. polygoni* from five hosts)
Zheng, R.-y., and Chen, G.-q. 1981. The genus *Erysiphe* in China. Sydowia 34:214-327. (key based on host family, descriptions, illustrations, host index)

Erysiphe includes many important powdery mildews, some of which have wide host ranges. Conidial states of Erysiphales can be separated into at least five morphological types. See *Oidium*. Members of *Erysiphe* and other Erysiphaceae are usually treated on a host or geographic basis. See Erysiphales. Spencer (1978) and Yarwood (1973) provide keys to genera and give numerous references. Hawksworth et al. (1983) list additional regional treatments. Important species of *Erysiphe* include: *Erysiphe betae* (Vanha) Weltzien, powdery mildew of sugar beet (CMI 151); *E. cichoracearum* DC., powdery mildew of lettuce, safflower, and various Asteraceae and Cucurbitaceae (CMI 152); *E. cruciferarum* Opiz ex Junell, powdery mildew of turnips and swedes (CMI 251); *E. graminis* DC., powdery mildew of wheat, barley, and other cereals and grasses (CMI 153, F. Can. 71); *E. heraclei* DC., powdery mildew of carrot, fennel, parsley, and other Apiaceae (CMI 154); *E. pisi* DC., powdery mildew of pea (CMI 155); *E. polygoni* DC., powdery mildew of *Polygonum* and *Rumex* (CMI 509); and *E. trifolii* Grev., powdery mildew of clover (CMI 156).

ERYTHRICIUM J. Eriksson & Hjortstam
    Aphyllophorales   3 spp.
Oniki, M., Ogoshi, A., and Araki, T. 1985. Development of the perfect state and taxonomic assessment of the citrus pink disease fungus, *Corticium salmonicolor*. Trans. Mycol. Soc. Jpn. 26:441-448. (in Japanese with English summary, description of teleomorph)

Erythricium salmonicolor (Berk. & Broome) Burdsall (syns. *Corticium salmonicolor* Berk. & Broome, *Phanerochaete salmonicolor* (Berk. & Broome) Juelich) causes pink disease of rubber, tea, and other tropical plants (CMI 511 as *C. salmonicolor*).

EUPENICILLIUM C. A. Ludw.
    Eurotiales   37 spp.
Pitt, J. I. 1974. A synoptic key to the genus *Eupenicillium* and to sclerotigenic *Penicillum* spp. Can. J. Bot. 52:2231-2236. (synoptic key to 36 European *Eupenicillium* spp. and 22 sclerotigenic *Penicillium* spp.)
Stolk, A. C., and Samson, R. A. 1983. The ascomycete genus *Eupenicillium* and *Penicillium* anamorphs. Stud. Mycol. 23:1-149. (monograph of *Eupenicillium* spp. and sclerotial *Penicillium* anamorphs, synoptic key using teleomorph and anamorph characters, keys to 33 *Eupenicillium* taxa, descriptions and illustrations)
Udagawa, S., and Horie, Y. 1973. Surface ornamentation of ascospores in *Eupenicillium* species. Antonie van Leeuwenhoek 39:313-319. (SEM of ascospores)
Species of *Eupenicillium* are teleomorphs of *Penicillium* spp.

EUROPHIUM   See *Ophiostoma*.

EUROTIUM Link:Fr.
    Eurotiales   19 spp.
Blaser, P. (1975)1976. Taxonomische und physiologische Untersuchungen ueber die Gattung *Eurotium* Link ex Fries. Sydowia 28:1-49. (in German with English summary, key to 19 species, descriptions, SEM illustrations)
Teleomorphs of the *Aspergillus glaucus* group belong to *Eurotium*.

EUTYPA Tul. & C. Tul.
    Diatrypales   40 spp.
Glawe, D. A., Dilley, M. A., and Moller, W. J. 1983. Isolation and identification of *Eutypa armeniacae* from *Malus domestica* in Washington State. Mycotaxon 18:315-318. (cultural and pathogenicity studies)
Glawe, D. A., and Rogers, J. D. 1982. Observations on the anamorphs of six species of *Eutypa* and *Eutypella*. Mycotaxon 14:334-346. (descriptions of cultures and anamorphs of four *Eutypa* spp.)
Glawe, D. A., and Rogers, J. D. 1984. Diatrypaceae in the Pacific Northwest. Mycotaxon 20:401-460. (key to four *Eutypa* spp., taxonomy, biology, host/fungus index)
Glawe, D. A., Skotland, C. B., and Moller, W. J. 1982. Isolation and identification of *Eutypa armeniacae* from diseased grapevines in Washington State. Mycotaxon 16:123-132. (cultural and pathogenicity studies)
Rappaz, F. 1984. Les especes santionees du genre *Eutypa* (Diatrypaceae, Ascomycetes). Etude taxonomicque et nomenclaturale. Mycotaxon 20:567-586. (in French with English summary, key, descriptions and illustrations)
No comprehensive reference exists. *Eutypa armeniacae* Hansf. & Carter

EUTYPA - cont.
causes *Cytosporina* dieback of apricots and other woody plants (CMI 436).

EUTYPELLA (Nitschke) Sacc.
Diatrypales   40 spp.
Glawe, D. A. 1983. Observations on the anamorph of *Eutypella parasitica*. Mycologia 75:742-743. (descriptions and illustrations of conidiogenesis)
Glawe, D. A., and Rogers, J. D. 1982. Observations on the anamorphs of six species of *Eutypa* and *Eutypella*. Mycotaxon 14:334-346. (descriptions of cultures and anamorphs of two *Eutypella* spp.)
Glawe, D. A., and Rogers, J. D. 1984. Diatrypaceae in the Pacific Northwest. Mycotaxon 20:401-460. (key to three *Eutypella* spp., taxonomy, biology, host/fungus index)
Rogers, J. D., and Glawe, D. A. 1983. *Diatrype whitmanensis* sp. nov. and the anamorphs of *Diatrype bullata* and *Eutypella sorbi*. Mycotaxon 18:73-80. (description and illustrations)
No comprehensive monograph exists. *Eutypella parasitica* Davidson & Lorenz causes a destructive disease of *Acer* spp.

EXOBASIDIUM Woronin
Exobasidiales   50 spp.
McNabb, R. F. R. 1962. The genus *Exobasidium* in New Zealand. Trans. R. Soc. N. Z. Bot. 1:259-268. (key to eight species, descriptions and illustrations)
Nannfeldt, J. A. 1981. *Exobasidium*, a taxonomic reassessment applied to the European species. Symb. Bot. Ups. 23:1-72. (key to 27 species, descriptions)
Nickerson, N. L. 1984. A previously unreported disease of cranberries caused by *Exobasidium perenne* sp. nov. Can. J. Plant Pathol. 6:218-220. (description, discussion, color photographs)
Savile, D. B. O. 1959. Notes on *Exobasidium*. Can. J. Bot. 37:641-656. (key to 11 taxa, descriptions)
Sundstrom, K.-R. 1964. Studies of the physiology, morphology, and serology of *Exobasidium*. Symb. Bot. Ups. 18:1-89. (delimits species based on a variety of characters)
Exobasidium species are widespread on Ericales especially in north temperate regions. Species include: *Exobasidium japonicum* Shirai, *Azalea* gall (CMI 780); *E. vaccinii* (Fuckel) Woronin, red leaf disease of *Vaccinium* (CMI 778); and *E. vexans* Massee, blister blight of tea (CMI 779). See *Arcticomyces* and *Muribasidiospora*.

EXSEROHILUM Leonard & Suggs
Hyphomycetes   17 spp.
Alcorn, J. L. 1983. Generic concepts in *Drechslera*, *Bipolaris* and *Exserohilum*. Mycotaxon 17:1-86. (includes list of accepted binomials)

EXSEROHILUM - cont.
Alcorn, J. L. 1986. A new homothallic *Setosphaeria* species and its *Exserohilum* anamorph. Trans. Br. Mycol. Soc. 86:313-317. (description and illustrations, isolated from *Dactyloctenium*)
Sivanesan, A. 1984. New species of *Exserohilum*. Trans. Br. Mycol. Soc. 83:319-329. (key to 16 species, descriptions and illustrations)
Within the *Helminthosporium*-like fungi, *Exserohilum* is a segregate characterized by conidia with a distinctly protuberant hilum and a *Setosphaeria* teleomorph. Ellis (1971, 1976) treats *Exserohilum* species under *Drechslera*. Species include *Exserohilum rostratum* (Drechs.) Leonard & Suggs, causing various cereal diseases, including foot rot of wheat (CMI 587 as *Drechslera rostrata* under *Setosphaeria rostrata*), and *E. turcicum* (Pass.) Leonard & Suggs, causing northern leaf blight of maize and sorghum (CMI 304 under *S. turcica* as *Trichometasphaeria turcica*.)

FABRAEA Sacc.
    Helotiales   1 sp.
Fabraea is a synonym of *Leptotrochila*. *Fabraea cincta* Sacc. & Scalia causes leaf spots on *Rubus pedatus* (F. Can. 81) and has not been transferred to an appropriate genus. See *Diplocarpon*.

FISTULINA Bull.:Fr.
    Aphyllophorales   2 spp.
Fistulina hepatica (Schaeff.:Fr.) Sibthorp, the beefsteak fungus, causes a brown heart rot of living oaks and chestnuts. See Aphyllophorales.

FOMES (Fr.) Fr.
    Aphyllophorales   2 spp.
Lowe, J. L. 1957. Polyporaceae in North America. The genus *Fomes*. State Univ. Coll. For. Syracuse Univ., Tech . Publ. 80:1-97. (key to 68 species formerly placed in *Fomes*, descriptions and illustrations)
Species of *Fomes* cause white rot of living or dead hardwoods. Gilbertson and Ryvarden (1986) provide a key to the two species retained in *Fomes*, *F. fasciatus* (Sw.:Fr.) Cooke and *F. fomentarius* (Fr.:Fr.) Kickx. See segregate genera *Fomitopsis, Heterobasidion, Phellinus, Rigidoporus*, and Aphyllophorales.

FOMITOPSIS P. Karst.
    Aphyllophorales   18 spp.
Carranza-Morse, J., and Gilbertson, R. L. 1986. Taxonomy of the *Fomitopsis rosea* complex (Aphyllophorales; Polyporaceae). Mycotaxon 25:469-486. (key to seven species, descriptions and illustrations)
Fomitopsis rosea (Albertini & Schwein.:Fr.) P. Karst. (syn. *Fomes*

FOMITOPSIS - cont.
roseus (Albertini & Schwein.:Fr.) Cooke) causes a brown top rot of conifers, especially Douglas fir (CMI 191). *Fomitopsis pinicola* (Swartz:Fr.) P. Karst. causes heart rot of living conifers and some hardwoods. See Aphyllophorales.

FOVEOSTROMA DiCosmo
    Coelomycetes   3 spp.
DiCosmo, F. 1978. A revision of *Corniculariella*. Can. J. Bot. 56:1665-1690. (key to seven *Corniculariella* spp., treats two *Foveostroma* spp., amends *Gelatinosporium*, descriptions and illustrations)
Funk, A. 1976. The genus *Dermea* and related conidial states on Douglas fir. Can. J. Bot. 54:2852-2856. (describes and illustrates four *Micropera* spp., now placed in *Foveostroma*) DiCosmo (1978) describes *Foveostroma* to replace the illegitimate genus *Micropera*. Funk (1981) discusses *Foveostroma abietinum* (Peck) DiCosmo and *F. boycei* (Dearn.) Funk both occurring on conifers with teleomorphs in *Dermea*. Sutton (1980) includes *F. drupacearum* (Lev.) DiCosmo, the anamorph of *Dermea cerasi* (Pers.:Fr.) Fr., which occurs on *Prunus* spp.

FULVIA Cif.
    Hyphomycetes   1 sp.
Fulvia fulva (Cooke) Cif. causes leaf mold of tomato (CMI 487; Ellis, 1971).

FUSARIUM Link:Fr.
    Hyphomycetes   50 spp.
Booth, C. 1971. The Genus *Fusarium*. Commonwealth Mycological Institute, Kew, Surrey, England. 237 pp. (key, descriptions and illustrations)
Booth, C. 1975. The present status of *Fusarium* taxonomy. Ann. Rev. Phytopathol. 13:83-93. (discusses problems in each section of *Fusarium*)
Fisher, N. L., Marasas, W. F. O., and Toussoun, T. A. 1983. Taxonomic importance of microconidial chains in *Fusarium* section *Liseola* and effects of water potential on their formation. Mycologia 75:693-698. (compares three species)
Gerlach, W., and Nirenberg, H. 1982. The genus *Fusarium*-a pictorial atlas. Mitt. Biol. Bundes. Land- & Forst. 209:1-409. (descriptions and illustrations)
Joffe, A. Z. 1986. *Fusarium* Species: Their Biology and Toxicology. Wiley and Sons, New York. 588 pp. (one chapter on taxonomy; keys to sections, species, and varieties; descriptions and illustrations)
Marasas, W. F. O., Nelson, P. E., and Toussoun, T. A. 1984. Toxigenic *Fusarium* Species: Identity and Mycotoxicology. Pennsylvania

FUSARIUM - cont.
State University Press, University Park. 328 pp. (companion volume to Nelson et al. 1983, emphasizes toxins)
Moss, M. O., and Smith, J. E., eds. 1984. The Applied Mycology of *Fusarium*. Cambridge University Press, Cambridge. 264 pp. (primarily non-taxonomic, one chapter by Booth includes overview of systematics)
Nelson, P. E., Toussoun, T. A., and Cook, R. J., eds. 1981. *Fusarium*: Diseases, Biology, and Taxonomy. Pennsylvania State University Press, University Park. 457 pp. (four chapters on taxonomy that include keys and some descriptions of both anamorphs and teleomorphs, discussion)
Nelson, P. E., Toussoun, T. A., and Marasas, W. F. O. 1983. *Fusarium* Species: An Illustrated Manual for Identification. Pennsylvania State University Press, University Park. 193 pp. (synoptic key to 30 species, descriptions and illustrations)
Nirenberg, H. 1976. Untersuchungen ueber die morphologische und biologische Differenzierung in der *Fusarium*-Sektion *Liseola*. Mitt. Biol. Bundes. Land- & Forst. 169:1-117. (in German)
Nirenberg, H. 1981. A simplified method for identifying *Fusarium* spp. occurring on wheat. Can. J. Bot. 59:1599-1609. (synoptic and dichotomous keys to eight species, descriptions and illustrations)
Seemuller, E. 1968. Untersuchungen ueber die morphologische und biologische Differenzierung in der *Fusarium*-Sektion *Sporotrichiella*. Mitt. Biol. Bundes. Land- & Forst. 127:1-93. (in German with English summary, descriptions and illustrations of five taxa, pathology)
Teetro-Barsch, G. H., and Roberts, D. W. 1983. Entomogenous *Fusarium* species. Mycopathologia 84:3-16. (pathogenicity tests on insects, biology)
Toussoun, T. A., and Nelson, P. E. 1975. Variation and speciation in the Fusaria. Ann. Rev. Phytopathol. 13:71-82. (review of classification systems)

Given the plethora of comprehensive accounts, identification of *Fusarium* species is now possible. Cultures must be studied carefully according to the conditions specified by the monographer and isolates often degenerate after they are transferred a few times, no longer forming macroconidia. References to all aspects of *Fusarium* are found in Domsch et al. (1980) who treat 27 species. The teleomorphs are placed in *Gibberella* and *Nectria*. Some *Fusarium* species have been transferred to *Microdochium*. *Fusarium* species causing diseases include: *Fusarium avenaceum* (Corda:Fr.) Sacc., various diseases on a variety of hosts (CMI 25); *F. coccophilum* (Desmaz.) Wollenw. & Reinking, on scale insects (CMI 715 under *Nectria flammea*, with key to *Nectria* and *Fusarium* species on scale insects); *F. culmorum* (W. G. Sm.) Sacc., on grains and other hosts (CMI 26); *F. decemcellulare* Brick, on *Theobroma cacao* and numerous

FUSARIUM - cont.
tropical crops (CMI 21 under *Calonectria rigidiuscula*); *F. equiseti* (Corda) Sacc., various diseases on a variety of hosts (CMI 571); *F. graminearum* Schwabe, various diseases of cereals and other hosts (CMI 384 under *Gibberella zeae*); *F. heterosporum* Nees:Fr., head blight of cereals (CMI 572); *F. larvarum* Fuckel, on scale insects (CMI 714 under *Nectria aurantiicola*); *F. lateritium* Nees, diseases of a variety of woody hosts (CMI 310 under *Gibberella baccata*); *F. moniliforme* Sheldon, on Poaceae and a variety of hosts (CMI 22 under *Gibberella fujikuroi*); *F. oxysporum* Schlechtend.:Fr., wilt pathogens on many crop plants (CMI 28, 211-220); *F. poae* (Peck) Wollenw., various diseases on a number of hosts (CMI 308, F. Can. 234); *F. redolens* Wollenw., a tracheid parasite causing wilts, damping off, and a cortical rot (CMI 27); *F. sambucinum* Fuckel, dieback of hops, root rot of cereals and other hosts, storage rot of potato (CMI 385 under *Gibberella pulicaris*); *F. semitectum* Berk. & Ravenel, storage rot of tropical crops (CMI 573); *F. solani* (Mart.) Sacc., various diseases on a variety of hosts (CMI 29); *F. sporotrichoides* Sherb. var. *sporotrichoides*, on various plants (F. Can. 235); *F. stilboides* Wollenw., scaly bark and collar rot of coffee (CMI 30); *F. subglutinans* (Wollenw. & Reinking) P. Nelson, Toussoun & Marasas, seedling blight, root, stalk, and kernel rot of maize, and pitch canker of southern pines (CMI 23 as *F. moniliforme* var. *subglutinans* under *Gibberella fujikuroi* var. *subglutinans*); *F. sulphureum* Schlechtend., potato tuber dry rot (CMI 574); *F. udum* E. J. Butler, *Fusarium* wilt of pigeon pea (CMI 575); and *F. xylarioides* Steyaert, tracheomycosis of coffee (CMI 24 under *Gibberella xylarioides*).

FUSICLADIUM Bonord.
    Hyphomycetes   40 spp.
Fusicladium species are anamorphs of *Apiosporina, Microcyclus*, and *Venturia*. Ellis (1971, 1976) treats five species. See also Sivanesan (1984).

FUSICOCCUM Corda
    Coelomycetes   50 spp.
Pennycook, S. R., and Samuels, G. J. 1985. *Botryosphaeria* and *Fusicoccum* species associated with ripe fruit rot of *Actinidia deliciosa* (kiwifruit) in New Zealand. Mycotaxon 24:445-458. (two *Botryosphaeria* spp., three *Fusicoccum* sp., descriptions and illustrations, discussion of taxonomy of *Fusicoccum* and *Dothiorella*)
Samuels, G. J., and Singh, B. 1986. *Botryosphaeria xanthocephala*, cause of stem canker in pigeon pea. Trans. Br. Mycol. Soc. 86:295-299. (descriptions and illustrations of *Botryosphaeria xanthocephala* and anamorph *F. cajani*)
No monographic account of this genus exists. Species in this genus have often been placed in *Dothiorella*. Sutton (1980) includes

FUSICOCCUM - cont.
Fusicoccum aesculi Corda, the anamorph of *Botryosphaeria dothidea* (Moug.:Fr.) Ces. & De Not., a ubiquitous pathogen, and presents a nomenclatural history of the genus.

GAEUMANNOMYCES Arx & Olivier
    Diaporthales   4 spp.
Walker, J. 1972. Type studies on *Gaeumannomyces graminis* and related fungi. Trans. Br. Mycol. Soc. 58:427-457. (description, illustration, synonymy)
Walker, J. 1975. Take-all diseases of Gramineae: A review of recent work. Rev. Plant Pathol. 54:113-144. (reviews biology and recent research, extensive bibliography)
Walker, J. 1980. Taxonomy of take-all fungi and related genera and species. Pages 15-75 in: Biology and Control of Take-all. M. J. C. Asher and P. J. Shipton, eds. Academic Press, New York. (key to related genera and 11 taxa in *Gaeumannomyces* and *Phialophora*, descriptions and illustrations of varieties of *G. graminis* and related species)
Walker, J. 1980. *Gaeumannomyces, Linocarpon, Ophiobolus*, and several other genera of scolecospored Ascomycetes and *Phialophora* conidial states, with a note on hyphopodia. Mycotaxon 11:1-129. (describes genus, discusses accepted species)
Species of *Gaeumannomyces* are parasites of roots, culms, and leaf sheaths of Poaceae and Cyperaceae. Originally these pathogens were placed in *Ophiobolus* but fundamental differences in ascus structure and biology necessitated their transfer to *Gaeumannomyces*. Walker (1980) provides a complete account of *Gaeumannomyces* species and related fungi. Diseases include those caused by: *Gaeumannomyces graminis* (Sacc.) Arx & Olivier var. *avenae* (E. M. Turner) Dennis, take-all of oats and other Poaceae (CMI 382); *G. graminis* var. *graminis*, crown sheath rot of rice (CMI 381); and *G. graminis* var. *tritici* Walker, take-all of cereals and grasses (CMI 383, F. Can. 37). See the anamorph genus *Phialophora* for organisms that cause similar diseases.

GANODERMA P. Karst.
    Aphyllophorales   50 spp.
Corner, E. J. H. 1983. Ad Polyporaceas I. *Amaroderma* and *Ganoderma*. Nova Hedw. Beih. 75:1-182. (keys to sections and species, descriptions and illustrations)
Steyaert, R. L. 1962. Genus *Ganoderma* (Polyporaceae). Taxa nova - 2. Bull. Jard. Bot. Brux. 32:89-104. (in French, descriptions)
Steyaert, R. L. 1980. Study of some *Ganoderma* species. Bull. Jard. Bot. Nat. Belg. 50:135-186. (refers to other papers by this author)
Despite the recent studies, some species of *Ganoderma* are poorly known and difficult to identify, especially those occurring in

GANODERMA - cont.
tropical areas. Pathogenic species include: *Ganoderma applanatum* (Pers.) Pat., on many broad-leaved and coniferous trees (CMI 443); *G. boninense* Pat., on Palmaceae, basal rot of *Elaeis guineensis* and *Cocos nucifera* (CMI 444); *G. lucidum* (Curtis:Fr.) P. Karst., butt rot and lethal root disease of many tree species (CMI 445); *G. philippii* (Bres. & Henn.) Bres., red root rot of rubber, tea, and others (CMI 446); *G. pseudoferrum* (Wakef.) Overh. & Steinmann, root rot of cacao, coffee, rubber, tea, and others; *G. tornatum* (Pers.) Bres., heart and butt rot (CMI 447); and *G. zonatum* Murrill, base rot of palms (CMI 448). See Aphyllophorales.

GELATINOSPORIUM Peck
    Coelomycetes   8 spp.
Species of *Gelatinosporium* are anamorphs of *Dermea* species causing cankers on conifers. Sutton (1980) distinguishes *Gelatinosporium* from anamorphs of other canker-causing discomycetes. Funk (1981) provides a synoptic table of eight species. See the similar genus *Foveostroma*.

GENICULOSPORIUM   See *Hypoxylon*.

GEOTRICHUM Link:Fr.
    Hyphomycetes   8 spp.
Arx, J. A. von, Miranda, L. Rodrigues de, Smith, M. T., and Yarrow, D. 1977. The genera of yeasts and the yeast-like fungi. Stud. Mycol. 14:1-42. (discusses species placed in *Geotrichum*, makes three new combinations)
Butler, E. E., and Petersen, L. J. 1972. *Endomyces geotrichum* a perfect state of *Geotrichum candidum*. Mycologia 64:365-374. (description and illustrations, sexuality, pathology)
Carmichael, J. W. 1957. *Geotrichum candidum*. Mycologia 49:820-830. (description, illustration, list of synonyms)
Gueho, E. 1979. Deoxyribonucleic acid base composition and taxonomy in the genus *Geotrichum* Link. Antonie van Leeuwenhoek 45:199-210. (results suggest three groups of species, correlated with biochemical characters)
Gueho, E., Tredick, J., and Phaff, H. J. 1985. DNA relatedness among species of *Geotrichum* and *Dipodascus*. Can. J. Bot. 63:961-966. (discusses taxonomic problems within these genera)
Sigler, L., and Carmichael, J. W. 1976. Taxonomy of *Malbranchea* and some other Hyphomycetes with arthroconidia. Mycotaxon 4:349-488. (lists taxonomic disposition of *Geotrichum* names)
Identification of these yeast-like fungi is not easy and no definitive work exists. A few species cause postharvest damage to fruits.

GERLACHIA   See *Microdochium*.

GIBBERELLA Sacc.
    Hypocreales   10 spp.
Booth, C. 1971. The Genus *Fusarium*. Commonwealth Mycological Institute, Kew, Surrey, England. 237 pp. (descriptions and illustrations under the *Fusarium* anamorph)
Booth, C., and Prior, C. 1984. A new *Gibberella* species on *Imperata* (Gramineae). Trans. Br. Mycol. Soc. 82:180-182. (description and illustration)
Booth, C., and Spooner, B. M. 1984. *Gibberella avenacea*, teleomorph of *Fusarium avenaceum*, from stems of *Pteridium aquilinum*. Trans. Br. Mycol. Soc. 82:178-180. (description and illustration)
Kuhlman, E. G. 1982. Varieties of *Gibberella fujikuroi* with anamorphs in *Fusarium* section *Liseola*. Mycologia 74:759-768. (describes four varieties with illustrations)
All *Gibberella* species have *Fusarium* anamorphs, thus *Gibberella* is included in literature on *Fusarium*. No monograph exists for the teleomorph. Domsch et al. (1980) include descriptions of six *Gibberella* species. Booth (1971) is also useful. Species causing diseases include: *Gibberella baccata* (Wallr.) Sacc., diseases of *Pinus* and other hosts (CMI 310); *G. cyanogena* (Desmaz.) Sacc., potato tuber dry rot (CMI 574); *G. fujikuroi* (Sawada) Ito, various diseases (CMI 22); *G. fujikuroi* var. *subglutinans* Edwards, various diseases (CMI 23); *G. gordonii* C. Booth, head blight of cereals (CMI 572 under *Fusarium heterosporum*); *G. intricans* Wollenw., various diseases (CMI 571 under *Fusarium equiseti*); *G. pulicaris* (Fr.:Fr.) Sacc., canker and dieback of hops, root rot of cereals and other hosts, storage rot of potato (CMI 385); *G. xylarioides* R. Heim & Saccas, tracheomycosis of coffee (CMI 24); and *G. zeae* (Schwein.) Petch, various diseases of cereals and other hosts (CMI 384).

GIBELLINA Pass. ex Roum.
    Polystigmatales   2 spp.
Glynne, M. B., Fitt, B. D. L., and Hornby, D. 1985. *Gibellina cerealis*, an unusual pathogen of wheat. Trans. Br. Mycol. Soc. 84:653-659. (describes, illustrates, and discusses the disease and its causal fungus)
Mueller and Arx (1962) present a description and illustration of *Gibellina cerealis* Pass., the cause of white foot rot of wheat (CMI 534).

GILBERTELLA Hesseltine
    Mucorales   1 sp.
Hesseltine, C. W. 1960. *Gilbertella* gen. nov. (Mucorales). Bull. Torrey Bot. Club 87:21-30. (description and illustration)
Gilbertella persicaria (Eddy) Hesseltine occurs on peach, mulberry, and other hosts (CMI 104), and *G. persicaria* var. *indica* M. D. Mehrotra & B. S. Mehrotra occurs on a variety of hosts (CMI 105). See Mucorales.

GLIOCLADIUM Corda
Hyphomycetes   13 spp.
Morquer, R., Viala, G., Rouch, J., Fayret, J., and Berge, G. 1963. Contribution a l'etude morphogenique du genre *Gliocladium*. Bull. Trimest. Soc. Mycol. Fr. 79:137-241. (in French, presents cultural characteristics of 12 species, descriptions and illustrations)
Seifert, K. 1985. A monograph of *Stilbella* and some allied hyphomycetes. Stud. Mycol. 27:1-235. (key to five *Gliocladium* spp., descriptions and illustrations including teleomorph)
Species of *Gliocladium* are anamorphs of hypocrealean fungi. Seifert (1985) treats five species having *Nectria* and *Sphaerostilbella* teleomorphs. Domsch et al. (1980) include four common species.

GLOEOCERCOSPORA Bain & Edgerton ex Deighton
Hyphomycetes   2 spp.
Deighton, F. C. 1971. Validation of the generic name *Gloeocercospora* and the specific names *G. sorghi* and *G. inconspicua*. Trans. Br. Mycol. Soc. 57:358-360. (describes *G. sorghi*)
Gloeocercospora sorghi Bain & Edgerton ex Deighton causes zonate leaf spot of sorghum and copper spot of turf grasses (CMI 300).

GLOEOSPORIDIELLA Petr.
Hyphomycetes   5 spp.
Arx, J. A. von 1970. A revision of the fungi classified as *Gloeosporium*. Bibl. Mycol. 24:1-203. (key to related genera, treats six species, descriptions and illustrations)
Rimpau, R. H. 1961. Untersuchungen ueber die Gattung *Drepanopeziza* (Kleb.) Hoehnel. Phytopathol. Z. 43:257-306. (in German with English summary, descriptions of two *Gloeosporium* spp., anamorphs of *Drepanopeziza*)
Sutton (1980) briefly describes and illustrates five species, some of which are anamorphs of *Drepanopeziza*.

GLOEOSPORIUM Desmaz. & Mont.
Coelomycetes   500 spp.
Arx, J. A. von 1970. A revision of the fungi classified as *Gloeosporium*. Bibl. Mycol. 24:1-203. (key to 29 related genera, lists 734 names in *Gloeosporium* and indicates their taxonomic disposition, many determined to belong to *Colletotrichum*, descriptions and illustrations)
In spite of work on these fungi, identification is often difficult. After examination of the type species, many species have been redisposed in other genera (Arx, 1970; Sutton, 1980).

GLOEOTINIA M. Wilson, M. Noble & E. Gray
Helotiales   1 sp.
Hardison, J. R. 1962. Susceptibility of Gramineae to *Gloeotinia temulenta*. Mycologia 54:201-216. (pathogenicity and host range)
Schumacher, T. 1979. *Phialea granigena*, an older name for *Gloeotinia temulenta*. Mycotaxon 8:125-126. (lists synonyms of *G. granigena*)
Wilson, M., Noble, M., and Gray, E. 1954. *Gloeotinia* - a new genus of the Sclerotiniaceae. Trans. Br. Mycol. Soc. 37:29-32. (synonymy and description)
Gloeotinia granigena (Quelet) Schumacher causes blind-seed disease of grass seed.

GLOMERELLA Schrenk & Spauld.
Polystigmatales   3 spp.
Glomerella cingulata (Stoneman) Spauld. & Schrenk, the teleomorph of *Colletotrichum gloeosporioides* (Penz.) Penz. & Sacc., causes anthracnose of many plants (CMI 315) and is one of the most ubiquitous plant pathogenic fungi. *Glomerella tucumanensis* (Speg.) Arx & E. Mueller causes a disease of *Saccharum* (CMI 133). See *Colletotrichum*.

GNOMONIA Ces. & De Not.
Diaporthales   50 spp.
Bolay, A. 1971. Contribution a la connaissance de *Gnomonia comari* Karsten (syn. *G. fructicola* [Arnaud] Fall). Etude taxonomique, phytopathologique et recherches sur sa croissance in vitro. Ber. Schweiz Bot. Ges. 81:398-482. (in French with English summary, key to ten species on herbaceous Rosaceae, descriptions and illustrations)
Monod, M., and Ziegler, P. 1983. Taxonomie numerique de 89 especes Europeennes de la famille des Gnomoniaceae. Mycol. Helv. 1:101-124. (in French with English summary, list of species, phylogenetic analysis)
Barr (1978) includes a key to 35 species of *Gnomonia* and its segregates in North America with descriptions and illustrations. Many of these fungi produce leaf spots and cankers of woody plants and berries. Pathogenic species include: *Gnomonia comari* P. Karst. causing diseases of strawberry plants (CMI 737); *G. manihotis* Punithalingam on *Manihot utilissima* (CMI 136 with a *Sporonema* anamorph); and *G. platani* Kleb. causing plane tree (*Platanus*) scorch. See *Stegophora*.

GODRONIA Moug. & Lev.
Helotiales   26 spp.
Groves, J. W. 1965. The genus *Godronia*. Can. J. Bot. 43:1195-1276. (key to 24 species, descriptions and illustrations, considers *Crumenula* a synonym)
Schlaepfer-Bernard, E. 1968. Beitrag zur Kenntnis der

GODRONIA - cont.
Discomycetengattungen *Godronia, Ascocalyx, Neogodronia,* und *Encoeliopsis.* Sydowia 22:1-56. (in German with English summary, key to related genera and 12 *Godronia* spp., descriptions and illustrations)
Smerlis, E. 1969. Pathogenicity of some species of *Godronia* occurring in Quebec. Plant Dis. Rep. 53:807-810. (reports hosts for seven species)
Godronia species cause cankers on woody plants. The anamorphs have been placed in *Chondropodiella, Fuckelia,* and *Topospora.* See *Ascocalyx.*

GOPLANA Racib.
    Uredinales   10 spp.
Ono, Y., and Hennen, J. F. 1983. Taxonomy of the Chaconiaceous genera (Uredinales). Trans. Mycol. Soc. Jpn. 24:369-402. (key to ten *Goplana* spp. according to host, descriptions)
The biology of *Goplana* species is incompletely known. See Cummins and Hiratsuka (1983).

GRAPHIOLA Poiteau
    Graphiolales   5 spp.
Cole, G. T. 1983. *Graphiola phoenicis*: a taxonomic enigma. Mycologia 75:93-116. (a thorough treatment with illustrations)
Oberwinkler, F., Bandoni, R. J., Blanz, P., Deml, G., and Kisimova-Horovitz, L. 1982. Graphiolales: Basidiomycetes parasitic on palms. Plant Syst. Evol. 140:251-277. (studies developmental morphology, places *Graphiola phoenicis* in Heterobasidiomycetes, illustrations)
Oberwinkler et al. (1982) discuss *Graphiola phoenicis* Poiteau and other *Graphiola* species, all of which occur on palms. Despite detailed study, correct placement of this group is still under discussion.

GRAPHIUM   See *Ophiostoma* and *Pesotum.*

GREENERIA Scribner & Viala
    Coelomycetes   1 sp.
Greeneria uvicola (Berk. & M. A. Curtis) Punithalingam causes bitter rot and ripe rot of grape (CMI 538). Sutton (1980) provides a description.

GREMMENIELLA   See *Ascocalyx.*

GRIFOLA S. F. Gray
    Aphyllophorales   1 sp.
Grifola frondosa (Dickson:Fr.) S. F. Gray (syn. *Polyporus frondosus* (Dicks.:Fr.) Fr.) causes a white rot and butt rot of living trees.

GRIFOLA - cont.
See Aphyllophorales.

GROVESINIA See *Cristulariella*.

GUIGNARDIA Viala & Ravaz
  Dothideales 40 spp.
Aa, H. A. van der 1973. Studies in *Phyllosticta* I. Stud. Mycol. 5:1-110. (includes *Guignardia* species with *Phyllosticta* anamorphs previously placed in *Phyllostictina*, descriptions and illustrations)
Bissett, J. 1986. A note on the typification of *Guignardia*. Mycotaxon 25:519-522. (*Discochora* is the correct name for *Guignardia* species with *Phyllosticta* anamorphs)
Bissett, J. 1986. *Discochora yuccae* sp. nov. with *Phyllosticta* and *Leptodothiorella* synanamorphs. Can. J. Bot. 64:1720-1726. (descriptions and illustrations, discussion of similar fungi on *Yucca*, table comparing morphology of 12 *Guignardia* spp. on Liliales, transfers most species to *Discochora*)
Punithalingam, E. 1974. Studies on Sphaeropsidales in culture II. Mycol. Pap. 136:1-63. (describes and illustrates nine species)
Guignardia species are mostly foliicolous saprophytes or weak parasites. Limits between this genus and *Botryosphaeria* are controversial. Sivanesan (1984) treats 22 species and their synanamorphs in *Phyllosticta* and *Leptodothiorella*. The anamorphs of *Guignardia* species were formerly placed in *Phyllostictina* but are now considered species of *Phyllosticta* for nomenclatural reasons. Important diseases are caused by: *Guignardia aesculi* (Peck) V. B. Stewart, leaf blotch of horse chestnut; *G. bidwellii* (Ellis) Viala & Ravaz, black rot of grapevine (CMI 710); *G. camelliae* (Cooke) E. J. Butler, copper blight of tea; *G. citricarpa* Kiely, citrus black spot (CMI 85); and *G. musae* Racib., freckle of banana (CMI 467).

GYMNOCONIA Lagerh.
  Uredinales 2 spp.
Gymnoconia nitens (Schwein.) F. Kern & Thurston causes orange rust of brambles (*Rubus* spp.) (CMI 201) and *G. peckiana* (Howe) Trott. causes orange rust of *Rubus*. See Cummins and Hiratsuka (1983).

GYMNOSPORANGIUM R. Hedw. ex DC.
  Uredinales 57 spp.
Grasso, V. 1972. [Considerations about nomenclature of some *Gymnosporangium*.] Ann. Accad. Ital. Sci. For. 21:359-367. (in Italian, discussion, host list)
Harada, Y. 1984. Pear and apple rusts in Japan, with special reference to their life cycles and host ranges. Rep. Tottori Mycol. Inst. 22:108-119. (includes table comparing aecial and telial characters of *G. asiaticum* and *G. yamadae*)

GYMNOSPORANGIUM - cont.
Kern, F. D. 1964. Lists and keys to cedar rusts of the world. Mem.
N. Y. Bot. Gard. 10:305-326. (list of *Gymnosporangium* names, keys
to species by aecia, telia, and host, illustrations)
Kern, F. D. 1973. Revised Taxonomic Account of *Gymnosporangium*.
Pennsylvania State University Press, University Park. 134 pp.
(key to 57 species, descriptions, illustrated glossary)
Parmelee, J. A. 1965. The genus *Gymnosporangium* in eastern Canada.
Can. J. Bot. 43:239-267. (key to ten species based on both telial
and aecial characters, descriptions and illustrations)
Parmelee, J. A. 1971. The genus *Gymnosporangium* in western Canada.
Can. J. Bot. 49:903-926. (key to 13 species, descriptions and
illustrations)
Wang, Y.-c., and Guo, L. 1984. [Taxonomic studies on *Gymnosporangium*
in China.] Acta Mycol. Sin. 4:24-34. (in Chinese, key to 15
Gymnosporangium spp., discussion of each species, illustrations)
Gymnosporangium contains rusts that occur mostly on Rosaceae in the
aecial stage and on Cupressaceae in the telial stage. Kern (1973)
provides a key to species with descriptions, illustrations, and
numerous references; Ziller (1974) treats 12 species, with a key
according to hosts, descriptions, and illustrations. Species causing
diseases include: *Gymnosporangium asiaticum* Miyabe ex Yamada,
Japanese pear rust (CMI 541); *G. clavariiforme* (Pers.) DC., European
hawthorn rust (CMI 542, F. Can. 115); *G. clavipes* (Cooke & Peck)
Cooke & Peck, quince rust (CMI 543, F. Can. 116); *G. confusum*
Plowr., medlar rust (CMI 544); *G. connersii* Parmelee,
Crataegus/Juniperus rust (F. Can. 28); *G. cornutum* Arth. ex F. Kern,
Malus/Sorbus rust (F. Can. 117); *G. fuscum* DC., European pear rust
(CMI 545, F. Can. 43); *G. gaeumanii* Zogg ssp. *albertense* Parmelee,
rust of *Juniperus communis* var. *depressa* (F. Can. 118); *G. globosum*
(Farl.) Farl., American hawthorn rust (CMI 546, F. Can. 135); *G.
inconspicuum* F. Kern, rust of *Amelanchier, Crataegus,* and *Juniperus*
(F. Can. 136); *G. juniperi-virginianae* Schwein., American
cedar-apple rust (F. Can. 137); *G. libocedri* (Henn.) F. Kern, rust
of apple and other Rosaceae (CMI 548); *G. nelsonii* Arth., rust of
Amelanchier, Malus, and *Juniperus* (F. Can. 138); *G. nidus-avis*
Thaxter, rust of *Amelanchier* and *Juniperus* (F. Can. 139); *G.
tremelloides* R. Hartig, European apple rust (CMI 549, F. Can. 119);
and *G. yamadae* Miyabe ex Yamada, Japanese apple rust (CMI 550). See
Cummins and Hiratsuka (1983).

HAPLOBASIDION Eriks.
 Hyphomycetes   3 spp.
Ellis, M. B. 1957. *Haplobasidion, Lacellinopsis* and *Lacellina*.
Mycol. Pap. 67:1-15. (key to three *Haplobasidion* spp.,
descriptions and illustrations)
Ellis, M. B. 1972. Dematiaceous Hyphomycetes XI. Mycol. Pap.
131:1-25. (additional information on *H. lelebae*, illustrations)

HAPLOBASIDION - cont.
Ellis (1971) provides a key to three species with descriptions and illustrations. *Haplobasidion musae* M. B. Ellis causes diamond leaf spot of banana (CMI 496) and *H. thalictri* Eriks. causes leaf spots on *Aquilegia* and *Thalictrum*.

HARKNESSIA Cooke
    Coelomycetes   24 spp.
Nag Raj, T. R., and DiCosmo, F. 1981. A monograph of *Harknessia* and *Mastigosporella* with notes on associated teleomorphs. Bibl. Mycol. 80:1-62. (synoptic key to 24 *Harknessia* spp., descriptions and illustrations)
Sutton, B. C. 1971. Coelomycetes. IV. The genus *Harknessia* and similar fungi on *Eucalyptus*. Mycol. Pap. 123:1-46. (key to 16 species, descriptions and illustrations)
Sutton (1980) includes a key, descriptions, and illustrations of 20 species. Nag Raj and DiCosmo (1981) suggest that *Harknessia* spp. have *Cryptosporella* teleomorphs.

HELICOBASIDIUM Pat.
    Auriculariales   10 spp.
Spaulding, P. 1961. Foreign diseases of forest trees of the world. U.S. Dep. Agric., Agric. Handb. 197:1-361. (lists five *Helicobasidium* spp., hosts, distribution)
Valder, P. G. 1958. The biology of *Helicobasidium purpureum* Pat. Trans. Br. Mycol. Soc. 41:283-308. (studies nutritional and physiological factors)
Helicobasidium purpureum Pat., the teleomorph of *Rhizoctonia crocorum* DC.:Fr., causes violet root rot of a number of plants. See Juelich (1984) under Aphyllophorales.

HELMINTHOSPORIUM Link:Fr.
    Hyphomycetes   20 spp.
Alcorn, J. L. 1983. Generic concepts in *Drechslera, Bipolaris* and *Exserohilum*. Mycotaxon 17:1-86. (reviews the basis for the delimitation of these genera of *Helminthosporium*-like fungi)
Ellis, M. B. 1961. Dematiaceous Hyphomycetes. III. Mycol. Pap. 82:1-55. (key to ten species, descriptions and illustrations)
Luttrell, E. S. 1963. Taxonomic criteria in *Helminthosporium*. Mycologia 55:643-674. (discussion of taxonomic characters using ten species in five genera as examples)
Misra, A. P. 1974. *Helminthosporium* Species Occurring on Cereals and other Gramineae. Thirut College of Agriculture, Dholi, Muzaffarpur, India. 289 pp. (descriptions, presents species based on host, many species are included under their old *Helminthosporium* names)
Sisterna, M. N. 1984. Hongos fitopatogenos. Conceptos sobre la segregacion del genero *Helminthosporium* en *Drechslera, Bipolaris*,

HELMINTHOSPORIUM - cont.
y *Exserohilum*. Rev. Fac. Agron. 60:117-120. (in Spanish, review of generic concepts)
Helminthosporium contains species that are mostly lignicolous and generally not plant pathogenic. Ellis (1971) provides a key to ten species with illustrations and descriptions. Most species previously placed in *Helminthosporium* in the broad sense, have been transferred to *Bipolaris, Curvularia, Drechslera*, and *Exserohilum* on the basis of their mode of conidium production, teleomorph, and biology. Domsch et al. (1980) give an overview of the characteristics differentiating the genera that now include former *Helminthosporium* spp. See also Alcorn (1983) for rationale. Species of *Helminthosporium* in the narrow sense have no known teleomorphs. *Helminthosporium oligosporum* (Corda) S. J. Hughes is found on *Tilia americana* (F. Can. 245), *H. solani* Durieu & Mont. causes silver scurf of potato (CMI 166, F. Can. 236), and *H. velutinum* Link:Fr. occurs on dead wood and bark of hardwoods (F. Can. 163).

HEMILEIA Berk. & Broome
   Uredinales 35 spp.
Fulton, R. H., ed. 1984. Coffee Rust in the Americas. American Phytopathological Society, St. Paul, MN. 120 pp. (history, epidemiology, host-parasite interaction, control of *H. vastatrix*)
Gopalkrishnan, K. S. 1951. Notes on the morphology of the genus *Hemileia*. Mycologia 43:271-283. (compares the morphology of 32 species, distinguishes three types of sori)
Schieber, E., and Zentmeyer, G. A. 1984. Coffee rust in the western hemisphere. Plant Dis. 68:89-93. (reviews the disease, illustrations)
Thirumalachar, M. J. 1947. Some noteworthy rusts-II. Mycologia 39:231-248. (discussion of *H. wrightiae*, description of *H. mysorensis* and comparison with other *Hemileia* spp. on Asclepiadaceae)
Hemileia vastatrix Berk. & Broome causes coffee leaf rust (CMI 1) and *H. coffeicola* Maubl. & Roger causes another coffee rust (CMI 2).

HENDERSONIA Berk.
   Coelomycetes 100 spp.
Fripp, Y. J., and Forrester, R. I. 1981. Variation in size of *Hendersonia* conidia on *Eucalyptus* species. Trans. Br. Mycol. Soc. 76:169-172. (defines three host-specific races)
Park, R. F., and Keane, P. J. 1984. Further *Mycosphaerella* species causing leaf diseases of *Eucalyptus*. Trans. Br. Mycol. Soc. 83:93-105. (describes and illustrates two *Hendersonia* anamorphs)
The generic name *Hendersonia* has been formally rejected in favor of *Stagonospora* (Sacc.) Sacc. Many species have not been transferred to a taxonomically appropriate genus. Grove (1937) gives descriptions of many species formerly placed in *Hendersonia*. Teleomorphs are

HENDERSONIA - cont.
placed in *Phaeosphaeria*. See *Wojnowicia*.

HENDERSONINA E. J. Butler
    Coelomycetes   1 sp.
Hendersonina sacchari E. J. Butler causes a disease of sugar cane and is included in Sutton (1980).

HENDERSONULA Speg.
    Coelomycetes   2 spp.
Hendersonula includes stromatic Coelomycetes, many of which are hyperparasites. Sutton (1980) treats *H. australis* Speg. and suggests that *H. toruloidea* Nattrass, the cause of diseases on numerous plants and humans (CMI 274), should be placed in another genus.

HERPOBASIDIUM Lind
    Auriculariales   3 spp.
Oberwinkler, F., and Bandoni, R. 1984. *Herpobasidium* and allied genera. Trans. Br. Mycol. Soc. 83:639-658. (key to related genera, most of which are fern parasites, key to three *Herpobasidium* spp., descriptions and illustrations)
Herpobasidium species are parasitic on ferns. See *Insolibasidium*.

HERPOTRICHIA Fuckel
    Pleosporales   17 spp.
Arx, J. A. von, and Mueller, E. 1984. Notes on some Ascomycetes. Sydowia 37:6-10. (generic treatment, disagrees with Barr, 1984)
Barr, M. E. 1984. *Herpotrichia* and its segregates. Mycotaxon 20:1-38. (separates North American species into five genera, key to species, descriptions and illustrations)
Bose, S. K. 1961. Studies on *Massarina* Sacc. and related genera. Phytopathol. Z. 41:151-213. (key to 12 species, descriptions and illustrations)
Freyer, K. von, and Aa, H. A. van der 1975. [*Pyrenochaeta parasitica* spec. nov., the conidial stage of *Herpotrichia parasitica* (Hartig) E. Rostrup (= *Trichosphaeria parasitica* Hartig)]. Eur. J. For. Pathol. 5:177-182. (in German with English summary, describes and illustrates anamorph of *H. parasitica*)
Pirozynski, K. A. 1972. Microfungi of Tanzania. I. Miscellaneous fungi on oil palm. Mycol. Pap. 129:1-39. (transfers three species to *Herpotrichia*, illustrations)
Samuels, G. J., and Mueller, E. 1978. Life-history studies of Brazilian Ascomycetes 4) Three species of *Herpotrichia* and their *Pyrenochaeta*-like anamorphs. Sydowia 31:157-168. (describes and illustrates three species)
Sivanesan, A. 1971. The genus *Herpotrichia*. Mycol. Pap. 127:1-37. (key to 18 species, descriptions and illustrations)
Subramanian, C. V., and Sekar, G. 1980. *Chaetosphaerulina yasudae* and

HERPOTRICHIA - cont.
its *Xenosporium* anamorph. Kavaka 8:73-77. (describes and illustrates teleomorph and anamorph)
The genus *Herpotrichia* in the broad sense includes species that are parasites, hyperparasites, and saprophytes. Barr (1984) restricts the genus and includes mostly species that are saprophytic on woody or herbaceous substrates. Sivanesan (1984) treats nine species. *Herpotrichia juniperi* (Duby) Petr. causes black snow mold of conifers (CMI 328; Funk, 1981, 1985). *Herpotrichia parasitica* (R. Hartig) Rostr., anamorph *Pyrenochaeta parasitica* Freyer & van der Aa, causes a decay of conifer needles. See *Neopeckia*.

HETEROBASIDION Bref.
    Aphyllophorales   2 spp.
Heterobasidion annosum (Fr.:Fr.) Bref. causes butt rot of living trees, especially conifers (CMI 192 as *Fomes annosus*). See Aphyllophorales.

HETEROPATELLA Fuckel
    Coelomycetes   5 spp.
Heteropatella antirrhini Buddin & Wakef. causes *Antirrhinum* shot hole. Sutton (1980) includes three species. No comprehensive treatment of this genus exists.

HORMOTHECA Bonord.
    Pleosporales   3 spp.
Corlett, M., and Barr, M. E. 1986. *Hormotheca* for species of *Coleroa* with hemispherical ascomata. Mycotaxon 25:255-257. (discussion of generic characters, transfers three *Coleroa* spp. to *Hormotheca*)
Corlett and Barr (1986) place three species in *Hormotheca*. These include: *Hormotheca plantaginis* (Ellis) Corlett & Barr, on *Plantago rugelii* (F. Can. 120 as *Coleroa plantaginis*); *H. robertiani* (Fr.:Fr.) Hoehn., on *Geranium robertianum* (F. Can. 93 as *C. robertiani*); and *H. rubicola* (Ellis & Everh.) Corlett & Barr, on *Rubus* spp. (F. Can. 92 as *C. rubicola*).

HYALODENDRON   See *Ophiostoma*.

HYALOTHYRIDIUM Tassi
    Coelomycetes   4 spp.
Latterell, F. M., and Rossi, A. E. 1984. A new species of *Hyalothyridium* associated with a leaf spot of maize in Latin America. Mycologia 76:506-514. (descriptions and illustrations, discussion of generic concept and disposition of species named in *Hyalothyridium*)
Hyalothyridium maydis Latterell & Rossi causes a leaf spot disease of maize in Latin America.

HYMENELLA Fr.
      Hyphomycetes   20 spp.
Gams, W. 1971. *Cephalosporium*-artige Schimmelpilze (Hyphomycetes). Gustav Fischer Verlag, New York. 262 pp. (in German with summary, glossary, and keys in English, treats *Hymenula cerealis*) The genus *Hymenella* replaces *Hymenula* which is a *nomen illegitimum*. "*Hymenula*" *cerealis* Ellis & Everh. (syn. *Cephalosporium graminearum* Nishikado & Ikata) causes leaf stripe of cereals and grasses (CMI 501). This species has not yet been transferred to an appropriate genus.

HYMENULA   See *Hymenella*.

HYPNOTHECA   See *Monochaetiellopsis*.

HYPOCREA   See *Trichoderma*.

HYPODERMA De Not.
      Rhytismatales   8 spp.
Cannon, P. F., and Minter, D. W. 1983. The nomenclatural history and typification of *Hypoderma* and *Lophodermium*. Taxon 32:572-583. (discusses nomenclature and taxonomy)
Cannon, P. F., and Minter, D. W. 1986. The Rhytismataceae of the Indian subcontinent. Mycol. Pap. 155:1-123. (key to genera of Rhytismataceae, key to four *Hypoderma* spp., descriptions and illustrations)
Darker, G. D. 1932. The Hypodermataceae of conifers. Contrib. Arnold Arbor., Harv. Univ. 1:1-131. (key to seven species, descriptions, illustrations, disposition of names)
Darker, G. D. 1967. A revision of the genera of the Hypodermataceae. Can. J. Bot. 45:1399-1444. (lists four *Hypoderma* spp. and synonyms, indicates disposition of species excluded from *Hypoderma*)
Hunt, R. S., and Ziller, W. G. 1978. Host-genus keys to the Hypodermataceae of conifer leaves. Mycotaxon 6:481-496. (keys only)
Suto, Y. 1983. A new species of *Hypoderma* causing dieback in *Chamaecyparis obtusa* and *Thuja orientalis*. Trans. Mycol. Soc. Jpn. 24:419-424. (describes *H. shimanense*)
Hypoderma rubi (Pers.:Fr.) DC. occurs on *Rubus* spp. and other herbaceous and woody hosts (CMI 781). See *Lophodermium*.

HYPODERMELLA Tubeuf
      Rhytismatales   20 spp.
Darker, G. D. 1967. A revision of the genera of the Hypodermataceae. Can. J. Bot. 45:1399-1444. (includes *H. laricis* and disposition of names)

HYPODERMELLA - cont.
Thyr, B. D., and Shaw, C. G. 1966. Ontogeny of the needle cast fungus, *Hypodermella arcuata*. Mycologia 58:192-200. (describes development of the fungus in host tissue, illustrations)
Most species in this genus have been transferred to other genera (Darker, 1967). See *Lophodermium*.

HYPODERMINA Hoehn.
  Coelomycetes   4 spp.
Hypodermina hartigii Hilitzer causes needle disease of spruce (CMI 794 under *Lirula macrospora*). See *Lirula*.

HYPOXYLON Bull.:Fr.
  Sphaeriales   120 spp.
Abe, Y. 1984. The tissue types of stromata in *Hypoxylon* and its allied genera. Trans. Mycol. Soc. Jpn. 25:399-411. (discusses and describes taxonomic significance of tissue types in *Daldinia, Hypoxylon*, and *Ustulina*.)
Dennis, R. W. G. 1963. Hypoxyloideae of Congo. Bull. Jard. Bot. Brux. 33:317-343. (key to 17 species, descriptions and illustrations)
Greenhalgh, G. N., and Chesters, C. G. C. 1968. Conidiophore morphology in some British members of the Xylariaceae. Trans. Br. Mycol. Soc. 51:57-82. (discusses anamorphs of Xylariaceae, describes and illustrates anamorphs of eight *Hypoxylon* spp.)
Jong, S. C., and Rogers, J. D. 1972. Illustrations and descriptions of conidial states of some *Hypoxylon* species. Wash. Agric. Exp. Stn. Tech. Bull. 71:1-51. (describes and illustrates anamorphs of 18 *Hypoxylon* spp.)
Martin, P. 1967. Studies in the Xylariaceae I. New and old concepts. J. S. Afr. Bot. 33:205-240. (key to genera of Xylariaceae, discusses generic charcters including anamorphs, illustrations)
Martin, P. 1967. Studies in the Xylariaceae II. *Rosellinia* and the Primocinerea section of *Hypoxylon*. J. S. Afr. Bot. 33:315-328. (key to species, descriptions and illustrations)
Martin, P. 1968. Studies in the Xylariaceae III. South African and foreign species of *Hypoxylon* sect. Entoleuca. J. S. Afr. Bot. 34:153-199. (descriptions and illustrations of 26 species with non-colored spores)
Martin, P. 1968. Studies in the Xylariaceae IV. *Hypoxylon*, sections Papillata and Annulata. J. S. Afr. Bot. 34:303-330. (key to 16 species, descriptions and illustrations)
Martin, P. 1968. Studies in the Xylariaceae: V. *Euhypoxylon*. J. S. Afr. Bot. 35:149-206. (key to 25 *Hypoxylon* spp., descriptions and illustrations)
Martin, P. 1976. Studies in the Xylariaceae: Supplementary note. J. S. Afr. Bot. 42:71-83. (nomenclature)
Miller, J. H. 1961. A Monograph of the World Species of *Hypoxylon*.

HYPOXYLON - cont.
University of Georgia Press, Athens. 158 pp. (key to 120 species, descriptions, illustrations, disposition of excluded taxa)
Petrini, L. E., and Mueller, E. 1986. Teleomorphs and anamorphs of European species of *Hypoxylon* (Xylariaceae, Sphaeriales) and allied genera. Mycol. Helv. 1:501-627. (in German with English, French, and Italian summaries, dichotomous and synoptic key to 25 species, descriptions and illustrations)
Petrini, L. E., and Rogers, J. D. 1986. A summary of the *Hypoxylon serpens* complex. Mycotaxon 26:401-436. (key to 12 taxa, provides illustrations, discussion, and information on teleomorph, anamorph, and cultural characters)
Petrini, L. E., and Whalley, A. J. S. 1984. *Hypoxylon macrocarpum* from Australia. Trans. Br. Mycol. Soc. 82:550. (compares with similar species)
Petrini, L., and Petrini, O. 1985. Xylariaceous fungi as endophytes. Sydowia 38:216-234. (keys to European xylariaceous endophytes based on cultural characters, key to ten anamorph genera, keys to species of eight anamorph genera, descriptions of 22 cultures, most of which are *Hypoxylon* spp.)
Rogers, J. D. 1975. A large-spored variety of *Hypoxylon uniapiculatum*. Mycologia 67:1061-1065. (description and illustration)
Rogers, J. D. 1979. The Xylariaceae: systematic, biological and evolutionary aspects. Mycologia 71:1-42. (overview of family, discussion of anamorphs)
Thind, K. S., and Waraitch, K. S. 1976. Xylariaceae of India - III. Indian J. Mycol. & Plant Pathol. 6:113-120. (seven species, descriptions and illustrations)
Whalley, A. J. S., Edwards, R. L., and Francis, S. M. 1983. *Hypoxylon gwyneddii* sp. nov. from Wales. Trans. Br. Mycol. Soc. 81:389-392. (description and illustration)
Whalley, A. J. S., and Greenhalgh, G. N. 1973. Numerical taxonomy of *Hypoxylon*. I. Comparison of classifications of the cultural and the perfect states. Trans. Br. Mycol. Soc. 61:435-454. (results of analysis agree with species concepts of Miller, 1961)
Whalley, A. J. S., and Greenhalgh, G. N. 1973. Numerical taxonomy of *Hypoxylon*. II. A key for the identification of British species of *Hypoxylon*. Trans. Br. Mycol. Soc. 61:455-459. (key to British species using teleomorph and anamorph morphology as well as biochemical and cultural characteristics)
Miller (1961) provides a comprehensive monograph of this genus. Many species are known to have *Geniculosporium* or *Nodulisporium* anamorphs. Species causing diseases include: *Hypoxylon mammatum* (Wahl.) J. H. Miller, canker of *Populus* and other trees (CMI 356); *H. mediterraneum* (De Not.) Ces. & De Not., charcoal disease of *Quercus suber* (CMI 359); *H. rubiginosum* (Pers.:Fr.) Fr., disease of

HYPOXYLON - cont.
Hevea and a weak pathogen of many hosts (CMI 357); and *H. serpens* (Pers.:Fr.) Kickx, wood rot of *Camellia sinensis* (CMI 358).

IDRIELLA P. Nelson & Wilhelm
    Hyphomycetes   15 spp.
Arx, J. A. von 1981. Notes on *Microdochium* and *Idriella*. Sydowia 34:30-38. (compares *Microdochium* and *Idriella*, key to five *Microdochium* spp. and 13 *Idriella* spp., describes cultural characteristics)
Crivelli, P., and Mueller, E. 1983. [*Dothiopeltis arunci* and its anamorph (Ascomycetes)]. Bot. Helv. 93:33-37. (in German with English summary, describes and illustrates anamorph and teleomorph)
Morgan-Jones, G. 1979. Notes on Hyphomycetes XXX. On three species of *Idriella*. Mycotaxon 8:402-410. (describes and illustrates three *Idriella* spp., discusses previously described species)
Nelson, P. E., and Wilhelm, S. 1956. An undescribed fungus causing a root rot of strawberry. Mycologia 48:547-551. (describes and illustrates *I. lunata*)
Idriella was established for *I. lunata* P. Nelson & Wilhelm, the cause of root rot of strawberry. Ellis (1971) includes this species and its *Trichocladium*-like synanamorph.

INONOTUS P. Karst.
    Aphyllophorales   40 spp.
Gilbertson, R. L. 1976. The genus *Inonotus* (Aphyllophorales: Hymenochaetaceae) in Arizona. Mem. N. Y. Bot. Gard. 28:67-85. (key to 11 species, descriptions and illustrations)
Pegler, D. N. 1964. A survey of the genus *Inonotus* (Polyporaceae). Trans. Br. Mycol. Soc. 47:175-195. (key to 40 species, descriptions and illustrations, indicates disposition of excluded species)
Some *Inonotus* species cause white heart rots of living trees. Pathogenic species include *Inonotus tomentosus* (Fr.:Fr.) Teng (syn. *Polyporus tomentosus* Fr.:Fr.) on conifers and causing stand-opening disease of spruce in Canada, and *I. hispidus* (Bull.:Fr.) P. Karst. causing heart rot of ash, oak, and other hardwoods (CMI 193). See *Phellinus* and Aphyllophorales.

INSOLIBASIDIUM Oberwinkler & Bandoni
    Auriculariales   1 sp.
Oberwinkler, F., and Bandoni, R. 1984. *Herpobasidium* and allied genera. Trans. Br. Mycol. Soc. 83:639-658. (key to related genera, description and illustration of *I. deformans* and its anamorph)
Insolibasidium deformans (Gould) Oberwinkler & Bandoni (syn. *Herpobasidium deformans* Gould), anamorph *Glomopsis lonicerae* Donk,

INSOLIBASIDIUM - cont.
is parasitic on *Lonicera*.

ISTHMIELLA Darker
    Rhytismatales   4 spp.
Species in this genus cause needle blight of conifers, especially *Abies* and *Picea*. Funk (1985) provides descriptions and information on three species. *Isthmiella faulii* (Darker) Darker causes needle blight of balsam fir (CMI 793). See Hunt and Ziller (1978) under *Hypoderma*.

ITERSONILIA Derx
    Hyphomycetes   2 spp.
Channon, A. G. 1963. Studies on parsnip canker. I. The causes of the diseases. Ann. Appl. Biol. 51:1-15. (description, illustration, discussion of the diseases)
Olive, L. S. 1952. Studies on the morphology and cytology of *Itersonilia perplexans* Derx. Bull. Torrey Bot. Club 79:126-138. (illustrations)
Sowell, G., Jr., and Korf, R. P. 1960. An emendation of the genus *Itersonilia* based on studies of morphology and pathogenicity. Mycologia 52:934-945. (emended to include *I. perplexans*, compares with similar genera including *Tilletiopsis*)
Itersonilia perplexans Derx causes a leaf spot and canker of parsnip. Channon (1963) differentiates *I. pastinacae* Channon, pathogenic on parsnip, from *I. perplexans*. See the similar genus *Tilletiopsis*.

KABATIELLA Bubak
    Hyphomycetes   15 spp.
Arx, J. A. von 1970. A revision of the fungi classified as *Gloeosporium*. Bibl. Mycol. 24:1-203. (describes and illustrates 20 *Kabatiella* spp.)
Hermanides-Nijhof, E. J. 1977. *Aureobasidium* and allied genera. Stud. Mycol. 15:141-177. (considers *Kabatiella* a synonym of *Aureobasidium*, descriptions and illustrations)
Kabatiella belongs to a group of fungi known as black yeasts. They are ubiquitous and generally saprophytic, however, *Kabatiella caulivora* (Kirchn.) Karakulin causes clover scorch and *K. zeae* Narita & Y. Hiratsuka causes eyespot of corn.

KABATINA R. Schneider & Arx
    Coelomycetes   3 spp.
Butin, H., and Schneider, R. 1976. *Kabatina populi* nov. spec. Phytopathol. Z. 85:39-42. (in German with English summary, description and illustration)
Hermanides-Nijhof, E. J. 1977. *Aureobasidium* and allied genera. Stud. Mycol. 15:141-177. (includes three *Kabatina* spp. in key)
Schneider, R., and Arx, J. A. von 1966. Zwei neue, als Erreger von

KABATINA - cont.
Zweigsterben nachgewiesene Pilze: *Kabatina thujae* n. g., n. sp.
und *K. juniperi* n. sp. Phytopathol. Z. 57:176-182. (in German
with English summary, descriptions and illustrations)
This genus belongs to the black yeasts. One species, *Kabatina thujae*
R. Schneider & Arx, causes diseases of Cupressaceae (CMI 489).
Sutton (1980) also includes *K. thujae* var. *juniperi* (R. Schneider &
Arx) Morelet on *Juniperus* spp.

KERNKAMPELLA Rajendren
    Uredinales   8 spp.
Laundon, G. F. 1975. Taxonomy and nomenclature notes on Uredinales.
Mycotaxon 3:133-161. (includes eight *Kernkampella* spp. with
illustrations, nomenclature, and discussion)
Tyagi, R. N. S. 1974. A critical account of the *Kernkampella*. Indian
J. Mycol. & Plant Pathol. 3:63-66. (discusses generic limits,
  does not consider *Kernkampella* distinct from *Ravenelia*)
Species of this genus are parasitic on Euphorbiaceae. See Cummins and
Hiratsuka (1983).

KHUSKIA H. Hudson
    Sphaeriales   1 sp.
Khuskia oryzae H. Hudson is the teleomorph of *Nigrospora oryzae*
(Berk. & Broome) Petch causing cob and stalk rot of maize and other
diseases of various hosts (CMI 311).

KORDYANA Racib.
    Exobasidiales   5 spp.
Kordyana species are tropical parasites, especially on members of the
Commelinaceae.

KRIEGERIA Bres.
    Auriculariales   1 sp.
Kriegeria eriophori Bres. (syn. *Xenogloea eriophori* Sydow & P. Sydow)
is parasitic on *Scirpus* and *Eriophorum* in Europe and North America.
See Juelich (1984).

KUEHNEOLA Magnus
    Uredinales   8 spp.
Bagyanarayana, G., and Rao, K. N. 1985. A new species of *Kuehneola* on
Gymnosporia montana from India. Can. J. Bot. 63:762-764.
(describes *K. ramacharii* on Celastraceae)
Kuehneola uredinis (Link) Arth. causes yellow stem rust of blackberry
and other *Rubus* spp. (CMI 202, F. Can. 307). See Uredinales,
especially Wilson and Henderson (1966).

LACELLINA Sacc.
    Hyphomycetes   4 spp.
Ellis, M. B. 1957. *Haplobasidion, Lacellinopsis* and *Lacellina*. Mycol. Pap. 67:1-15. (key to three *Lacellina* spp., descriptions and illustrations)
Ellis (1971) includes three species with brief descriptions and illustrations.

LACELLINOPSIS C. V. Subramanian
    Hyphomycetes   5 spp.
Ellis, M. B. 1957. *Haplobasidion, Lacellinopsis* and *Lacellina*. Mycol. Pap. 67:1-15. (key to three *Lacellinopsis* spp., descriptions and illustrations)
Rao, D. 1973. Studies on *Lacellinopsis* Subram. Proc. Ind. Acad. Sci. Sect. B. 77:1-4. (key to five species)
Ellis (1971) includes three species with brief descriptions and illustrations.

LACHNELLULA P. Karst.
    Helotiales   35 spp.
Baral, H. O. 1984. [Taxonomical and ecological studies on the European species of the genus *Lachnellula* Karsten growing on conifers.] Beitr. Kenntn. Pilz. Mittleuropas 1:143-156. (in German with English summary, key to 21 species, descriptions and illustrations)
Dennis, R. W. G. 1962. A reassessment of *Belonidium* Mont. & Dur. Persoonia 2:171-191. (discusses generic concept, lists 19 *Lachnellula* spp.)
Dharne, C. G. 1964. Taxonomic investigations on the discomycetous genus *Lachnellula* Karst. Phytopath. Z. 53:101-144. (key to 17 species, cultural characteristics, descriptions and illustrations)
Galan, R., and Moreno, G. 1985. Dos especies descritas pro J. L. Grelet poco conocidas, *Calycellina albida* (Grelet & Crozals) Galan y Moreno comb. nov. y *Lachnellula robusta* Grelet ex Baral & Matheis, en Espana Peninsular. Crypt. Mycol. 6:21-28. (in Spanish with English summary, descriptions and illustrations)
Manners, J. G. 1953. Studies on larch canker I. The taxonomy and biology of *Trichoscyphella wilkommii* (Hart.) Nannf. and related species. Trans. Br. Mycol. Soc. 36:362-374. (taxonomy and pathology of *T. hahniana* and *T. wilkommii*, both now placed in *Lachnellula*)
Oguchi, T. 1980. Morphological and cultural studies on *Lachnellula* spp. in Hokkaido I. Species on *Abies*. Trans. Mycol. Soc. Jpn. 21:435-447. (includes six species, cultural characteristics)
Oguchi, T. 1981. Morphological and cultural studies on *Lachnellula* spp. in Hokkaido II. Species on *Larix*. Trans. Mycol. Soc. Jpn. 22:165-172. (includes five species, cultural characteristics)

LACHNELLULA - cont.
Oguchi, T. 1981. Pathogenicity of *Lachnellula* spp. collected from conifers in Hokkaido. Trans. Mycol. Soc. Jpn. 22:377-382. (presents pathogenicity of 12 species on three potted host plant species)
Raitviir, A. 1970. Synopsis of the Hyaloscyphaceae. Acad. Sci. Estonian SSR, Inst. Zool. Bot. Tartu. 115 pp. (key to 22 *Lachnellula* spp., describes *L. kamtschatica*)
Dharne (1964) provides a monographic study and Funk (1981) includes descriptions and a table comparing ten *Lachnellula* spp. Larch canker is caused by *Lachnellula wilkommii* (R. Hartig) Dennis (CMI 450 as *Trichoscyphella wilkommii*).

LAETICORTICIUM   See *Corticium*.

LAETIPORUS Murrill
  Aphyllophorales   3 spp.
The common species *Laetiporus sulphureus* (Bull.:Fr.) Murrill (syn. *Polyporus sulphureus* (Bull.:Fr.) Fr.) causes a brown heart rot of many trees (CMI 441). This edible fungus is commonly known as sulphur shelf or chicken-of-the-woods. See Aphyllophorales.

LAETISARIA Burdsall
  Aphyllophorales   1 sp.
Burdsall, H. H., Jr. 1979. *Laetisaria* (Aphyllophorales, Corticiaceae), a new genus for the teleomorph of *Isaria fuciformis*. Trans. Br. Mycol. Soc. 72:419-422. (descriptions and illustrations)
Kaplan, J. D., and Jackson, N. 1983. Red thread and pink patch diseases of turfgrasses. Plant Dis. 67:159-162. (describes complex of species involved)
Stalpers, J. A., and Loerakker, W. M. 1982. *Laetisaria* and *Limonomyces* species (Corticiaceae) causing pink diseases in turf grasses. Can. J. Bot. 60:529-537. (key to species on grasses, taxonomy, nomenclature, disease symptoms, illustrations)
Members of this genus and *Limonomyces* cause red thread diseases of turf grasses.

LASIODIPLODIA Ellis & Everh.
  Coelomycetes   1 sp.
Lasiodiplodia theobromae (Pat.) Griff. & Maubl. is apparently the correct name for the omnivorous tropical pathogen previously placed in *Botryodiplodia* (CMI 519 as *Botryodiplodia theobromae*; Sutton, 1980). Sivanesan (1984) refers to *Botryosphaeria rhodina* (Cooke) Arx as the teleomorph of *L. theobromae* (as *Botryodiplodia theobromae*).

LAUROBASIDIUM Juelich
   Exobasidiales   1 sp.
Laurobasidium lauri (Geyler) Juelich is parasitic on *Laurus*. See Juelich (1984).

LECANOSTICTA Sydow
   Coelomycetes   3 spp.
Evans, H. C. 1984. The genus *Mycosphaerella* and its anamorphs *Cercoseptoria, Dothistroma* and *Lecanosticta* on pines. Mycol. Pap. 153:1-102. (treats *L. acicola* and teleomorph *Mycosphaerella dearnessii*, extensive treatment, descriptions and illustrations)
Lecanosticta acicola (Thuem.) H. Sydow is the anamorph of *Mycosphaerella dearnessii* Barr (syn. *Scirrhia acicola* (Desmaz.) Siggers), the cause of *Lecanosticta* blight or brown spot needle blight of pine (CMI 367). See Sutton (1980), Sivanesan (1984), and Funk (1985).

LEIOSPHAERELLA Hoehn.
   Sphaeriales   8 spp.
Fisher, P. J., and Petrini, O. 1983. Two new Pyrenomycetes from submerged wood. Trans. Br. Mycol. Soc. 81:396-398. (new *Leiosphaerella* sp., description and illustrations)
Katumoto, K. 1981. Notes on some plant-inhabiting Ascomycotina from western Japan (2). Trans. Mycol. Soc. Jpn. 22:37-46. (descriptions of two new species and illustrations)
These fungi occur on living or dead plant parts but do not seem to be pathogenic. Mueller and Arx (1962) treat five species.

LEPTEUTYPA Petr.
   Sphaeriales   4 spp.
Boesewinkel, H. J. 1983. New records of the three fungi causing Cypress canker in New Zealand, *Seiridium cupressi* (Guba) comb. nov. and *S. cardinale* on *Cupressocyparis* and *S. unicorne* on *Cryptomeria* and *Cupressus*. Trans. Br. Mycol. Soc. 80:544-547. (recognizes three *Seiridium* spp. causing Cypress canker, descriptions, illustrations, and table comparing the three species)
Nag Raj, T. R., and Kendrick, B. 1985. *Ellurema* gen. nov., with notes on *Lepteutypa cisticola* and *Seiridium canariense*. Sydowia 38:178-193. (*L. indica* transferred to *Ellurema*, description and illustrations, key to four accepted *Lepteutypa* spp., status of five others indicated)
Shoemaker, R. A., and Mueller, E. 1965. Types of the pyrenomycete genera *Hymenopleella* and *Lepteutypa*. Can. J. Bot. 43:1457-1460. (describes and illustrates the type species *L. fuckelii*)
Swart, H. J. 1973. The fungus causing cypress canker. Trans. Br. Mycol. Soc. 61:71-82. (describes and illustrates *L. cupressi*, discusses anamorph and disease epidemic)

LEPTEUTYPA - cont.
Lepteutypa cupressi (Nattrass, Booth, & Sutton) Swart causes cypress canker (CMI 325 as *Rhynchosphaeria cupressi*). The anamorph is *Seiridium cupressi* (Guba) Boesewinkel (syn. *Cryptostictus cupressi* Guba); some consider this a synonym of *Seiridium unicorne* (Cooke & Ellis) B. C. Sutton (Sutton, 1980).

LEPTODOTHIORELLA Hoehn.
    Coelomycetes   3 spp.
Aa, H. A. van der 1973. Studies in *Phyllosticta* I. Stud. Mycol. 5:1-110. (descriptions and illustrations of species associated with *Guignardia*)
Punithalingam, E. 1974. Studies on Sphaeropsidales in culture. II. Mycol. Pap. 136:1-63. (includes three *Guignardia* spp. with *Leptodothiorella* spermatial states, descriptions and illustrations)
This genus accommodates the spermatial state of *Botryosphaeria* and *Guignardia*. Sutton (1980) includes the *Leptodothiorella* state of *Guignardia sansevieriae* Punithalingam.

LEPTOGRAPHIUM Lagerberg & Melin
    Hyphomycetes   30 spp.
Bertagnole, C. L., Woo, J. Y., and Partridge, A. D. 1983. Pathogenicity of five *Verticicladiella* species to lodgepole pine. Can. J. Bot. 61:1861-1867. (results support five species)
Davidson, R. W. 1978. Staining fungi associated with *Dendroctonus adjunctus* in pines. Mycologia 70:35-40. (describes a new species *L. pyrinum*)
Gambogi, P., and Lorenzini, G. 1977. Conidiophore morphology in *Verticicladiella serpens*. Trans. Br. Mycol. Soc. 69:217-223. (descriptions and illustrations, now placed in *Leptographium*)
Harrington, T. C., and Cobb, F. W., Jr. 1986. Varieties of *Verticicladiella wageneri*. Mycologia 78:562-567. (descriptions and illustrations of three morphological and apparently host-specific variants, this species is now placed in *Leptographium*)
Jooste, W. J. 1978. *Leptographium reconditum* sp. nov. and observations on conidiogenesis in *Verticicladiella*. Trans. Br. Mycol. Soc. 70:152-155. (illustrations of *L. reconditum* and *V. procera*)
Kendrick, W. B. 1962. The *Leptographium* complex. *Verticicladiella* Hughes. Can. J. Bot. 40:771-797. (describes and illustrates seven *Verticicladiella* spp., most of which are now placed in *Leptographium*)
Kendrick, W. B. 1980. The generic concept in Hyphomycetes - a reappraisal. Mycotaxon 11:339-364. (compares generic concepts of *Leptographium, Phialocephala* and *Verticicladiella*, illustrations)
Kendrick, W. B., and Molnar, A. C. 1965. A new *Ceratocystis* and its

LEPTOGRAPHIUM - cont.
Verticicladiella imperfect state associated with the bark beetle
Dryocoetes confusus on *Abies lasiocarpa*. Can. J. Bot. 43:39-43.
(description and illustrations)
Wingfield, M. J. 1985. Reclassification of *Verticicladiella* based on conidial development. Trans. Br. Mycol. Soc. 85:81-93. (reduces *Verticicladiella* to synonymy with *Leptographium*, transfers all species, illustrations)
Wingfield, M. J., and Marasas, W. F. O. 1981. *Verticicladiella alacris*, a synonym of *V. serpens*. Trans. Br. Mycol. Soc. 76:508-510. (descriptions and illustrations, now placed in *Leptographium*)
Wingfield, M. J., and Marasas, W. F. O. 1983. Some *Verticicladiella* species, including *V. truncata* sp. nov., associated with root diseases of pine in New Zealand and South Africa. Trans. Br. Mycol. Soc. 80:231-236. (discusses three species associated with *P. strobus* in New Zealand, description and illustrations of *V. truncata*)
No comprehensive treatment of *Leptographium* exists. The genus includes fungi previously placed in *Verticicladiella* (Wingfield, 1985), many of which are anamorphs of *Ophiostoma* species. Species are often associated with bark beetles, blue stain of timber, and root disease of conifers.

LEPTOMELANCONIUM Petr.
    Coelomycetes   3 spp.
Hunt, R. S. 1985. *Leptomelanconium pinicola* comb. nov. and associated needle blight of pines. Can. J. Bot. 63:1157-1159. (discusses generic concept)
Morgan-Jones, G. 1971. An addition to the genus *Leptomelanconium* Petrak. Can. J. Bot. 49:1011-1013. (key to three species, descriptions and illustrations)
Sutton (1980) includes three species and rejects the placement by Morgan-Jones (1971) of *Gloeocoryneum cinereum* (Dearn.) Weindlmayr in *Leptomelanconium*.

LEPTOSPHAERIA Ces. & De Not.
    Pleosporales   100 spp.
Boerema, G. H., and Loerakker, W. M. 1981. *Phoma piskorzii* (Petrak) comb. nov., the anamorph of *Leptosphaeria acuta* (Fuckel) P. Karst. Persoonia 11:311-315. (description and illustration)
Eriksson, O. 1967. On graminicolous pyrenomycetes from Fennoscandia 2. Phragmosporous and scolecosporous species. Ark. Bot. 6:381-440. (key to 31 phragmosporous species and eight scolecosporous species in segregate genera)
Holm, L. 1952. Taxonomical notes on Ascomycetes. II. The herbicolous Swedish species of the genus *Leptosphaeria* Ces. et De Not. Sven. Bot. Tidskr. 46:18-46. (key to 62 species,

LEPTOSPHAERIA - cont.
descriptions)
Holm, L. 1957. Etudes taxonomiques sur les Pleosporacees. Symb. Bot. Ups. 14:1-188. (in French, key to 35 *Leptosphaeria* spp., descriptions)
Hosford, R. M., Jr. 1978. Effects of wetting period on resistance to leaf spotting of wheat by *Leptosphaeria microscopica* with conidial stage *Phaeoseptoria urvilleana*. Phytopathology 68:908-912. (describes and illustrates both stages, presents table comparing several *Leptosphaeria* spp. associated with wheat leaf spotting)
Koponen, H., and Makela, K. 1975. *Leptosphaeria* s. lat. (*Keissleriella, Paraphaeosphaeria, Phaeosphaeria*) on Gramineae in Finland. Ann. Bot. Fenn. 12:141-160. (includes descriptions and illustrations of 13 species in these three genera)
Lucas, M. T. 1963. Culture studies on Portugese species of *Leptosphaeria*. I. Trans. Br. Mycol. Soc. 46:361-367. (four species)
Lucas, M. T. 1968. Culture studies on Portugese species of *Leptosphaeria*. II. Trans. Br. Mycol. Soc. 51:411-415. (four species)
Lucas, M. T., and Webster, J. 1967. Conidial states of British species of *Leptosphaeria*. Trans. Br. Mycol. Soc. 50:85-121. (describes conidial states and cultural characteristics of 21 species with illustrations)
Mueller, E. 1950. Die schweizerischen Arten der Gattung *Leptosphaeria* und ihrer Verwandten. Sydowia 4:185-319. (in German, keys to 116 species, descriptions and illustrations, includes related genera)
Mueller, E. 1951. Neue, alpine Arten der Gattung *Leptosphaeria*. Sydowia 5:49-55. (in German, five new *Leptosphaeria* spp. from Switzerland)
Shoemaker, R. A. 1984. Canadian and some extralimital *Leptosphaeria* species. Can. J. Bot. 62:2688-2729. (key to 61 species mostly from Canada, key to allied genera, descriptions and illustrations)
Smiley, R. W., and Fowler, M. C. 1984. *Leptosphaeria korrae* and *Phialophora graminicola* associated with *Fusarium* blight of *Poa pratensis* in New York. Plant Dis. 68:440-442. (describes and illustrates fungi, production of ascomata, and disease)
Walker, J., and Smith, A. M. 1972. *Leptosphaeria normari* and *L. korrae* spp. nov., two long-spored pathogens of grasses in Australia. Trans. Br. Mycol. Soc. 58:459-466. (discusses generic concepts, compares with similar species, descriptions and illustrations)
Despite several major works, identification of species in this and related genera is difficult. Holm (1957) and Eriksson (1967) recognize genera segregated from *Leptosphaeria* and Shoemaker (1984) provides a key separating 14 genera allied with *Leptosphaeria*. See

LEPTOSPHAERIA - cont.
Entomodesmium, Nodulisporium, Paraphaeosphaeria, and *Phaeosphaeria.*
Sivanesan (1984) treats 33 *Leptosphaeria* spp. with keys based on anamorphs placed in *Coniothyrium, Diplodina, Phaeoseptoria, Phoma, Scolecosporiella, Septoria,* and *Stagonospora.* Plant pathogenic species include: *Leptosphaeria bicolor* D. Hawksworth, W. Kaiser, & Ndimande, leaf scorch of sugarcane (CMI 771); *L. bondari* Bitancourt & Jenk., areolate spot of citrus in South America; *L. coniothyrium* (Fuckel) Sacc., rose stem canker, cane blight of raspberry and other *Rubus* spp. (CMI 663); *L. maculans* (Desmaz.) Ces. & De Not., blackleg of cabbage and diseases of other Brassicaceae (CMI 331 with anamorph *Phoma lingam*); and *L. taiwanensis* Yen & Chi, leaf blight of sugarcane (CMI 506).

LEPTOSPHAERULINA McAlpine
    Dothideales   7 spp.
Graham, J. H., and Luttrell, E. S. 1961. Species of *Leptosphaerulina* on forage plants. Phytopathology 51:680-693. (key to six species, descriptions, illustrations and pathology)
Irwin, J. A. G., and Davis, R. D. 1985. Taxonomy of some *Leptosphaerulina* spp. on legumes in Eastern Australia. Aust. J. Bot. 33:233-237. (discusses taxonomic characters, includes three species, considers *L. briosiana* and *L. australis* to be synonyms of *L. trifolii*)
Roux, C. 1986. *Leptosphaerulina chartarum* sp. nov., the teleomorph of *Pithomyces chartarum.* Trans. Br. Mycol. Soc. 86:319-323. (descriptions and illustrations)
Leptosphaerulina species cause diseases on Fabaceae and other plants e.g. *L. trifolii* (Rostr.) Petr. causing a leaf spot of legumes and other plants (CMI 146). Graham and Luttrell (1961) provide a useful account as does Barr (1972), who treats five species with references for other species. The only known anamorph belongs in *Pithomyces.*

LEPTOSTROMA Fr.
    Coelomycetes   200 spp.
Minter, D. W. 1980. *Leptostroma* on pine needles. Can. J. Bot. 58:906-917. (key to 11 taxa, descriptions and illustrations)
Minter (1980) places the anamorphs of *Lophodermium* and *Meloderma* in *Leptostroma.* See also *Ploioderma.*

LEPTOTROCHILA P. Karst.
    Helotiales   14 spp.
Schuepp, H. 1959. Untersuchungen ueber *Pseudopezizoideae* sensu Nannfeldt. Phytopathol. Z. 36:213-269. (in German, key to 14 *Leptotrochila* spp., many transferred from *Fabraea*, descriptions and illustrations)
Schuepp (1959) provides a monographic account of this genus and includes *Fabraea* as a synonym. Sutton (1980) includes a description

LEPTOTROCHILA - cont.
and illustration of *Sporonema phacidioides* Desmaz., the anamorph of *Leptotrochila medicaginis* (Fuckel) Schuepp (syn. *Pyrenopeziza medicaginis* Fuckel), a pathogen on leaves of *Medicago*.

LEUCOSTOMA (Nitschke) Hoehn.
    Diaporthales   10 spp.
Dhanvantari, B. N. 1982. Relative importance of *Leucostoma cincta* and *L. persoonii* in perennial canker of peach in southwestern Ontario. Can. J. Plant Pathol. 4:221-225. (pathology)
Hubbes, M. 1960. Systematische und physiologishe Untersuchungen an Valsaceen auf Weiden. Phytopathol. Z. 39:65-93. (in German with English summary, describes and illustrates two *Leucostoma* spp. including pathology)
Kern, H. 1961. Physiologische und systematische Untersuchungen in der Gattung *Leucostoma*. Phytopathol. Z. 40:303-314. (in German with English summary, key to groups)
Urban, Z. 1958. Revise ceskoslovenskych zastpcu rodu *Valsa, Leucostoma*, a *Valsella*. Rozpr. Cesk. Akad. Ved., Rada Mat. Prir. Ved. 68:1-101. (in Czechoslovakian, keys to teleomorphs, detailed descriptions of both teleomorphs and anamorphs, host lists)
Urban (1958) provides the best keys and descriptions but species limits are still unclear and identification is difficult. See the related genus *Valsa*.

LEVEILLULA Arnaud
    Erysiphales   7 spp.
Braun, U. 1980. The genus *Leveillula* - a preliminary study. Nova Hedw. 32:565-583. (key to seven species, descriptions and illustrations, species delimited by conidial characters)
Durrieu, G., and Rostam, S. 1984. Specificite parasitaire et systematique de quelques *Leveillula* (Erysiphaceae). Crypt. Mycol. 5:279-292. (in French with English summary, five species, results of inoculation experiments)
Braun (1980) recognizes seven species in *Leveillula*. *Leveillula taurica* (Lev.) Arnaud, the cause of powdery mildew of cotton and guar (CMI 182), is a variable species. See *Erysiphe* and Erysiphales.

LIBERTELLA Desmaz.
    Coelomycetes   20 spp.
Hinds, T. E. 1981. *Cryptosphaeria* canker and *Libertella* decay of aspen. Phytopathology 71:1137-1145. (description and biology of the disease)
Messner, K., and Sutton, B. C. 1982. *Libertella blepharis*, pathogenic on apple trees of the variety McIntosh. Mycotaxon 14:325-333. (typification, descriptions and illustrations)
Anamorphs of *Cryptosphaeria* and *Diatrypella* have been placed in

LIBERTELLA - cont.
Libertella but Glawe and Rogers (1984 under *Diatrypella*) discuss the difficulty in assigning these anamorphs to form-genera. Sutton (1980) treats three species stating that 70 species have been described and that a revision of *Libertella* species and their teleomorphs is needed. *Libertella blepharis* A. L. Smith causes a disease of apple as does the *Libertella* anamorph of *Cryptosphaeria populina* (Pers.) Sacc. on aspen (Funk, 1981).

LIDOPHIA J. Walker & Sutton
    Pleosporales   1 sp.
Walker, J. 1980. *Gaeumannomyces, Linocarpon, Ophiobolus*, and several other genera of scolecospored Ascomycetes and *Phialophora* conidial states, with a note on hyphopodia. Mycotaxon 11:1-129. (description, discussion of related genera)
Walker, J., and Sutton, B. C. 1974. *Dilophia* Sacc. and *Dilophospora* Desm. Trans. Br. Mycol. Soc. 62:231-241. (transfers *Dilophia graminis* to *Lidophia*, description and illustration, discusses four other *Dilophia* spp.)
Walker and Sutton (1974) proposed *Lidophia* to replace the invalid genus *Dilophia* Sacc. *Dilophospora* is a possible anamorph.

LIMONOMYCES   See *Laetisaria.*

LINOCARPON Sydow & P. Sydow
    Diaporthales   5 spp.
Kobayashi, T. 1970. Taxonomic studies of Japanese Diaporthaceae with special reference to their life-histories. Bull. Gov. For. Exp. Stn. (Jpn.) 226:1-242. (keys to genera and species of Diaporthaceae, key to two *Linocarpon* spp., descriptions and illustrations)
Pirozynski, K. A. 1972. Microfungi of Tanzania. I. Miscellaneous fungi on oil palm. Mycol. Pap. 129:1-39. (describes and illustrates *L. cajani* and *L. elaeidis*)
Walker, J. 1980. *Gaeumannomyces, Linocarpon, Ophiobolus*, and several other genera of scolecospored Ascomycetes and *Phialophora* conidial states, with a note on hyphopodia. Mycotaxon 11:1-129. (describes genus, describes and illustrates the type species *Linocarpon pandani*, discusses other names)
Most species occur on members of the Palmaceae as either endophytes or saprophytes. Barr (1978) provides a key to three species.

LINOSPORA Fuckel
    Diaporthales   4 spp.
Kojwang, H. O., and Kurkela, T. 1984. *Linospora ceuthocarpa* on aspen (*Populus tremula*) in Finland. Karstenia 24:33-40. (biology, histology of perithecial development and infected leaves)
Walker, J. 1980. *Gaeumannomyces, Linocarpon, Ophiobolus*, and several

LINOSPORA - cont.
other genera of scolecospored Ascomycetes and *Phialophora* conidial states, with a note on hyphopodia. Mycotaxon 11:1-129. (discusses genus and disposition of six names)
Barr (1978) includes four species and discusses three other names. Most species occur on fallen leaves.

LIRULA Darker
  Rhytismatales   7 spp.
Darker, G. D. 1967. A revision of the genera of the Hypodermataceae. Can. J. Bot. 45:1399-1444. (includes seven taxa in *Lirula* with synonymy)
Hunt, R. S., and Ziller, W. G. 1978. Host-genus keys to the Hypodermataceae of conifer leaves. Mycotaxon 6:481-496. (keys only)
Lirula species cause needle blight of conifers, including *Lirula macrospora* (R. Hartig) Darker, needle disease of spruce (CMI 794 with anamorph *Hypodermina hartigii* Hilitzer), and *L. nervisequia* (DC.:Fr.) Darker, needle cast of fir (CMI 783). Funk (1985) treats five species. See *Lophodermium*.

LOPHODERMELLA Hoehn.
  Rhytismatales   8 spp.
Millar, C. S. 1986. *Lophodermella* species on pines. Pages 45-55 in: Proc. Recent Research on Conifer Needle Diseases. G. W. Peterson, ed. U.S. Dep. Agric., For. Serv., Gen. Tech. Rep. GTR-WO 50. (describes eight species, discusses biology)
Funk (1985) presents a table comparing seven *Lophodermella* spp. that occur on pines. *Lophodermella conjuncta* (Darker) Darker causes needle blight of pines (CMI 658) and *L. sulcigena* (Rostr.) Hoehn. causes Swedish pine cast (CMI 562). See *Lophodermium*.

LOPHODERMIUM Chev.
  Rhytismatales   50 spp.
Cannon, P. F., and Minter, D. W. 1983. The nomenclatural history and typification of *Hypoderma* and *Lophodermium*. Taxon 32:572-583. (discusses nomenclature and taxonomy)
Cannon, P. F., and Minter, D. W. 1986. The Rhytismataceae of the Indian subcontinent. Mycol. Pap. 155:1-123. (key to genera of Rhytismataceae, key to 23 *Lophodermium* spp., descriptions and illustrations)
Darker, G. D. 1932. The Hypodermataceae of conifers. Contrib. Arnold Arbor., Harv. Univ. 1:1-131. (classic monograph but outdated)
Darker, G. D. 1967. A revision of the genera of the Hypodermataceae. Can. J. Bot. 45:1399-1444. (lists 21 *Lophodermium* spp. and disposition of names for genera in the family)
Hunt, R. S., and Ziller, W. G. 1978. Host-genus keys to the Hypodermataceae of conifer leaves. Mycotaxon 6:481-496. (keys to

LOPHODERMIUM - cont.
species in many genera)
Minter, D. W. 1981. *Lophodermium* on pines. Mycol. Pap. 147:1-54. (dichotomous and synoptic keys to 16 species, descriptions including anamorphs, illustrations)
Minter, D. W. 1986. Some members of the Rhytismataceae (Ascomycetes) on conifer needles from Central and North America. Pages 71-106 in: Proc. Recent Research on Conifer Needle Diseases. G. W. Peterson, ed. U.S. Dep. Agric., For. Serv., Gen. Tech. Rep. GTR-WO 50. (describes and illustrates 27 species in 14 genera)
Minter, D. W., and Millar, C. S. 1980. Ecology and biology of three *Lophodermium* species on secondary needles of *Pinus sylvestris*. Eur. J. For. Pathol. 10:169-181. (presents data on seasonal occurrence and other ecological factors)
Minter, D. W., and Millar, C. S. 1980. A study of three pine inhabiting *Lophodermium* species in culture. Nova Hedw. 32:361-368. (descriptions in culture)
Minter, D. W., and Sharma, M. P. 1982. Three species of *Lophodermium* from the Himalayas. Mycologia 74:702-711. (descriptions and illustrations)
Minter, D. W., Staley, J. M., and Millar, C. S. 1978. Four species of *Lophodermium* on *Pinus sylvestris*. Trans. Br. Mycol. Soc. 71:295-301. (key to species, descriptions and illustrations, distinctions based on stromatic lines and relationship of ascocarp to substrate)
Roux, C., and Lundquist, J. E. 1984. Needle disease found on pines in South Africa, caused by *Lophodermium australe, L. seditiosum*, and *L. indianum*. Plant Dis. 68:628. (new host and distribution records)
Sharma, M. P., and Sharma, R. 1983. The genus *Lophodermium* Chev. (Phacidiales) in India. Bibl. Mycol. 91:157-169. (includes ten species)
Staley, J. M. 1975. The taxonomy of *Lophodermium* on pines, with special reference to problems in North American christmas tree plantations. Mitt. Bundesforschungsanst. Forst Holzwirtsch 108:79-85. (discusses taxonomic problems in *Lophodermium*, neotypifies *L. pinastri*)
Stephan, B. R. 1973. [Studies on the variability of *L. pinastri* I. Variant forms in culture.] Eur. J. For. Pathol. 3:6-12. (in German with English summary, recognizes three cultural types)
Tehon, L. R. 1935. A monographic rearrangement of *Lophodermium*. Ill. Biol. Monogr. 13:1-151. (outdated but useful, comprehensive treatment)
Species of *Lophodermium* and related genera on conifers have been well studied but those on angiosperms have received little attention. Funk (1985) provides a table comparing ten species on conifers. Species causing diseases include: *Lophodermium agathidis* Minter & Hettige on *Agathis australis* (CMI 784); *L. australe* Dearn., needle

LOPHODERMIUM - cont.
cast of pines (CMI 563); *L. baculiferum* Mayr, needle cast of pines (CMI 795); *L. canberrianum* Stahl ex Minter & Millar, needle cast of pines (CMI 564); *L. conigenum* (Brunaud) Hilitzer, needle cast of pines (CMI 565); *L. durilabrum* Darker, needle disease of *Pinus* spp. (CMI 796); *L. indianum* S. Singh & Minter, needle cast of pines (CMI 787); *L. juniperinum* (Fr.:Fr.) De Not., on *Juniperus* spp. (CMI 797); *L. kumaunicum* Minter & P. Sharma, needle cast of pines (CMI 786); *L. maculare* (Fr.:Fr.) De Not., on *Vaccinium uliginosum* (F. Can. 196); *L. mangiferae* Koorders, leaf spot of mango (CMI 798); *L. melaleucum* (Fr.:Fr.) De Not., on *Vaccinium* spp. (F. Can. 195); *L. nitens* Darker, needle cast of pines (CMI 566); *L. orientale* Minter, needle cast of pines (CMI 788); *L. pinastri* (Schrad. ex Hook.) Chev., needle cast of pines (CMI 567); *L. pini-excelsae* Ahmad, needle cast of pines (CMI 785); *L. seditiosum* Minter, Staley & Millar, severe needle cast of pines (CMI 568); *L. sphaerioides* (Albertini & Schwein.:Fr.) Duby, on leaves of *Ledum groenlandicum* (CMI 197); and *L. vagulum* N. Wilson & N. F. Robertson, leaf spot of Chinese rhododendrons (CMI 789). See the anamorph *Leptostroma*.

LOPHOMERUM Ouellette & Magasi
    Rhytismatales   7 spp.
Darker, G. D. 1967. A revision of the genera of the Hypodermataceae. Can. J. Bot. 45:1399-1444. (includes generic description, lists six *Lophomerum* spp. and synonyms)
Ouellette, G. B., and Magasi, L. P. 1966. *Lophomerum*, a new genus of Hypodermataceae. Mycologia 58:275-280. (describes genus and three species)
This genus is segregated from *Lophodermium*. Funk (1985) treats two species on *Abies*. *Lophomerum ponticum* Minter causes leaf spot of rhododendron (CMI 790). See *Lophodermium*.

LOPHOPHACIDIUM Lagerberg
    Helotiales   2 spp.
Corlett, M., and Shoemaker, R. A. 1984. *Lophophacidium dooksii* n. sp., a phacidiaceous fungus on needles of white pine. Can. J. Bot. 62:1836-1840. (descriptions and illustrations)
DiCosmo, F., Nag Raj, T. R., and Kendrick, W. B. 1984. A revision of the Phacidiaceae and related anamorphs. Mycotaxon 21:1-234. (a complete monographic treatment of the family, description and illustration of *L. hyperboreum*)
Funk (1985) and DiCosmo et al. (1984) provide a description of *Lophophacidium hyperboreum* Lagerberg, the cause of snow-blight of spruce.

MACROPHOMA (Sacc.) Berl. & Vogl.
    Coelomycetes   600 spp.
Petrak, F., and Sydow, H. 1927. Die Gattungen der Pyrenomyzeten,

MACROPHOMA - cont.
Sphaeropsideen und Melanconieen. II. Beih. Rep. Spec. Nov.
Regni Veg. 42:1-551. (in German, includes about 200 *Macrophoma*
names, descriptions)
Zachos, D. G., and Tzavella-Klonari, K. 1979. Recherches sur
l'identite et la position systematique du champignon qui provoque
la maladie des olives attribuee au champignon *Macrophoma* ou
Sphaeropsis dalmatica. Ann. Inst. Phytopathol. Benaki 12:59-71.
(in French, descriptions and illustrations)
Sutton (1980) indicates that over 600 species have been described in
this heterogeneous genus. After studying the type, *Macrophoma
sapinea* (Fr.) Petr., he considers *Macrophoma* to be a synónym of
Sphaeropsis thus the numerous *Macrophoma* names must be transferred
to appropriate genera. Petrak and Sydow (1927) indicate the generic
placement for many species.

MACROPHOMINA Petr.
   Coelomycetes 1 sp.
Dhingra, O. D., and Sinclair, J. B., eds. 1977. An Annotated
Bibliography of *Macrophomina phaseolina*, 1905-1975. Univ. Fed.
Vicosa, Brazil and Univ. Illinois, Champaign/Urbana. 244 pp.
(905 citations from world literature)
Goidanich, G. 1947. [A revision of the genus *Macrophomina* Petrak.
Typical species: *Macrophomina phaseolina* (Tassi) G. Goid. n.
comb. nec. *M. phaseoli* (Maubl.) Ashby.] Ann. Sper. Agrar.
1:449-461. (in Italian with English summary, descriptions and
illustrations)
Punithalingam, E. 1982. Conidiation and appendage formation in
Macrophomina phaseolina (Tassi) Goid. Nova Hedw. 36:249-290.
(description and illustrations, reviews taxonomy, compares with
Tiarosporella paludosa)
Macrophomina includes the ubiquitous *M. phaseolina* (Tassi) Goidanich,
the cause of charcoal rot and ashy stem blight of many hosts (CMI
275). This is a synanamorph of *Rhizoctonia bataticola* (Taub.)
Briton-Jones. Dhingra and Sinclair (1977) report many host and
distribution records.

MAGNAPORTHE Krause & R. Webster
   Diaporthales 3 spp.
Arx, J. A. von 1977. On *Yukonia caricis* and some other Ascomycetes.
Rev. Mycol. 41:265-270. (describes and illustrates *M. salvinii*)
Barr, M. E. 1977. *Magnaporthe, Telimenella*, and *Hyponectria*
(Physosporellaceae). Mycologia 69:952-966. (discusses generic
characters)
Krause, R. A., and Webster, R. K. 1972. The morphology, taxonomy, and
sexuality of the rice stem rot fungus, *Magnaporthe salvinii*
(*Leptosphaeria salvinii*). Mycologia 64:103-114. (describes and
illustrates development and morphology, determines

MAGNAPORTHE - cont.
heterothallism)
Ou, S. H. 1985. Rice Diseases. Second Edition. Commonwealth Mycological Institute, Kew, Surrey, England. 380 pp. (describes and discusses *M. salvinii* and its pathology as the cause of stem rot of rice)
Punter, D., Reid, J., and Hopkin, A. A. 1984. Notes on sclerotium-forming fungi from *Zizania aquatica* (wildrice) and other hosts. Mycologia 76:722-732. (describes and illustrates *Sclerotium oryzae*, its alternate state *Nakataea sigmoidea*, and its teleomorph *Magnaporthe salvinii*)
Scott, D. B., and Deacon, J. W. 1983. *Magnaporthe rhizophila* sp. nov., a dark mycelial fungus with a *Phialophora* conidial state, from cereal roots in South Africa. Trans. Br. Mycol. Soc. 81:77-81. (describes, illustrates, and compares *M. rhizophila* with *Gaeumannomyces graminis*)
Tsuda, M., and Ueyama, A. 1978. [Formation of ascigerous stage of *Magnaporthe salvinii* in culture by the crossing of Japanese isolates.] Trans. Mycol. Soc. Jpn. 19:425-431. (in Japanese with English summary, describes teleomorph, compares with ascigerous states of *Pyricularia* spp.)
Magnaporthe contains species parasitic on Poaceae including *M. salvinii* (Cattaneo) Krause & R. Webster, the cause of stem rot of rice (CMI 344 as *Leptosphaeria salvinii*). The anamorph of *M. salvinii* is *Nakataea sigmoidea* (Cavara) K. Hara and the sclerotial state is *Sclerotium oryzae* Cattaneo. Other *Magnaporthe* anamorphs are placed in *Pyricularia*.

MARASMIELLUS Murrill
    Agaricales   200 spp.
Latterell, F. M., and Rossi, A. E. 1984. A *Marasmiellus* disease of maize in Latin America. Plant Dis. 68:728-731. (describes disease called "borde blanco")
Pegler, D. N. 1977. A preliminary agaric flora of East Africa. Kew Bull. Add. Ser. 6:1-615. (key to 14 *Marasmiellus* spp., descriptions and illustrations)
Pegler, D. N. 1983. Agaric flora of the Lesser Antilles. Kew Bull. Add. Ser. 9:1-668. (key to 26 *Marasmiellus* spp., descriptions and illustrations)
Sabet, K. A., Ashour, W. A., Samra, A. S., and Abdel-Azim, O. F. 1970. Root-rot of maize caused by *Marasmiellus inoderma*. Trans. Br. Mycol. Soc. 54:123-126. (describes fungus and disease, illustrations)
Singer, R. 1973. The genera *Marasmiellus, Crepidotus*, and *Simocybe* in the neotropics. Nova Hedw. Beih. 44:1-517. (key to 134 neotropical *Marasmiellus* spp., descriptions and illustrations)
Species of *Marasmiellus* are lignicolous on living or dead wood or on other plant material. This genus is segregated from *Marasmius* based

MARASMIELLUS - cont.
on pileus cuticle morphology. Singer (1973) presents a comprehensive treatment of the genus including *Marasmiellus inoderma* (Berk.) Singer causing a root rot of maize and sugarcane, and *M. semiustus* (Berk. & M. A. Curtis) Singer, another pathogenic species. See *Marasmius*.

MARASMIUS Fr.
    Agaricales   350 spp.
Gilliam, M. S. 1976. The genus *Marasmius* in the northeastern United States and adjacent Canada. Mycotaxon 4:1-144. (a comprehensive monograph with description and illustrations of *M. oreades*)
Ikediugwu, F. E. O. 1984. Leaf blight of *Ficus pumila* caused by a basidiomycete. Trans. Br. Mycol. Soc. 83:501-505. (describes *Marasmius* sp. and the disease, similar to threadblight of rubber)
Lucas, L. T., Warren, T. B., Woodhouse, W. W., Jr., and Seneca, E. D. 1971. *Marasmius* blight, a new disease of American beachgrass. Plant Dis. Rep. 55:582-585. (describes and illustrates *Marasmius* sp. and disease)
Pegler, D. N. 1977. A preliminary agaric flora of East Africa. Kew Bull. Add. Ser. 6:1-615. (key to 42 *Marasmius* spp., descriptions and illustrations)
Pegler, D. N. 1983. Agaric flora of the Lesser Antilles. Kew Bull. Add. Ser. 9:1-668. (key to 41 *Marasmius* spp., descriptions and illustrations)
Singer, R. 1976. Marasmieae (Basidiomycetes-Tricholomataceae). Flora Neotropica, Monograph 17. New York Botanical Garden, Bronx. 347 pp. (key to over 200 *Marasmius* spp., descriptions and illustrations)
Sivanesan, A., and Waller, J. M. 1986. Sugarcane diseases. Phytopathol. Pap. 29:1-88. (describes and illustrates three *Marasmius* spp. that cause diseases on sugarcane)
Most species of *Marasmius* are saprophytic. *Marasmius oreades* (Bolton:Fr.) Fr. causes fairy rings on turf grasses. Pegler (1983) provides a key to 41 species with descriptions and illustrations. Singer (1976) treats over 200 *Marasmius* species. See *Crinipellis* and *Marasmiellus*.

MARAVALIA Arth.
    Uredinales   31 spp.
Cummins, G. B. 1950. The genus *Scopella* of the Uredinales. Bull. Torrey Bot. Club 77:204-213. (key to 13 species, descriptions and illustrations)
Laundon, G. F. 1964. *Angusia* (Uredinales). Trans. Br. Mycol. Soc. 47:327-329. (describes genus and one species)
Ono, Y. 1984. A monograph of *Maravalia* (Uredinales). Mycologia 76:892-911. (key to 31 species, descriptions, host list)
Ono (1984) provides the most recent and comprehensive account of this

MARAVALIA - cont.
tropical rust genus. *Scopella* and *Angusia* are considered synonyms of *Maravalia*.

MARSSONINA Magnus
Coelomycetes   70 spp.

Arx, J. A. von 1970. A revision of the fungi classified as *Gloeosporium*. Bibl. Mycol. 24:1-203. (key to related genera, discusses relationship of *Gloeosporium* and *Marssonina*, describes *Marssonina* spp. previously placed in *Gloeosporium*)

Chuandao, L. 1984. [Two specialized forms of *Marssonina populi* (Lib.) Magn.]. J. Nanjing Inst. For. 4:10-17. (in Chinese with English summary, divides *M. populi* into two formae speciales based on germination characteristics and host range, illustrations)

Gremmen, J. 1965. Three poplar-inhabiting *Drepanopeziza* species and their life-history. Nova Hedw. 9:170-176. (key to three species of *Drepanopeziza* and their *Marssonina* anamorphs, descriptions)

Gremmen, J. 1965. [The *Marssonina* disease of poplar. The occurrence of *Marssonina brunnea* (Ellis & Everh.) Magn. in the Netherlands.] Ned. Bosbouw Tijdschr. 37:196-198. (in Dutch with English summary, anamorph of *Drepanopeziza punctiformis*)

Lentz, P. L. 1950. The genus *Marssonina* on *Quercus* and *Castanea*. Mycologia 42:259-264. (discusses three species, describes and illustrates *M. ochroleuca*)

Rimpau, R. H. 1961. Untersuchungen ueber die Gattung *Drepanopeziza* (Kleb.) Hoehnel. Phytopathol. Z. 43:257-306. (in German with English summary, key to five *Marssonina* spp., describes conidial states and cultural characteristics of species on *Populus, Ribes,* and *Salix*)

Spiers, A. G. 1983. Studies of *Marssonina* and *Drepanopeziza* species pathogenic to poplars. Soil Conserv. Centre, Aokautere, New Zealand. Publ. 4. 41 pp. (recognizes three *Drepanopeziza-Marssonina* holomorphs, key to three *Marssonina* spp., discussion, pathology)

Spiers, A. G., and Hopcroft, D. H. 1983. Ultrastructure of conidial and microconidial ontogeny of *Marssonina* species pathogenic to poplars. Can. J. Bot. 61:3529-3532. (describes annellidic conidiogeny in three *Marsoonina* spp.)

Vegh, I., and Velastegui, J. 1983. Etude comparee de trois especes de *Marssonina* salicicoles. Crypt. Mycol. 4:345-352. (in French with English summary, presents table comparing three species)

Marssonina is in need of critical study and revision. The teleomorphs have been placed in *Diplocarpon* and *Drepanopeziza*. Sutton (1980) treats two species. Many species of *Marssonina* cause leaf spots including: *Marssonina brunnea* (Ellis & Everh.) Magn. on *Populus* spp. (F. Can. 13); *M. castagnei* (Desmaz. & Mont.) Magn. on leaves of *Populus alba* (F. Can. 14); *M. fragariae* (Lib.) Kleb., strawberry leaf scorch (CMI 486 under *Diplocarpon earliana*); *M. ochroleuca*

MARSSONINA - cont.
(Berk. & M. A. Curtis) Lentz, chestnut leaf spot; *M. populi* (Lib.)
Magn. on *Populus* spp. (F. Can. 15); and *M. rosae* (Lib.) Died., black
spot of roses (CMI 485 under *D. rosae*). See *Microdochium*.

MASSARIA De Not.
   Pleosporales 20 spp.
Barr, M. E. 1979. On the Massariaceae in North America. Mycotaxon
 9:17-37. (key to four *Massaria* spp., descriptions and
 illustrations)
Punithalingam, E. 1969. Studies on Sphaeropsidales in culture.
 Mycol. Pap. 119:1-24. (describes and illustrates *M. indica* and
 its *Hyalotropsis* anamorph)
Shoemaker, R. A., and LeClair, P. M. 1975. Type studies of *Massaria*
 from the Wehmeyer collection. Can. J. Bot. 53:1568-1598.
 (accounts for 65 names in *Massaria*, discusses generic concepts)
Massaria species are saprophytic or weakly parasitic on woody
branches. *Massaria inquinans* (Tode:Fr.) De Not. is found on branches
of *Acer*, especially *A. saccharum* (F. Can. 104); *M. lantanae* (G.
Otth) Shoemaker & LeClair occurs on *Viburnum* spp. (F. Can. 105); and
M. zanthoxyli (Peck) Petr. is reported most commonly on *Zanthoxylum
americanum* (F. Can. 106).

MASTIGOSPORIUM Riess
   Hyphomycetes 5 spp.
Austwick, P. K. C. 1954. *Mastigosporium deschampsiae* Jorstad in Great
 Britain. Trans. Br. Mycol. Soc. 37:161-165. (compares with *M.
 album*, both on *Deschampsia*, illustrations)
Bollard, E. G. 1950. Studies on the genus *Mastigosporium*. I.
 General account of the species and their host ranges. Trans. Br.
 Mycol. Soc. 33:250-264. (describes and illustrates four taxa,
 host range, inoculation studies, biology and cultural
 characteristics)
Gunnerbeck, E. 1971. Studies on foliicolous Deuteromycetes. I. The
 genus *Mastigosporium* in Sweden. Sven. Bot. Tidskr. 65:39-52.
 (describes and illustrates five species, excludes three species)
Hughes, S. J. 1951. Studies on microfungi. III. *Mastigosporium,
 Camposporium*, and *Ceratophorum*. Mycol. Pap. 36:1-43. (describes
 and illustrates *M. album*)
Makela, K. 1970. The genus *Mastigosporium* Riess in Finland.
 Karstenia 11:5-22. (describes and illustrates three species, host
 studies)
Species of *Mastigosporium* cause leaf spots of grasses. Gunnerbeck
(1971) provides the most complete treatment of the genus.

MELAMPSORA Castagne
   Uredinales 80 spp.
Hiratsuka, N., and Kaneko, S. 1982. A taxonomic revision of

MELAMPSORA - cont.
Melampsora on willows in Japan. Rep. Tottori Mycol. Inst. 20:1-32. (key to 14 species, descriptions and illustrations)
Hubbes, M., Jeng, R. S., and Zsuffa, L. 1983. *Melampsora* rust in poplar plantations across southern Ontario. Plant Dis. 67:217-218. (describes *M. medusae*, compares with *M. occidentalis* in western Canada)
Kaneko, S., and Hiratsuka, N. 1984. Some criteria in taxonomy of melampsoraceous rust species. Rep. Tottori Mycol. Inst. 22:141-147. (considers urediniospore germ pore, basidiospores, and isolated teliospores in uredinia to be useful taxonomic characters at the species level)
Pinon, J. 1973. [Poplar rusts in France - classification and distribution.] Eur. J. For. Pathol. 3:221-228. (in French with English summary, key to eight species based on urediniospores and paraphyses, illustrations)
Melampsora occurs primarily in temperate regions. No monographic treatment exists except Sydow and Sydow (Vol. III, 1915). Ziller (1974) includes descriptions and illustrations of seven species occurring on western trees. See Cummins and Hiratsuka (1983). Important species include: *Melampsora hirculi* Lindr. on *Saxifraga* spp. (F. Can. 141); *M. larici-populina* Kleb., leaf rust of poplars (CMI 479); *M. lini* (Ehrenb.) Desmaz., flax rust (CMI 51); *M. medusae* Thuem., poplar rust (CMI 480); and *M. ricini* Noronha, castorbean rust (CMI 171).

MELAMPSORIDIUM Kleb.
    Uredinales   6 spp.
Hiratsuka, N. 1958. Revision of Taxonomy of the Puccinastreae, with Special Reference to Species of the Japanese Archipelago. Kasai, Tokyo. 167 pp. (key to four *Melampsoridium* spp., hosts, distribution, references)
Kaneko, S., and Hiratsuka, N. 1981. [Classification of the *Melampsoridium* species based on the position of urediniospore germ pores.] Trans. Mycol. Soc. Jpn. 22:463-473. (in Japanese with English summary, key to four species, illustrations)
Kaneko, S., and Hiratsuka, N. 1982. Taxonomic significance of the urediniospore germ pores in the puccinastraceous and melampsoraceous rust fungi. Trans. Mycol. Soc. Jpn. 23:201-210. (discussion of taxonomic criteria)
Kaneko, S., and Hiratsuka, N. 1983. A new species of *Melampsoridium* on *Carpinus* and *Ostrya*. Mycotaxon 18:1-4. (description and illustrations, comparison with two other species)
Roll-Hansen, F., and Roll-Hansen, H. 1981. *Melampsoridium* on *Alnus* in Europe. *M. alni* conspecific with *M. betulinum*. Eur. J. For. Pathol. 11:77-87. (conclusions based on morphology and inoculation experiments, illustrations)
Singh, S., and Pandy, P. C. 1972. New *Melampsoridium* on *Magnolia*.

MELAMPSORIDIUM - cont.
Trans. Br. Mycol. Soc. 58:342-344. (description and illustrations)
The telial states of five *Melampsoridium* spp. occur on betulaceous hosts. See Cummins and Hiratsuka (1983).

MELANCONIS Tul. & C. Tul.
    Diaporthales   26 spp.
Jensen, J. D. 1984. *Melanconis marginalis* from northern Idaho. Mycotaxon 20:275-281. (defines related but distinct populations in Idaho as *M. marginalis* and *M. alni*, illustrations)
Kobayashi, T. 1971. [Physiology of *Melanconis* spp. causing dieback disease of broad-leaved trees and variation in the size of conidia produced on various media.] J. Jpn. For. Soc. 53:57-67. (in Japanese with English summary, presents table comparing six species with anamorph and cultural characteristics)
Wehmeyer, L. E. 1941. A revision of *Melanconis, Pseudovalsa, Prosthecium*, and *Titania*. Univ. Mich. Stud. Sci. Ser. 14:1-161. (key to 27 species, descriptions and illustrations including conidial states)
The most comprehensive account of *Melanconis* is provided by Wehmeyer (1941). Barr (1978) discusses the genus and includes *M. stilbostoma* (Fr.) Tul. on *Betula* spp. *Melanconis juglandis* (Ellis & Everh.) Groves causes dieback of *Juglans*. The anamorphs are placed in *Melanconium*.

MELANCONIUM Link:Fr.
    Coelomycetes   50 spp.
Sutton, B. C. 1964. *Melanconium* Link ex Fries. Persoonia 3:193-198. (typifies genus, describes and illustrates *M. atrum*)
Melanconium species are anamorphs of *Melanconis*. Sutton (1980) includes the type species *Melanconium atrum* Link, on *Fagus sylvatica*, and includes several names in other genera. A revision is needed. Funk (1981) treats *Melanconium bicolor* Nees and *M. sphaeroideum* Link:Fr. Grove (1937) describes 11 *Melanconium* spp. arranged by hosts. See *Phaeocytostroma*.

MELANODOTHIS R. Arnold
    Dothideales   1 sp.
Melanodothis caricis R. Arnold is parasitic on ovaries of *Carex* spp. in Asia and North America (F. Can. 30). See Barr (1972) and Sivanesan (1984).

MELANOPSICHIUM G. Beck
    Ustilaginales   6 spp.
Spooner, B. M. 1985. *Melanopsichium* (Ustilaginales), a genus new to the British Isles. Trans. Br. Mycol. Soc. 85:540-544. (describes and illustrates *M. nepalense* on *Polygonum*, reviews other species)

MELANOPSICHIUM - cont.
Species of this smut genus cause galls on *Polygonum*. Species also occur on Lauraceae and Poaceae.

MELANOTAENIUM de Bary
    Ustilaginales   25 spp.
Whitehead, M. D., and Thirumalachar, M. J. 1953. Notes on two smuts reported from the United States. Bull. Torrey Bot. Club 80:498-499. (description and discussion of *M. euphorbiae*)
Zambettakis, C., and Joly, P. 1972. Application de traitements numeriques a la systematique des Ustilaginales. I. Le genre *Melanotaenium*. Bull. Trimest. Soc. Mycol. Fr. 88:193-208. (in
    French, synoptic key to 25 species, key to species by host)
Melanotaenium selaginellae Henn. & Nyman and *M. oreophilum* Sydow are parasites of *Selaginella* in Europe. *Melanotaenium euphorbiae* (L. Lenz) Whitehead & Thirumalachar occurs on *Euphorbia* spp. in Louisiana.

MELODERMA Darker
    Rhytismatales   1 sp.
Funk (1985) includes *Meloderma desmazierii* (Duby) Darker, the cause of needle blight of pines infecting mostly *Pinus monticola* in western North America (CMI 569) and indicates that the anamorph is *Leptostroma strobicola* Hilitzer. See *Leptostroma* and *Lophodermium*.

MEMNONIELLA   See *Stachybotrys*.

MERIA Vuill.
    Hyphomycetes   3 spp.
Barron, G. L. 1977. The Nematode-destroying Fungi. Topics in
    Mycobiology I. Canad. Biol. Publ. Ltd., Guelph, Ontario. 140
    pp. (describes and illustrates *M. coniospora*)
Meria laricis Vuill. causes a needle cast of *Larix* spp. in North America and Europe (Funk, 1985). See *Rhabdocline*.

METACOLEROA Petr.
    Pleosporales   1 sp.
Metacoleroa dickiei (Berk. & Broome) Petr. occurs on living and dead leaves of *Linnaea borealis* in North America (F. Can. 114; Barr, 1968).

MICROASCUS   See *Scopulariopsis*.

MICROCYCLUS Sacc.
    Dothideales   18 spp.
Anahosur, K. H. (1970)1971. Ascomycetes of Coorg (India) III.
    Sydowia 24:169-172. (describes spermatial state of *M. indicus*)
Holliday, P. 1970. South American leaf blight (*Microcyclus ulei*) of

MICROCYCLUS - cont.
Hevea brasiliensis. Phytopathol. Pap. 12:1-31. (complete account
of the disease and the pathogen, illustrations)
Sivanesan, A. 1970. *Parmulariopsella buseracearum* gen. et sp. nov.
and *Microcyclus placodisci* sp. nov. Trans. Br. Mycol. Soc.
55:509-514. (description and illustration)
Sivanesan, A. 1975. New Ascomycetes and some revisions. Trans. Br.
Mycol. Soc. 65:19-27. (describes and illustrates two species)
Parasitic species of *Microcyclus* include *M. ulei* (Henn.) Arx causing
South American leaf blight of rubber (CMI 225). Anamorphs are placed
in *Aposphaeria, Fusicladium*, and *Pazschkeella*. Mueller and Arx
(1962) treat 11 *Microcyclus* spp. and Sivanesan (1984) includes two
species.

MICRODIPLODIA Allesch.
    Coelomycetes   27 spp.
Zambettakis, C. 1954. Recherches sur la systematique des
Sphaeropsidales-Phaeodidymae. Bull. Trimest. Soc. Mycol. Fr.
70:219-350. (in French, key to 23 genera, describes and lists
synonyms of 38 *Microdiplodia* spp., nomenclature generally not
accepted)
This genus is not well known. Grove (1937) includes 19 species.

MICRODOCHIUM Sydow
    Hyphomycetes   7 spp.
Arx, J. A. von 1981. Notes on *Microdochium* and *Idriella*. Sydowia
34:30-38. (compares *Microdochium* and *Idriella*, key to five
Microdochium spp. and 13 *Idriella* spp., describes cultural
characteristics)
Arx, J. A. von 1985. Notes on *Monographella* and *Microdochium*. Trans.
Br. Mycol. Soc. 83:373-374. (transfers two *Fusarium* spp. to
Microdochium)
Galea, V. J., Price, T. V., and Sutton, B. C. 1986. Taxonomy and
biology of the lettuce anthracnose fungus. Trans. Br. Mycol.
Soc. 86:619-628. (describes and illustrates *Microdochium
panattonianum*)
Gams, W., and Mueller, E. 1980. Conidiogenesis of *Fusarium nivale* and
Rhynchosporium oryzae and its taxonomic implications. Neth. J.
Plant Pathol. 86:45-53. (describes annellidic conidiogenesis)
Harris, O. C. 1985. *Microdochium fusarioides* sp. nov. from oospores
of *Phytophthora syringae*. Trans. Br. Mycol. Soc. 84:358-361.
(description and illustrations)
Moline, H. E., and Pollack, F. G. 1976. Conidiogenesis of *Marssonina
panattoniana* and its potential as a serious postharvest pathogen
of lettuce. Phytopathology 66:669-674. (describes and
illustrates the disease and causal fungus with annellidic
conidiogenesis, now placed in *Microdochium*)
Mouchacca, J., and Samson, R. A. 1973. Deux nouvelles especes du

MICRODOCHIUM - cont.
genre *Microdochium* Sydow. Rev. Mycol. 37:267-275. (in French, describes and illustrates two species)
Parkinson, V. O. 1980. Cultural characteristics of the rice leaf scald fungus, *Rhynchosporium oryzae*. Trans. Br. Mycol. Soc. 74:509-514. (describes morphology, illustrations, now placed in *Microdochium*)
Parkinson, V. O., Sivanesan, A., and Booth, C. 1981. The perfect state of the rice leaf scald fungus and the taxonomy of both the perfect and imperfect states. Trans. Br. Mycol. Soc. 76:59-69. (compares *Monographella* spp. and their *Gerlachia* anamorphs which are now placed in *Microdochium*, descriptions and illustrations)
Samuels, G. J., and Hallett, I. C. 1983. *Microdochium stoveri* and *Monographella stoveri*, new combinations for *Fusarium stoveri* and *Micronectriella stoveri*. Trans. Br. Mycol. Soc. 81:473-483. (descriptions and illustrations)
Sutton, B. C., and Hodges, C. S., Jr. 1976. Eucalyptus microfungi: *Microdochium* and *Phaeoisaria* species from Brazil. Nova Hedw. 27:215-229. (key to eight *Microdochium* spp., descriptions and illustrations)
Sutton, B. C., Pirozynski, K. A., and Deighton, F. C. 1972. *Microdochium* Syd. Can. J. Bot. 50:1899-1907. (describes and illustrates five species)
Thomas, M. D. 1984. Dry-season survival of *Rhynchosporium oryzae* in rice leaves and stored seeds. Mycologia 76:1111-1113. (biology, now placed in *Microdochium*)
Species of *Microdochium* are separated from *Fusarium* on the basis of their annellidic or sympodially proliferating, holoblastic conidiogenous cells and non-hypocrealean teleomorphs belonging to *Monographella*. *Microdochium* is an earlier name for *Gerlachia*. The taxonomy of this genus is presented by Arx (1985), Harris (1985), and Samuels and Hallett (1983). *Microdochium nivale* (Fr.) Samuels & Hallett is the anamorph of *Monographella nivalis* (Schaffnit) E. Mueller which causes snow mold of turf (CMI 309 as *Micronectriella nivalis*). *Microdochium oryzae* (Hashioka & Yokogi) Samuels & Hallett (syn. *Rhynchosporium oryzae* Hashioka & Yokogi) causes leaf scald of rice (CMI 729 under *Monographella albescens*). *Microdochium panattonianum* (Berl.) Sutton, Galea & Price in Galea, Price & Sutton (1986) (syn. *Marssonina panattoniana* (Berl.) Magnus) is the cause of lettuce anthracnose.

MICRONECTRIELLA   See *Monographella*.

MICROPERA   See *Foveostroma*.

MICROSPHAERA Lev.
    Erysiphales   60 spp.
Braun, U. 1981. Miscellaneous notes on the Erysiphaceae (II). Feddes

MICROSPHAERA - cont.
Repert. 92:499-513. (descriptions and illustrations of seven *Microsphaera* spp.)
Braun, U. 1982. Descriptions of new species and combinations in *Microsphaera* and *Erysiphe*. Mycotaxon 14:369-374. (describes and illustrates six taxa near *M. penicillata*)
Braun, U. 1983. Descriptions of new species and combinations in *Microsphaera* and *Erysiphe*. (III). Mycotaxon 16:417-424. (describes and illustrates six taxa in *Microsphaera*)
Braun, U. 1983. Descriptions of new species and combinations in *Microsphaera* and *Erysiphe*. (IV). Mycotaxon 18:113-129. (key to six *Microsphaera* spp. on *Berberis*, descriptions)
Braun, U. 1984. Descriptions of new species and combinations in *Microsphaera* and *Erysiphe*. (V). Mycotaxon 19:375-383. (describes and illustrates three *Microsphaera* spp.)
Braun, U. 1984. Descriptions of new species and combinations in *Microsphaera* and *Erysiphe*. (VI). Mycotaxon 20:491-498. (describes and illustrates two *Microsphaera* spp.)
Braun, U. 1984. A short survey of the genus *Microsphaera* in North America. Nova Hedw. 39:211-243. (keys to 47 species by morphology and by host family, notes, illustrations)
Braun, U. 1985. The *Erysiphe-Microsphaera* complex on Fabaceae. Zbl. Mikrobiol. 140:393-417. (key to 29 species in the complex, descriptions and illustrations)
Nomura, Y., and Tanda, S. 1983. [Notes on the powdery mildew fungus on *Deutzia* in Japan.] Trans. Mycol. Soc. Jpn. 24:201-204. (in Japanese with English summary, describes and illustrates *M. deutziae*)
Sivanesan, A. 1971. A new *Microsphaera* species on begonia. Trans. Br. Mycol. Soc. 56:304-306. (description and illustration)
Yu, Y.-n., and Lai, Y.-q. 1982. [Taxonomic studies on the genus *Microsphaera* of China. IV. New and known species of *Microsphaera* on family Fagaceae.] J. North-eastern For. Inst., China 4:24-36. (in Chinese with English summary, key to four *Microsphaera* spp., descriptions and illustrations)
Yu, Y.-n., and Lai, Y.-q. 1983. [Taxonomic studies on the genus *Microsphaera* of China. V. New and known species of *Microsphaera* on family Caprifoliaceae.] Acta Mycol. Sin. 2:89-95. (in Chinese with English summary, key to five species, descriptions and illustrations)
Microsphaera species cause powdery mildew diseases of many hosts including: *Microsphaera alphitoides* Griffon & Maubl., oak mildew; *M. grossulariae* (Wallr.) Lev., European mildew of gooseberry and currant (CMI 252); and *M. penicillata* (Wallr.:Fr.) Lev., powdery mildew of alder and lilac (CMI 183). See *Erysiphe* and Erysiphales.

MICROSPHAEROPSIS Hoehn.
    Coelomycetes   10 spp.
Laundon, G. 1984. *Diplodia pittospororum* and *Diplodia pittospori*. Trans. Br. Mycol. Soc. 83:164-166. (transfers *D. pittospororum* to *Microsphaeropsis*, description and illustrations)
Morgan-Jones, G. 1974. Concerning some species of *Microsphaeropsis*. Can. J. Bot. 52:2575-2579. (describes and illustrates *M. centaureae*, transfers three *Coniothyrium* spp. to *Microsphaeropsis*)
Sutton, B. C. 1971. Coelomycetes. IV. The genus *Harknessia* and similar fungi on *Eucalyptus*. Mycol. Pap. 123:1-46. (describes and illustrates two *Microsphaeropsis* spp., discusses generic concept)
Sutton, B. C. 1974. Miscellaneous Coelomycetes on *Eucalyptus*. Nova Hedw. 25:161-172. (describes and illustrates three *Microsphaeropsis* spp.)
Sutton, B. C. 1980. *Microsphaeropsis clidemiae* sp. nov., associated with leaf lesions on *Clidemia hirta*. Trans. Br. Mycol. Soc. 74:645-647. (description and illustration)
Sutton (1980) includes nine *Microsphaeropsis* species, many of which were formerly placed in *Coniothyrium*. *Microsphaeropsis centaureae* Morgan-Jones causes a disease of *Centaurea* spp., *M. clidemiae* Sutton causes a leaf spot of *Clidemia hirta*, and *M. pittospororum* (Sacc.) Laundon is a weak pathogen of *Pittosporum*. Many species of *Microsphaeropsis* occur on *Eucalyptus*.

MOESZIOMYCES K. Vanky
    Ustilaginales   2 spp.
Rao, K. V. S., and Thakur, R. P. 1983. *Tolyposporium penicillariae*, the causal agent of pearl millet smut. Trans. Br. Mycol. Soc. 81:597-603. (describes and illustrates cultural characteristics, considered a synonym of *M. bullatus* by Vanky, 1986)
Vanky, K. 1977. *Moesziomyces*, a new genus of Ustilaginales. Bot. Notiser 130:131-135. (describes *Moesziomyces* for four *Tolyposporium* spp., illustrations)
Vanky, K. 1986. The genus *Moesziomyces* (Ustilaginales). Nord. J. Bot. 6:67-74. (*Moesziomyces* emended to include only two species, three original species considered synonyms of *M. bullatus*)
This genus contains species transferred from *Tolyposporium* including *Moesziomyces bullatus* (J. Schroet.) K. Vanky, pearl millet smut (CMI 77 as *Tolyposporium penicillariae*).

MONASCUS Tiegh.
    Pezizales   3 spp.
Bridge, P. D., and Hawksworth, D. L. 1985. Biochemical tests as an aid to the identification of *Monascus* species. Lett. Appl. Microbiol. 1:25-29. (describes use of API ZYM enzyme testing strips)

MONASCUS - cont.
Hawksworth, D. L., and Pitt, J. I. 1983. A new taxonomy for *Monascus* species based on cultural and microscopical characters. Aust. J. Bot. 31:51-61. (key to three species, descriptions and illustrations)
Monascus ruber Tiegh. causes a light red discoloration of silage. Some species are used in the production of Asian fermented foods.

MONILIA Bonord.
    Hyphomycetes   2 spp.
Arx, J. A. von 1981. On *Monilia sitophila* and some families of Ascomycetes. Sydowia 34:13-29. (restricts *Monilia* to anamorphs of *Monilinia*, transfers *Monilia sitophila* to *Chrysonilia*, discusses related genera)
Korf, R. P., and Kohn, L. M. 1979. Later starting point blues. I. *Monilia fructigena*. Mycotaxon 9:521-522. (nomenclature)
Monilia species are anamorphs of *Monilinia* which are parasitic on Rosaceae and Ericaceae. See *Monilinia* and *Moniliophthora*.

MONILINIA Honey
    Helotiales   30 spp.
Batra, L. R. 1979. First authenticated North American record of *Monilinia fructigena*, with notes on related species. Mycotaxon 8:476-484. (describes three species with geographical information)
Batra, L. R. 1983. *Monilinia vaccinii-corymbosi* (Sclerotiniaceae): its biology on blueberry and comparison with related species. Mycologia 75:131-152. (key to ten species on Ericaceae, descriptions and illustrations, information on biology, ecology, and pathogenicity)
Batra, L. R., and Harada, Y. 1986. A field record of apothecia of *Monilinia fructigena* in Japan and its significance. Mycologia 78:913-917. (describes micromorphology and cultural characteristics of Japanese specimens)
Boesewinkel, H. J., and Corbin, J. B. 1970. A new record of brown rot, *Sclerotinia (Monilinia) laxa*, in New Zealand. Plant Dis. Rep. 54:504-506. (compares with *M. fructicola*, description and illustration)
Byrde, R. J. W., and Willetts, H. J. 1977. The Brown Rot Fungi of Fruit: Their Biology and Control. Pergamon Press, Oxford. 171 pp. (a complete account of fruit-rotting *Monilinia* spp., one chapter on taxonomy and nomenclature, presents table comparing three species)
Gjaerum, H. B. 1969. Some fruit inhabiting Sclerotinias in Norway. Friesia 9:18-28. (describes and illustrates seven species)
Harada, Y. 1977. Studies on the Japanese species of *Monilinia* (Sclerotiniaceae). Bull. Fac. Agric. Hirosaki Univ. 27:30-109. (in English and Japanese, key to nine species, discusses

MONILINIA - cont.
taxonomic characters especially in culture, illustrations)
Penrose, L. J., Tarran, J., and Wong, A.-L. 1976. First record of *Sclerotinia laxa* Aderh. & Ruhl. in New South Wales: Differentiation from *S. fructicola* (Wint.) Rehm by cultural characteristics and electrophoresis. Aust. J. Agric. Res. 27:547-556. (illustrations)
Sonoda, R. M., Ogawa, J. M., and Manji, B. T. 1982. Use of interactions of cultures to distinguish *Monilinia laxa* from *M. fructicola*. Plant Dis. 66:325-326. (cultures on oatmeal agar, illustrations)
Willetts, H. J. 1969. Cultural characteristics of the brown rot fungi (*Sclerotinia* spp.). Mycologia 61:332-339. (descriptions and illustrations)
Willetts, H. J., Byrde, R. J. W., Fielding, A. H., and Wong, A.-L. 1977. The taxonomy of the brown rot fungi (*Monilinia* spp.) related to their extracellular cell wall-degrading enzymes. J. Gen. Microbiol. 103:77-83. (identifies species using enzymes)
Willetts, H. J., and Harada, Y. 1984. A review of apothecial production by *Monilinia* fungi in Japan. Mycologia 76:314-325. (presents table comparing nine species)
Monilinia species occur primarily on Ericaeae and Rosaceae including three important species on apples, pears, and stone fruits: *Monilinia fructicola* (Wint.) Honey, brown rot of stone fruits (CMI 616 as *Sclerotinia fructicola*, F. Can. 38); *M. fructigena* (Aderhold & Ruhland) Honey, brown rot of stone fruits (CMI 617 as *Sclerotinia fructigena*); and *M. laxa* (Aderhold & Ruhland) Honey, brown rot, blossom wilt, and twig blight of stone fruit (CMI 619 as *Sclerotinia laxa*). Anamorphs are placed in *Monilia*.

MONILIOPHTHORA H. Evans, Stalpers, R. A. Samson & Benny
    Hyphomycetes   1 sp.
Evans, H. C. 1981. Pod rot of cacao caused by *Moniliophthora* (*Monilia*) *roreri*. Phytopathol. Pap. 24:1-44. (describes fungus and disease, illustrations)
Evans, H. C., Stalpers, J. A., Samson, R. A., and Benny, G. L. 1978. On the taxonomy of *Monilia roreri*, an important pathogen of *Theobroma cacao* in South America. Can. J. Bot. 56:2528-2532. (describes new basidiomycete genus for *M. rorei* based on conidiogeny and presence of dolipore septa, descriptions and illustrations)
Moniliophthora roreri (Cif.) H. Evans et al. causes frosty pod rot or *Monilia* rot of cacao (CMI 226 as *Monilia roreri*). Based on the mode of conidiogeny and the presence of dolipore septa, Evans et al. (1978) reveal that this fungus is a basidiomycete and must be excluded from *Monilia*, a hyphomycete genus that includes anamorphs of the ascomycete *Monilinia*. Evans (1981) provides a monograph of the disease and the causal fungus.

MONOCHAETIA (Sacc.) Allesch.
  Coelomycetes   14 spp.
Graves, A. A., and Witcher, W. 1971. *Monochaetia* canker of Arizona cypress and redcedar in South Carolina. Plant Dis. Rep. 55:810-813. (description and pathogenicity)
Guba, E. F. 1961. Monograph of *Monochaetia* and *Pestalotia*. Harvard University Press, Cambridge, MA. 342 pp. (key to 41 *Monochaetia* spp., descriptions and illustrations)
Nag Raj, T. R. 1985. Redisposals and redescriptions in the *Monochaetia-Seiridium*, *Pestalotia-Pestalotiopsis* complexes. I. The correct name for the type species of *Pestalotiopsis*. Mycotaxon 22:43-51. (key to seven genera)
Nag Raj, T. R. 1985. Redisposals and redescriptions in the *Monochaetia-Seiridium*, *Pestalotia-Pestalotiopsis* complexes. III. *Monochaetia ilicina* (Sacc.) comb. nov. Mycotaxon 22:64-70. (presents table comparing *M. ilicina* and *M. saccardiana*)
Nag Raj, T. R. 1985. Redisposals and redescriptions in the *Monochaetia-Seiridium*, *Pestalotia-Pestalotiopsis* complexes. IV. On *Monochaetia miersi*. Mycotaxon 22:71-75. (places *M. miersi* in *Pestalotiopsis*)
Nag Raj, T. R. 1986. Redisposals and redescriptions in the *Monochaetia-Seiridium*, *Pestalotia-Pestalotiopsis* complexes. V. *Monochaetia alnea* and *M. berberidicola*. Mycotaxon 26:187-198. (treats two *Monochaetia* spp., transfers *M. berberidicola* to *Seimatosporium*, descriptions and illustrations)
Sutton, B. C. 1969. Forest microfungi. III. The heterogeneity of *Pestalotia* de Not. section *Sexloculatae* Klebahn sensu Guba. Can. J. Bot. 47:2083-2094. (generic concept expanded, *M. karstenii* and *M. nattrassii* new combinations)
Yokoyama, T. 1975. A new species of *Monochaetia* with arthroconidia. Trans. Br. Mycol. Soc. 65:499-503. (describes and illustrates *M. dimorphospora*)
Sutton (1980) includes a key to ten taxa with descriptions and illustrations. He treats some *Monochaetia* taxa in *Seimatosporium* or *Seiridium*.

MONOCHAETIELLA Castellani
  Coelomycetes   1 sp.
Castellani, E. 1943. *Monochaetiella*, un nuovo genere rappresentante un termine di passagio tra i Melanconiale e gli Sferopsidali. Nuovo Giornale Bot. Ital. 21:1-3. (in Italian, description and illustration)
Monochaetiella hyparrheniae Castellani causes leaf lesions on *Hyparrhenia* spp. (Sutton, 1980). See *Monochaetiellopsis*.

MONOCHAETIELLOPSIS Sutton & DiCosmo
Coelomycetes   2 spp.
Sutton, B. C., and DiCosmo, F. 1977. A revision of *Monochaetiella* (Deuteromycotina). Can. J. Bot. 55:2535-2543. (discussion of *Monochaetiella* and transfer of two species to *Monochaetiellopsis*, key, descriptions and illustrations)
Monochaetiellopsis themedae (Kandaswamy & Sundaram) Sutton & DiCosmo and *M. cymbopogonis* (Punithalingam & Sarwar) Sutton & DiCosmo cause leaf lesions on members of the Poaceae (Sutton, 1980). The teleomorphs belong to *Hypnotheca*.

MONOGRAPHELLA Petr.
Sphaeriales   4 spp.
Arx, J. A. von 1985. Notes on *Monographella* and *Microdochium*. Trans. Br. Mycol. Soc. 83:373-374. (transfers *Sphaerella opuntiae* and *Plectosphaerella [Venturia] cucumerina* to *Monographella*, discusses anamorphs)
Mueller, E. 1977. Die systematische Stellung des "Schneeschimmels". Rev. Mycol. 41:129-134. (in German with French summary, discusses generic concept, describes *M. nivalis*)
Mueller, E., and Samuels, G. J. 1984. *Monographella maydis* sp. nov. and its connection to the tar-spot disease of *Zea mays*. Nova Hedw. 40:113-121. (presents table comparing five taxa, describes teleomorph and *Microdochium* anamorph)
Parkinson, V. O., Sivanesan, A., and Booth, C. 1981. The perfect state of the rice leaf scald fungus and the taxonomy of both the perfect and imperfect states. Trans. Br. Mycol. Soc. 76:59-69. (describes and illustrates *M. albescens* and its anamorph, presents table comparing this species with two varieties of *M. nivalis*)
Samuels, G. J., and Hallett, I. C. 1983. *Microdochium stoveri* and *Monographella stoveri*, new combinations for *Fusarium stoveri* and *Micronectriella stoveri*. Trans. Br. Mycol. Soc. 81:473-483. (descriptions and illustrations)
Subramanian, C. V., and Bhat, D. J. 1978. Developmental morphology of Ascomycetes. III. *Monographella nivalis*. Rev. Mycol. 42:293-307. (includes taxonomy of teleomorph and anamorph, illustrations)
Monographella species have anamorphs in *Microdochium*. Mueller and Samuels (1984) present a table comparing four species as well as an overview of the generic taxonomy and anamorph characters.
Monographella species have previously been placed in *Calonectria, Micronectriella, Nectria, Plectosphaerella, Sphaerella*, and *Venturia*. *Monographella albescens* (Thuem.) Parkinson, Sivanesan & C. Booth causes leaf-scald disease of rice (CMI 729); *M. maydis* E. Mueller & Samuels is associated with *Phyllachora maydis* Maubl. in tar spot of maize; *M. nivalis* (Schaffnit) E. Mueller causes snow mold of turf and foot rot and head blight of cereals (CMI 309 as *Micronectriella nivalis*); and *M. stoveri* (C. Booth) Samuels &

MONOGRAPHELLA - cont.
Hallett (syn. *Micronectriella stoveri* C. Booth) is associated with *Mycosphaerella musicola* in Sigatoka leaf spot of banana.

MONOSTICHELLA Hoehn.
    Coelomycetes   9 spp.
Arx, J. A. von 1970. A revision of the fungi classified as *Gloeosporium*. Bibl. Mycol. 24:1-203. (describes and illustrates ten *Monostichella* spp.)
Morgan-Jones, G. 1971. Conidium ontogeny in Coelomycetes. I. Some amerosporous species which possess annellides. Can. J. Bot. 49:1921-1929. (discusses and illustrates three *Monostichella* spp.)
Morgan-Jones, G. 1971. Conidium ontogeny in Coelomycetes. II. Some Melanconiales which possess phialides. Can. J. Bot. 49:1931-1937. (discusses and illustrates three *Monostichella* spp.)
Sutton (1980) includes three species that have been transferred from *Gloeosporium* (Arx, 1970). *Monostichella salicis* (Westendorp) Arx causes leaf spots on *Salix* spp. and is the anamorph of *Drepanopeziza salicis* (Tul.) Hoehn.

MURIBASIDIOSPORA Kamat & Rajendren
    Exobasidiales   3 spp.
Rajendren, R. B. 1968. *Muribasidiospora* - a new genus of the Exobasidiaceae. Mycopathol. Mycol. Appl. 36:218-222. (description and illustrations of *M. indica*, transfers two *Exobasidium* spp. to *Muribasidiospora*)
Rajendren, R. B. 1970. *Muribasidiospora indica* in culture. Mycopathol. Mycol. Appl. 41:287-292. (description of disease and of fungus in culture, illustrations)
Muribasidiospora indica Kamat & Rajendren causes leaf spots on *Rhus*. See McNabb and Talbot (1973).

MYCENA (Pers.) Roussel
    Agaricales   200 spp.
Pegler, D. N. 1983. Agaric flora of the Lesser Antilles. Kew Bull. Add. Ser. 9:1-668. (describes and illustrates *M. citricolor*)
Mycena citricolor (Berk. & M. A. Curtis) Sacc. (syn. *Omphalia flavida* Maubl. & Rangel) is parasitic on many tropical hosts causing American leaf spot disease of coffee, as well as other diseases. The mycelium of *M. citricolor* is luminescent. Pegler (1983) provides a description and illustration of this species and the numerous saprophytic *Mycena* species.

MYCOCENTROSPORA Deighton
    Hyphomycetes   10 spp.
Constantinescu, O. 1978. Polymorphism of *Mycocentrospora acerina*

MYCOCENTROSPORA - cont.
conidia. Rev. Mycol. 42:105-112. (descriptions and illustrations)
Deighton, F. C. 1971. Studies on *Cercospora* and allied genera. III. *Centrospora*. Mycol. Pap. 124:1-13. (describes and illustrates four species as *Centrospora*, later transferred to *Mycocentrospora*)
Deighton, F. C. 1983. Studies on *Cercospora* and allied genera. VIII. Further notes on *Cercoseptoria* and some new species and redispositions. Mycol. Pap. 151:1-13. (discusses two *Mycocentrospora* spp.)
Laundon, G. F. 1970. Records of fungal plant diseases in New Zealand. N. Z. J. Bot. 8:51-66. (describes and illustrates *M. camelliae* as *Cercoseptoria theae* and *M. acerina* as *Centrospora*)
Pollack, F. G., and Ellett, C. W. 1974. *Mycocentrospora verrucosa*, the cause of foliar shot-hole of *Euonymus*. Mycologia 66:170-173. (description and illustration)
Wall, C. J., and Lewis, B. G. 1980. Infection of carrot leaves by *Mycocentrospora acerina*. Trans. Br. Mycol. Soc. 75:163-165. (pathology, biology, epidemiology)

Mycocentrospora is a segregate of *Cercospora* and replaces *Centrospora* Neerg., a later homonym. Ellis (1971, 1976) briefly describes and illustrates four species. *Mycocentrospora acerina* (R. Hartig) Deighton (syn. *Centrospora acerina* (R. Hartig) Newhall) causes diseases on many hosts, especially members of the Apiaceae, including licorice rot of carrot (CMI 537). *Mycocentrospora verrucosa* Pollack & Ellett causes a foliar shot hole of *Euonymus*.

MYCOLEPTODISCUS Ostazeski
    Hyphomycetes   6 spp.
Sutton, B. C., and Alcorn, J. L. 1985. Undescribed species of *Crinitospora* gen. nov., *Massariothea, Mycoleptodiscus* and *Neottiosporina* from Australia. Trans. Br. Mycol. Soc. 84:437-445. (describes and illustrates *Mycoleptodiscus lunatus* from lesions of *Carpobrotus*)
Sutton, B. C., and Hodges, C. S., Jr. 1976. *Eucalyptus* microfungi: *Mycoleptodiscus* species and *Pseudotracylla* gen. nov. Nova Hedw. 27:693-700. (key to four *Mycoleptodiscus* spp., descriptions and illustrations)
Vanev, S. G. 1983. *Mycoleptodiscus minimus* (Berk. & Curt.) Vanev, comb. nov. Proc. K. Ned. Akad. Wet., Ser. C. 86:433-435. (describes and illustrates *M. minimus* on *Ilex*, compares with four other species)

Mycoleptodiscus indicus (Sahni) Sutton occurs on a variety of hosts and is usually associated with large spreading lesions.

MYCOPAPPUS Redhead & G. P. White
  Hyphomycetes   2 spp.
Redhead, S. A., and White, G. P. 1985. *Mycopappus*, a new genus of leaf pathogens, and two parasitic *Anguillospora* species. Can. J. Bot. 63:1429-1435. (describes and illustrates two *Mycopappus* spp.)
This genus was described for two species on *Acer* and *Alnus* previously placed in *Cercosporella*.

MYCOSPHAERELLA Johan.
  Dothideales   500 spp.
Arx, J. A. von 1949. Beitraege zur Kenntnis der Gattung *Mycosphaerella*. Sydowia 3:28-100. (in German, describes and illustrates 20 species)
Arx, J. A. von 1983. *Mycosphaerella* and its anamorphs. Proc. K. Ned. Akad. Wet., Ser. C. 86:15-54. (key to 23 anamorph genera, overview of *Mycosphaerella*, descriptions and illustrations)
Evans, H. C. 1984. The genus *Mycosphaerella* and its anamorphs *Cercoseptoria*, *Dothistroma* and *Lecanosticta* on pines. Mycol. Pap. 153:1-102. (describes and illustrates three *Mycosphaerella* spp., their anamorphs, and the diseases they cause)
Ganapathi, A., and Corbin, J. B. 1979. *Colletogloeum nubilosum* sp. nov., the imperfect state of *Mycosphaerella nubilosa* on *Eucalyptus* in New Zealand. Trans. Br. Mycol. Soc. 72:237-244. (descriptions and illustrations including cultural characteristics)
Katumoto, K. 1983. Notes on some plant-inhabiting Ascomycotina from western Japan. Trans. Mycol. Soc. Jpn. 24:259-269. (describes and illustrates two *Mycosphaerella* spp.)
Kessler, K. J., Jr. 1984. *Mycosphaerella juglandis*, causal agent of a leaf spot of *Juglans nigra*. Mycologia 76:362-366. (describes and illustrates teleomorph and anamorph, *Cylindrosporium juglandis*)
Kessler, K. J., Jr. 1985. *Mycosphaerella* leaf spot of black walnut. Plant Dis. 69:1092-1094. (describes and illustrates teleomorph and anamorph, pathology)
Meredith, D. S. 1970. Banana leaf spot disease (Sigatoka) caused by *Mycosphaerella musicola* Leach. Phytopathol. Pap. 11:1-147. (complete account of the disease)
Park, R. F., and Keane, P. J. 1982. Three *Mycosphaerella* species from leaf diseases of *Eucalyptus*. Trans. Br. Mycol. Soc. 79:95-100. (descriptions and illustrations)
Park, R. F., and Keane, P. J. 1982. Leaf diseases of *Eucalyptus* associated with *Mycosphaerella* species. Trans. Br. Mycol. Soc. 79:101-115. (pathology of three species, illustrations)
Park, R. F., and Keane, P. J. 1984. Further *Mycosphaerella* species causing leaf diseases of *Eucalyptus*. Trans. Br. Mycol. Soc. 83:93-105. (describes and illustrates six species)
Patton, R. F., and Spear, R. N. 1983. Needle cast of European larch

MYCOSPHAERELLA - cont.
caused by *Mycosphaerella laricina* in Wisconsin and Iowa. Plant Dis. 67:1149-1153. (description, illustration, and pathology)
Sivanesan, A. 1985. The teleomorph of *Asperisporium pongamiae*. Trans. Br. Mycol. Soc. 84:363-367. (describes and illustrates *M. pongamiae*)
Sivanesan, A. 1985. The teleomorph of *Cercosporidium henningsii*. Trans. Br. Mycol. Soc. 84:551-555. (compares *M. henningsii* with *M. manihotis*)
Sivanesan, A. 1985. Teleomorphs of *Cercospora sesami* and *Cercoseptoria sesami*. Trans. Br. Mycol. Soc. 85:397-404. (describes and illustrates *M. sesami* and *M. sesamicola*)
Tomilin, B. A. 1979. [Classification Key of the Fungi of the Genus *Mycosphaerella* Johans.] Nauka, Leningrad. 319 pp. (in Russian, keys to 658 species based on host, descriptions and illustrations)

Mycosphaerella is a large genus with many plant pathogenic species that primarily cause leaf spots. Despite several major contributions, identification of species is still difficult. Arx (1983) has initiated work to clarify relationships among species based on anamorphs. Anamorphs are placed in many genera, including *Cercospora* and its segregate genera; *Cladosporium, Dothistroma, Hendersonia, Lecanosticta, Phloeospora, Phyllosticta, Ramularia*, and *Septoria*; and an *Asteromella* microconidial state. Sivanesan (1984) includes keys to over 60 species based on their anamorphs with descriptions and illustrations. Barr (1972) describes and illustrates 64 species that occur in North America; Mueller and Arx (1962) include 12 species. See the related genus *Didymella*. Species causing diseases include: *Mycosphaerella arachidis* Deighton, early leaf spot of groundnut (CMI 411); *M. berkeleyi* W. A. Jenkins, late leaf spot of groundnut (CMI 412); *M. brassicicola* (Duby) Lindau, ring spot of *Brassica* spp. (CMI 468); *M. carinthiaca* Jaap, mid-vein spot of clover; *M. citri* Whiteside, greasy spot of *Citrus* spp. (CMI 510); *M. dearnessii* Barr, *Lecanosticta* or brown spot needle blight of pines (CMI 367 as *Scirrhia acicola*); *M. fijiensis* Morelet, black leaf streak of banana (CMI 413); *M. fragariae* (Tul.) Lindau, leaf spot or white spot of strawberry (CMI 708); *M. graminicola* (Fuckel) Sanderson, speckled leaf blotch of wheat (F. Can. 244 as *Septoria tritici*); *M. grossulariae* (Fr.) Lindau, *Ribes* leaf spot; *M. holci* Tehon, on many Poaceae (CMI 584); *M. linicola* Naumov, Pasmo disease of flax (CMI 709); *M. macrospora* (Kleb.) Jorst., leaf spot or blotch of iris (CMI 435); *M. musicola* R. Leach, Sigatoka of banana (CMI 414); *M. pini* Rostr., *Dothistroma* blight or red-band needle blight of pines (CMI 368 as *Scirrhia pini*); *M. pinodes* (Berk. & Broome) Vestergren, foot rot and leaf, stem, and pod spot of pea (CMI 340); *M. rubi* Roark, *Rubus* leaf spot; *M. sentina* (Fr.) J. Schroet., pear leaf speck; and *M. ulmi* Kleb., elm leaf spot.

MYCOVELLOSIELLA Rangel
  Hyphomycetes   43 spp.
Deighton, F. C. 1974. Studies on *Cercospora* and allied genera. V. *Mycovellosiella* Rangel, and a new species of *Ramulariopsis*. Mycol. Pap. 137:1-73. (defines genus, treats 35 taxa, host list)
Deighton, F. C. 1979. Studies on *Cercospora* and allied genera. VII. New species and redispositions. Mycol. Pap. 144:1-56. (treats nine *Mycovellosiella* spp.)
Rai, B., and Kamal, A. 1985. New hyphomycetes from India. Trans. Br. Mycol. Soc. 85:566-570. (describes and illustrates three *Mycovellosiella* spp.)
Sutton, B. C., and Shahjahan, A. K. M. 1981. A comparison of the symptoms and causal agents of narrow brown leaf spot and white leaf streak of rice. Nova Hedw. 35:197-205. (compares *M. oryzae* with *Cercospora oryzae*, descriptions and illustrations)
Deighton (1974, 1979) provides the most complete account of the genus, a segregate of *Cercospora*. Ellis (1971, 1976) includes six species. *Mycovellosiella oryzae* (Deighton & Shaw) Deighton causes white leaf streak of rice. Other species causing diseases include: *Mycovellosiella cajani* (Henn.) Rangel ex Trott., minor leaf spot of pigeon pea (CMI 628); *M. concors* (Casp.) Deighton, leaf spot of *Solanum* spp. (CMI 724); *M. koepkei* (Krueger) Deighton, yellow spot of sugarcane (CMI 417 as *Cercospora koepkei*); *M. nattrassii* Deighton, leaf spot of *Solanum* (CMI 629); *M. phaseoli* (Drummond) Deighton, floury spot or mancha harinosa of *Phaseolus vulgaris* (CMI 870); and *M. vaginae* (Krueger) Deighton, leaf sheath lesions of sugarcane (CMI 725).

MYRIANGIUM Mont. & Berk.
  Dothideales   7 spp.
Arx, J. A. von 1963. Die Gattungen der Myriangiales. Persoonia 2:421-475. (in German, key to seven species, descriptions, reviews genera in the Myriangiales)
Miller, J. H. 1940. The genus *Myriangium* in North America. Mycologia 32:587-600. (key to four species, descriptions)
Petch, T. 1924. Studies in entomogenous fungi. V. *Myriangium*. Trans. Br. Mycol. Soc. 10:45-80. (key to four species on insects, descriptions and illustrations)
Species of *Myriangium* are saprophytic, parasitize scale insects, or are parasitic on woody plants.

MYRIOGENOSPORA Atk.
  Clavicipitales   2 spp.
Luttrell, E. S., and Bacon, C. W. 1977. Classification of *Myriogenospora* in the Clavicipitaceae. Can. J. Bot. 55:2090-2097. (describes fungal development, compares with *Balansia* spp., illustrations)
Rykard, D. M., Luttrell, E. S., and Bacon, C. W. 1982. Development of

MYRIOGENOSPORA - cont.
the conidial state of *Myriogenospora atramentosa*. Mycologia 74:648-654. (placed in *Ephelis*, descriptions and illustrations)
Sivanesan, A., and Waller, J. M. 1986. Sugarcane diseases. Phytopathol. Pap. 29:1-88. (includes *M. aciculispora*)
Myriogenospora atramentosa (Berk. & M. A. Curtis) Diehl is parasitic on sugarcane and other grasses and has an *Ephelis* conidial state.

MYROTHECIUM Tode:Fr.
  Hyphomycetes 16 spp.
DiCosmo, F., Michaelides, J., and Kendrick, B. 1980. *Myrothecium tongalense* anam.-sp. nov. Mycotaxon 12:219-224. (isolated from calcified green algae)
Nguyen, T. H., Mathur, S. B., and Neergard, P. 1973. Seed-borne species of *Myrothecium* and their pathogenic potential. Trans. Br. Mycol. Soc. 61:347-354. (describes and illustrates three species, pathology)
Rao, V., and Hoog, G. S. de 1983. A new species of *Myrothecium*. Persoonia 12:99-101. (describes and illustrates *M. bisetosum* from rotten bark)
Tulloch, M. 1972. The genus *Myrothecium* Tode ex Fr. Mycol. Pap. 130:1-42. (key to 13 species, descriptions and illustrations)
Species of *Myrothecium* are cellulolytic and produce antibiotics and mycotoxins. Tulloch (1972) provides a complete account of the genus. Ellis (1971, 1976) treats 14 species and Domsch et al. (1980) include a key to three species. *Myrothecium roridum* Tode:Fr. is pathogenic on many hosts including *Antirrhinum, Coffea, Gossypium,* and *Viola* (CMI 253).

MYXOFUSICOCCUM See *Pseudophacidium*.

NAEMACYCLUS Fuckel
  Rhytismatales 1 sp.
DiCosmo et al. (1983, 1984 under *Cyclaneusma*) include two *Naemacyclus* species causing needle casts of pine in *Cyclaneusma*. For nomenclatural reasons they retain only *Naemacyclus fimbriatus* (Schwein.) DiCosmo et al. (syn. *Lasiostictis fimbriata* (Schwein.) Baeumler) in *Naemacyclus*.

NAKATAEA K. Hara
  Hyphomycetes 3 spp.
Ou, S. H. 1985. Rice diseases. Second Edition. Commonwealth Mycological Institute, Kew, Surrey, England. 380 pp. (describes and illustrates *N. sigmoidea* and discusses disease)
Shearer, C. A., and Crane, J. L. 1979. Illinois fungi. XI. *Nakataea serpens* sp. nov., an aero-aquatic Hyphomycete. Trans. Br. Mycol. Soc. 73:370-372. (description and illustration)
Ellis (1971) includes *Nakataea sigmoidea* Hara, the anamorph of

NAKATAEA - cont.
Magnaporthe salvinii (Cattaneo) Krause & Webster, cause of stem rot of rice (CMI 344 as *Leptosphaeria salvinii*).

NECTRIA (Fr.) Fr.
Hypocreales   200 spp.
Booth, C. 1959. Studies of Pyrenomycetes: IV. *Nectria* (Part I.) Mycol. Pap. 73:1-115. (key to British species divided into ten groups, descriptions including cultural characteristics and anamorphs, illustrations, host list)
Chen, S.-c., and Zhang, J.-z. 1984. [A new parasitic fungus on tungoil trees]. Scientia Silvae Sin. 20:156-159. (in Chinese with English summary, *N. aleuritidia* sp. nov, description and illustrations, table comparing this species with *N. galligena* and *N. ditissima*)
Cotter, H. V. T., and Blanchard, R. O. 1981. Identification of the two *Nectria* taxa causing bole cankers on American beech. Plant Dis. 65:332-334. (differentiation of *N. galligena* and *N. coccinea* var. *faginata* by ascospore length)
Dehorter, B., and Perrin, R. 1983. Production in vitro de peritheces du *Nectria ditissima*, agent du chancre du hetre (*Fagus sylvatica*). I. Influence du milieu de culture et de la temperature. Application a la realisation d'infections artificielles du hetre. Can. J. Bot. 61:1941-1946. (in French with English summary, describes conditions favorable for in vitro production of *N. ditissima* ascomata, reports heterothallism)
Dingley, J. M. 1951. The Hypocreales of New Zealand II. The genus *Nectria*. Trans. R. Soc. N. Z. Bot. 79:177-202. (key to 31 species, descriptions and illustrations)
Dingley, J. M. 1957. Life-history studies of New Zealand species of *Nectria* Fr. Trans. R. Soc. N. Z. Bot. 84:467-477. (describes and illustrates the conidial states of 26 species)
Doebbeler, P. 1979. Moosbewohnende Ascomyceten III. Einige neue arten der gattungen *Nectria, Epibryon* and *Punctillum*. Mitt. Bot. Muenchen 15:193-221. (in German with English summary, describes and illustrates two *Nectria* spp. on bryophytes)
Doyle, A. F. 1978. Some secondary metabolites from *Nectria* species. Mycologia 70:355-362. (lists metabolites and pigments from 11 species)
Flack, N. J., and Swinburne, T. R. 1977. Host range of *Nectria galligena* Bres. and the pathogenicity of some northern Ireland isolates. Trans. Br. Mycol. Soc. 68:185-192. (summarizes host range, inoculation experiments with *N. coccinea, N. ditissima*, and *N. galligena* on various hosts)
Hawksworth, D. L., and Minter, D. W. 1980. New and interesting microfungi from the 1978 Exeter Foray. Trans. Br. Mycol. Soc. 74:567-577. (desribes and illustrates *Nectria boothii* from *Oenanthes crocata*)

NECTRIA - cont.
Kar, A. K., and Gupta, S. K. 1980. *Nectria* species from West Bengal. Indian Phytopathol. 33:547-550. (describes and illustrates three species)
Lohman, M. L., and Watson, A. J. 1943. Identity and host relations of *Nectria* species associated with diseases of hardwoods in the eastern states. Lloydia 6:77-108. (key to five taxa based on cultural characters, descriptions and illustrations)
Perrin, R. 1976. Clef de determination des *Nectria* d'Europe. Bull. Trimest. Soc. Mycol. Fr. 92:335-347. (in French, key to about 40 species by host)
Rossman, A. Y. 1983. The phragmosporous species of *Nectria* and related genera. Mycol. Pap. 150:1-164. (key to 53 species mostly in *Nectria*, descriptions and illustrations)
Samuels, G. J. 1973. The myxomyceticolous species of *Nectria*. Mycologia 65:401-420. (key to five species, descriptions and illustrations)
Samuels, G. J. 1976. Perfect states of *Acremonium*. The genera *Nectria, Actiniopsis, Ijuhya, Neohenningsia, Ophiodictyon*, and *Peristomialis*. N. Z. J. Bot. 14:231-260. (synoptic key to 12 *Nectria* spp. from New Zealand and their *Acremonium* anamorphs, descriptions and illustrations)
Samuels, G. J. 1976. A revision of the fungi formerly classified as *Nectria* subg. *Hyphonectria*. Mem. N. Y. Bot. Gard. 26:1-126. (key to 33 species, descriptions and illustrations)
Samuels, G. J. 1977. *Nectria consors* and its *Volutella* conidial state. Mycologia 69:255-262. (description and illustrations)
Samuels, G. J. 1978. Some species of *Nectria* having *Cylindrocarpon* imperfect states. N. Z. J. Bot. 16:73-82. (descriptions and illustrations)
Samuels, G. J. 1985. Four new species of *Nectria* and their *Chaetopsina* anamorphs. Mycotaxon 22:13-32. (descriptions and illustrations, discussion of *Chaetopsina* and related genera)
Samuels, G. J., and Dumont, K. P. 1982. The genus *Nectria* (Hypocreaceae) in Panama. Caldasia 13:379-423. (synoptic key to 32 species, descriptions)
Samuels, G. J., and Rogerson, C. T. 1984. *Nectria atrofusca* and its anamorph, *Fusarium staphyleae*, a parasite of *Staphylea trifolia* in eastern North America. Brittonia 36:81-85. (descriptions and illustrations)
Samuels, G. J., and Rossman, A. Y. 1979. Conidia and classification of the nectrioid fungi. Pages 167-182 in: The Whole Fungus: The Sexual-asexual Synthesis. Vol. 1. B. Kendrick, ed. National Museums of Canada, Ottawa. (discusses *Nectria* anamorphs)
Seifert, K. 1985. A monograph of *Stilbella* and some allied hyphomycetes. Stud. Mycol. 27:1-235. (key to *Nectria* spp. with synnematous anamorphs, descriptions and illustrations)
Subramanian, C. V., and Bhat, D. J. 1984. Developmental morphology of

NECTRIA - cont.
Ascomycetes XI. *Nectria kera.* Crypt. Mycol. 5:135-145.
(description and illustrations of developmental morphology of *N. kera* sp. nov. from *Cocos nucifera* and of its *Cylindrocarpon* anamorph)
Tilak, S. T., and Talde, U. K. 1979. Contribution to our knowledge of Ascomycetes of India. Indian J. Mycol. & Plant Pathol. 9:17-21. (descriptions and illustrations of two *Nectria* spp.)
Nectria is a large genus that contains plant pathogenic, saprophytic, and hyperparasitic species. Despite the recent work, no comprehensive account exists. Domsch et al. (1980) treat five species. The anamorphs belong to *Acremonium, Cylindrocarpon, Cylindrocladiella, Fusarium, Gliocladium, Rhizostilbella, Tubercularia, Verticillium, Volutella,* and *Zythiostroma.* Important diseases caused by *Nectria* species include: *Nectria aurantiicola* Berk. & Broome, on scale insects (CMI 714); *N. cinnabarina* (Tode:Fr.) Fr., coral spot fungus on many hosts (CMI 531); *N. coccinea* (Pers.:Fr.) Fr., beech bark disease (CMI 532); *N. coccinea* var. *faginata* Lohman, Watson & Ayers, beech bark disease (CMI 533); *N. flammea* (Tul.) Dingley, on scale insects (CMI 715); *N. fuckeliana* C. Booth, a wound parasite causing dieback of Pinaceae (CMI 624); *N. galligena* Bres. in Strasser, root rot, storage rot, black rot of strawberry, grapes and a wide range of hosts (CMI 147); *N. macrospora* (Wollenw.) Ouellette, canker of balsam fir and other Pinaceae (CMI 623); *N. mauritiicola* (Henn.) Seifert & Samuels, violet, red, or stinking root rot of many tropical hosts (CMI 391 as *Sphaerostilbe repens*; see Seifert, 1985); *N. radicicola* Gerlach & Nilsson, root plate rot of *Narcissus*, black rot of strawberry, black spot of grapes, and various other rots (CMI 148); and *N. rigidiuscula* Berk. & Broome, on *Theobroma cacao* and various tropical crops (CMI 21 as *Calonectria rigidiuscula*).

NECTRIELLA Nitschke
    Hypocreales   20 spp.
Alfieri, S. A., Jr., and Samuels, G. J. 1979. *Nectriella pironii* sp. nov. and its *Kutilakesa*-like anamorph, a parasite of ornamental shrubs. Mycologia 71:1178-1185. (descriptions, illustrations, biology)
Hawksworth, D. L. 1982. A new species of *Nectriella* with ornamented spores from Iceland, with a key to the lichenicolous species. Nova Hedw. 35:755-762. (discusses genus, description and illustration, key to six species)
Samuels, G. J., Rogerson, C. T., Rossman, A. Y., and Smith, J. D. 1984. *Nectria tuberculariformis, Nectriella muelleri, Nectriella* sp., and *Hyponectria sceptri*: low-temperature tolerant, alpine boreal fungal antagonists. Can. J. Bot. 62:1896-1903. (descriptions and illustrations)
Nectriella species differ from *Nectria* in having ascocarps immersed

NECTRIELLA - cont.
in the substrate. Species occur on many substrates but most are parasites of lichens or phanerogamic plants. *Nectriella pironii* Alfieri & Samuels, anamorph *Kutilakesa pironii* Alfieri, causes stem galls and cankers on *Aphelandra, Clerodendrum, Codiaeum*, and other hosts.

NEMATOSPORA Peglion
    Endomycetales   2 spp.
Batra, L. R. 1973. Nematosporaceae (Hemiascomycetidae): Taxonomy, Pathogenicity, Distribution, and Vector Relations. U.S. Dep. Agric., Tech. Bull. 1469. 71 pp. (key to two species, descriptions, illustrations, biology)
Nematospora coryli Peglion causes yeast spot of bean and other seeds and cotton stain or internal boll rot (CMI 184). See *Ashbya*.

NEOCOSMOSPORA E. F. Sm.
    Hypocreales   5 spp.
Cannon, P. F., and Hawksworth, D. L. 1984. A revision of the genus *Neocosmospora* (Hypocreales). Trans. Br. Mycol. Soc. 82:673-688. (key to five species, descriptions, illustrations, pathology)
Neocosmospora species are most commonly isolated from soil; *N. vasinfecta* E. F. Sm. causes stem and root rots, mainly of Fabaceae, as discussed by Cannon and Hawksworth (1984). Four taxa have *Acremonium* anamorphs. Domsch et al. (1980) provide brief descriptions of five species.

NEOHENDERSONIA Petr.
    Coelomycetes   2 spp.
Sutton, B. C. 1975. Coelomycetes. V. *Coryneum*. Mycol. Pap. 138:1-224. (transfers *Coryneum congoense* to *Neohendersonia*, description and illustration)
Sutton, B. C., and Pollack, F. G. 1974. Microfungi on *Cercocarpus*. Mycopathol. Mycol. Appl. 52:331-351. (describes and illustrates *N. kickxii*)
Sutton (1980) includes two species of *Neohendersonia*, *N. congoensis* (Torrend) Sutton on stems of *Aloe* and *Agave*, and *N. kickxii* (Westendorp) Sutton & Pollack on branches and bark of *Fagus sylvatica*.

NEOKELLERMANIA Punithalingam
    Coelomycetes   2 spp.
Punithalingam, E. 1981. New microfungi from cereals and grasses. II. Nova Hedw. 34:67-95. (key to two species, descriptions and illustrations)
Neokellermania species cause leaf spots on Poaceae.

NEOPECKIA Sacc.
   Pleosporales   1 sp.
Barr, M. E. 1984. *Herpotrichia* and its segregates. Mycotaxon
   20:1-38. (separates North American species into five genera, key
   to species, descriptions and illustrations)
Neopeckia coulteri (Peck) Sacc. (syn. *Herpotrichia coulteri* (Peck)
Bose) causes brown felt blight of pine (CMI 327 as *Herpotrichia coulteri*; Funk 1981, 1985). The anamorph is *Pyrenochaeta*-like (Barr, 1984).

NEOPYCNODOTHIS   See *Cytoplea*.

NEOTTIOSPORINA C. V. Subramanian
   Coelomycetes   9 spp.
Sutton, B. C., and Alcorn, J. L. 1974. *Neottiosporina*. Aust. J. Bot.
   22:517-530. (key to six species, descriptions and illustrations)
Sutton, B. C., and Alcorn, J. L. 1985. Undescribed species of
   Crinitospora gen. nov., *Massariothea, Mycoleptodiscus* and
   Neottiosporina from Australia. Trans. Br. Mycol. Soc.
   84:437-445. (describes and illustrates *N. cylindrica*)
Sutton, B. C., and Marasas, W. F. O. 1976. Observations on
   Neottiosporina and *Tiarosporella*. Trans. Br. Mycol. Soc.
   67:69-76. (describes new species on pine, illustrations)
Sutton (1980) includes a key to eight *Neottiosporina* spp. with brief descriptions and illustrations. Several species are associated with leaf lesions on various hosts.

NEOVOSSIA Koern.
   Ustilaginales   5 spp.
Khanna, A., and Payak, M. M. 1968. Teliospore morphology of some smut
   fungi. II. Light microscopy. Mycologia 60:655-662. (compares
   teliospore morphology and wall structure of *N. brachypodii, N.
   indica,* and *N. horrida.*)
Khanna, A., Payak, M. M., and Mehta, S. C. 1966. Teliospore
   morphology of some smut fungi. I. Electron microscopy. Mycologia
   58:562-569. (teliospore morphology of *N. indica* and *T. caries*)
Singh, R. A., and Pavgi, M. S. 1973. Development of sorus in kernel
   bunt of rice. Riso 22:243-250. (describes disease, mode of
   infection, and development, illustrations)
Singh, R. A., Whitehead, M. D., and Pavgi, M. S. 1979. Taxonomy of
   Neovossia horrida (Ustilaginales). Sydowia 32:305-308. (evidence
   suggests retaining in *Neovossia*)
Tullis, E. C., and Johnson, A. G. 1952. Synonymy of *Tilletia horrida*
   and *Neovossia barclayana*. Mycologia 44:773-788. (based on
   teliospore morphology and cross inoculations)
Neovossia species parasitize members of the Poaceae. Controversy exists over the generic limits of *Neovossia* and *Tilletia*. Vanky (1985) provides a key to two *Neovossia* species with descriptions and

NEOVOSSIA - cont.
illustrations. Duran and Fischer (1962) include four species as *Tilletia* spp. Important diseases include *Neovossia barclayana* Bref. (syn. *N. horrida* (Takahashi) Padwick & Azmatullah Khan), rice kernel smut (CMI 75 as *Tilletia barclayana*) and *N. indica* (Mitra) Mundk., karnal bunt of wheat (CMI 748 as *Tilletia indica*). See Ustilaginales.

NEWINIA Thaung
    Uredinales   2 spp.
Eboh, D. O. 1983. A new species of *Newinia* from Nigeria. Mycologia 75:316-318. (describes and illustrates *N. kigeliae* on *Kigelia africana*, compares with *N. heterophragmae*)
Thaung, M. M. 1973. A new genus of rusts from Burma. Mycologia 65:702-704. (describes and illustrates *N. heterophragmae* on *Heterophragma sulfureum*)
Species of *Newinia* occur on members of the Bignoniaceae.

NIGROSPORA Zimmermann
    Hyphomycetes   4 spp.
Hudson, H. J. 1963. The perfect state of *Nigrospora oryzae*. Trans. Br. Mycol. Soc. 46:355-360. (reviews genus, describes and illustrates the teleomorph *Khuskia oryzae*)
Ellis (1971) provides a key to four species with brief descriptions and illustrations. *Nigrospora oryzae* (Berk. & Broome) Petch causes diseases of maize, rice, and other hosts (CMI 311 under *Khuskia oryzae*).

NODULISPORIUM G. Preuss
    Hyphomycetes   14 spp.
Deighton, F. C. 1985. Some species of *Nodulisporium*. Trans. Br. Mycol. Soc. 85:391-395. (includes seven species with descriptions and illustrations)
Jong, S. C., and Rogers, J. D. 1972. Illustrations and descriptions of conidial states of some *Hypoxylon* species. Wash. Agric. Exp. Stn. Tech. Bull. 71:1-51. (describes and illustrates anamorphs of 18 *Hypoxylon* spp., most of which belong in *Nodulisporium*)
Petrini, L., and Petrini, O. 1985. Xylariaceous fungi as endophytes. Sydowia 38:216-234. (keys to European xylariaceous endophytes, key to eight species with *Nodulisporium* anamorphs, descriptions of cultures and ecological notes)
Whalley, A. J. S., and Greenhalgh, G. N. 1975. Numerical taxonomy of *Hypoxylon*. III. Comparison of the cultural states of some *Hypoxylon* species with *Nodulisporium* species. Trans. Br. Mycol. Soc. 64:229-233. (supports placement of *Hypoxylon* anamorphs near *Nodulisporium*)
Nodulisporium species are anamorphs of *Hypoxylon* and other Xylariaceae. Ellis (1971) treats two species.

NODULOSPHAERIA Rabenh.
Pleosporales   40 spp.
Holm, L. 1957. Etudes taxonomiques sur les Pleosporacees. Symb. Bot. Ups. 14:1-188. (in French, key to 18 *Nodulosphaeria* spp., descriptions and illustrations)
Holm, L. 1961. Taxonomical notes on Ascomycetes. IV. Notes on *Nodulosphaeria* Rbh. Sven. Bot. Tidskr. 55:63-80. (describes and illustrates 13 species, host list for 30 species)
Holm, L., and Mueller, E. 1963. Ueber eine neue Art aus der Gattung *Nodulosphaeria* Rbh. Sydowia 16:57-59. (in German, describes and illustrates *N. valesiaca* from onions)
Shoemaker, R. A. 1984. Canadian and some extralimital *Nodulosphaeria* and *Entodesmium* species. Can. J. Bot. 62:2730-2753. (key to 22 *Nodulosphaeria* spp., descriptions and illustrations)
Nodulosphaeria is a segregate of *Leptosphaeria*. Species occur on herbaceous dicotyledonous plants especially in the Apiaceae and Asteraceae. Species are specialized according to host genus.

NUMMULARIA   See *Biscogniauxia*.

NYSSOPSORA Arth.
Uredinales   9 spp.
Kakishima, M., Sato, T., and Sato, S. 1984. Notes on two rust fungi, *Pileolaria klugkistiana* and *Nyssopsora cedrelae*. Trans. Mycol. Soc. Jpn. 25:355-359. (determines life cycles by host inoculations, illustrations)
Many *Nyssopsora* species occur on Araliaceae including *N. clavellosa* (Berk.) Arth. on *Aralia nudicaulis* (F. Can. 221). See Cummins and Hiratsuka (1983).

OCOTOMYCES H. Evans & Minter
Rhytismatales   1 sp.
Evans, H. C., and Minter, D. W. 1985. Two remarkable new fungi on pine from Central America. Trans. Br. Mycol. Soc. 84:57-78. (descriptions and illustrations)
Ocotomyces parasiticus H. Evans & Minter, anamorph *Uyucamyces parasiticus* H. Evans & Minter, causes a canker disease of young pine branches and stems.

OIDIUM Link:Fr.
Hyphomycetes   50 spp.
Bhagyanarayana, G., and Ramachar, P. 1983. Nomenclatural changes in the genus *Oidium*. Curr. Sci. 52:170-171. (discusses three species)
Boesewinkel, H. J. 1977. Identification of Erysiphaceae by conidial characteristics. Rev. Mycol. 41:493-507. (key to 34 New Zealand species based on conidial characters)

OIDIUM - cont.
Boesewinkel, H. J. 1979. Erysiphaceae of New Zealand. Sydowia 32:13-56. (key to 40 species based on conidial characters, 228 host records, descriptions)
Boesewinkel, H. J. 1980. The morphology of the imperfect states of powdery mildews (Erysiphaceae). Bot. Rev. 46:167-224. (key to species of Erysiphaceae based on morphology of conidial states, discussion)
Braun, U. 1980. Morphological studies in the genus *Oidium*. Flora 170:77-90. (relates type of anamorph to teleomorph of Central European species)
Braun, U. 1982. Morphological studies in the genus *Oidium*. (II). Zbl. Mikrobiol. 137:138-152. (describes and illustrates *Oidium* anamorph of 21 species of Erysiphales)
Gorter, G. J. M. A., and Eicker, A. 1984. New South African records of Erysiphaceae from Transvaal. South African J. Bot. 3:38-42. (describes and illustrates anamorphs of six species)
Hammett, K. R. W. 1977. Taxonomy of Erysiphaceae in New Zealand. N. Z. J. Bot. 15:687-711. (key to 15 taxa on 66 hosts based on conidia, descriptions and illustrations)
Jaarsveld, A. B. van 1984. Powdery mildew fungi in South Africa. Phytophylactica 16:155-166. (compares conidial states from 68 plant hosts)
Quinn, J. A., and Powell, C. C. 1981. Identification and host range of powdery mildew of begonia. Plant Dis. 65:68-70. (compares host range of races of *Oidium begoniae* with other powdery mildews on begonia, illustrations)
Weresub, L. K. 1973. *Oidium* (Fungi) nom. cons. prop. Taxon 22:696-701. (reviews use of name *Oidium*, proposes conservation)
Yarwood, C. E. 1973. Pyrenomycetes: Erysiphales. Pages 71-86 in: The Fungi Vol. IVA. G. C. Ainsworth, F. K. Sparrow, and A. S. Sussman, eds. Academic Press, New York. (presents key to species based on conidial characters, discusses conidial states)
Yarwood, C. E. 1978. History and taxonomy of powdery mildews. Pages 1-38 in: The Powdery Mildews. D. M. Spencer, ed. Academic Press, New York. (key to genera and species based on conidial characters, discusses taxonomic characters)
Zaracovitis, C. 1965. Attempts to identify powdery mildew fungi by conidial characters. Trans. Br. Mycol. Soc. 48:553-558. (conidia of 28 species classified in three groups, illustrations)
Oidium is the name used for conidial states of many species in the Erysiphaceae. *Ovulariopsis* is the name used for conidial states of *Phyllactinia* spp., many of which are included in the references listed above. *Oidium heveae* Steinm. causes a powdery mildew of rubber (CMI 508). See *Erysiphe* and Erysiphales, especially Zheng (1985).

OLIVEA Arth.
    Uredinales   7 spp.
Ono, Y., and Hennen, J. F. 1983. Taxonomy of the Chaconiaceous genera (Uredinales) Trans. Mycol. Soc. Jpn. 24:369-402. (key to seven *Olivea* spp. according to host, descriptions)
Species of *Olivea* occur on Euphorbiaceae, Lamiaceae, and Verbenaceae. *Olivea tectonae* (T. S. Ramakrishnan & K. Ramakrishnan) Mulder causes teak rust (CMI 365).

OLPIDIUM (A. Braun) J. Schroet.
    Chytridiales   25 spp.
Sparrow, F. K. 1960. Aquatic Phycomycetes. Second Edition. University of Michigan Press, Ann Arbor. 1187 pp. (key to 21 *Olpidium* spp., descriptions and illustrations, discusses all names)
Most species of *Olpidium* are parasitic on algae, aquatic fungi, rotifers, and some flowering plants. *Olpidium brassicae* (Woronin) Dang occurs on the roots of many hosts and is a vector of lettuce big vein virus.

OMPHALIA   See *Mycena*.

OOSPORA Wallr.
    Hyphomycetes   2 spp.
Sigler, L., and Carmichael, J. W. 1976. Taxonomy of *Malbranchea* and some other Hyphomycetes with arthroconidia. Mycotaxon 4:349-488. (disposes of four *Oospora* spp.)
Sigler and Carmichael (1976) discuss the systematics of *Oospora*. Many described species require proper generic placement. *Oospora citri-aurantii* (Ferraris) Sacc. occurs on citrus fruits. See *Polyscytalum*.

OPHIOBOLUS Riess
    Pleosporales   3 spp.
Holm, L. 1957. Etudes taxonomiques sur les Pleosporacees. Symb. Bot. Ups. 14:1-188. (in French, treats three Swedish *Ophiobolus* spp., descriptions)
Mueller, E. 1952. Die schweizerischen Arten der Gattung *Ophiobolus* Riess. Ber. Schweiz Bot. Ges. 62:307-339. (in German, key to 20 species in Switzerland, descriptions and illustrations)
Shoemaker, R. A. 1976. Canadian and some extralimital *Ophiobolus* species. Can. J. Bot. 54:2365-2404. (key to 31 species, descriptions and illustrations)
Walker, J. 1980. *Gaeumannomyces, Linocarpon, Ophiobolus*, and several other genera of scolecospored Ascomycetes and *Phialophora* conidial states, with a note on hyphopodia. Mycotaxon 11:1-129. (describes type species of *Ophiobolus* and disposition of 60 names)

OPHIOBOLUS - cont.
Ophiobolus species are mostly saprophytic on herbaceous stems. Walker (1980) defines the genus in a narrow sense, indicates the disposition of *Ophiobolus* names, and discusses related genera. Shoemaker (1976) recognizes a broad generic concept and provides a key to species. The cause of take-all disease of cereals is now placed in *Gaeumannomyces*.

OPHIODOTHELLA (Henn.) Hoehn.
  Polystigmatales   7 spp.
Swart, H. J. 1982. Australian leaf-inhabiting fungi XV. *Ophiodothella longispora* sp. nov. Trans. Br. Mycol. Soc. 79:566-568. (descriptions and illustrations)
Ophiodothella atromaculans (Henn.) Hoehn. is parasitic on leaves of *Lonchocarpus* and *O. longispora* Swart occurs on leaves of *Eucalyptus*. No comprehensive account of this genus exists.

OPHIOSTOMA Sydow & P. Sydow
  Ophiostomatales   25 spp.
Hoog, G. S. de, and Scheffer, R. J. 1984. *Ceratocystis* versus *Ophiostoma*: a reappraisal. Mycologia 76:292-299. (summary of evidence for recognizing *Ophiostoma*, transfers 14 *Ceratocystis* spp. without *Chalara* anamorphs to *Ophiostoma*)
Robinson-Jeffrey, R. C., and Davidson, R. W. 1968. Three new *Europhium* species with *Verticicladiella* imperfect states on blue-stained pine. Can. J. Bot. 46:1523-1527. (key to four *Europhium* spp., now placed in *Ophiostoma*, descriptions and illustrations)
Samuels, G. J., and Mueller, E. 1978. Life-history studies of Brazilian Ascomycetes 5. Two new species of *Ophiostoma* and their *Sporothrix* anamorphs. Sydowia 31:169-179. (descriptions and illustrations of two species, discussion of distinction between *Ceratocystis* and *Ophiostoma*)
Ophiostoma (syn. *Europhium*) is distinguished from *Ceratocystis* on the basis of anamorphic states and biochemical characteristics (de Hoog and Scheffer, 1984). *Ophiostoma* species have anamorphs in *Acremonium, Graphium, Gabarnaudia, Hyalodendron, Leptographium, Pesotum, Phialocephala, Phialographium, Sporothrix*, and related genera. *Ophiostoma ulmi* (Buisman) Nannf. causes dutch elm disease (CMI 361 as *Ceratocystis ulmi*). See *Ceratocystis* for additional references.

OPHIOVALSA Petr.
  Diaporthales   10 spp.
Glawe, D. A., and Jensen, J. D. 1986. *Ophiovalsa* in the Pacific Northwest. Mycotaxon 25:645-655. (key to four species, descriptions and illustrations)
Petrak, F. 1965. Ueber die Gattung *Cryptospora* Tul. Sydowia

OPHIOVALSA - cont.
19:268-278. (replaces *Cryptospora* with *Ophiovalsa*, discusses nine species)
Species of *Ophiovalsa* occur on branches of deciduous trees and shrubs. Barr (1978) includes *O. suffusa* (Fr.) Petr., anamorph *Disculina vulgaris* (Fr.:Fr.) Sutton (syn. *D. neesii* (Corda) Hoehn.; Sutton, 1980), and briefly discusses the other described species.

OVULARIA Sacc.
    Hyphomycetes   13 spp.
Arx, J. A. von 1983. *Mycosphaerella* and its anamorphs. Proc. K. Ned. Akad. Wet., Ser. C. 86:15-54. (key to anamorph genera, discusses generic characters)
Deighton, F. C. 1984. *Tretovularia*, a new hyphomycetous genus. Trans. Br. Mycol. Soc. 82:743-745. (transfers *O. villiana* to *Tretovularia*, description and illustration)
Ondrej, M., and Zavrel, H. 1972. [Funde von parasitischen Pilzen (Fungi imperfecti) der Gattung *Ovularia* Sacc. aus Maehren und Schlesien]. Acta Mus. Silesiae, Ser. A 21:141-150. (in Czechoslovakian with German summary, describes and illustrates 13 species, host list)
Rao, R. 1968. A new species of *Ovularia* from India. Mycopathol. Mycol. Appl. 34:47-48. (describes and illustrates *O. indica*)
Ovularia is related to *Ramularia* (Arx, 1983). These species are anamorphs of *Mycosphaerella*. *Ovularia karelii* Petr. causes leaf spot of *Onobrychis heliobrychis* (CMI 863).

OVULARIOPSIS   See *Oidium*.

PAECILOMYCES Bainier
    Hyphomycetes   31 spp.
Brown, A. H. S., and Smith, G. 1957. The genus *Paecilomyces* Bainier and its perfect stage *Byssochlamys* Westling. Trans. Br. Mycol. Soc. 40:17-89. (key to 29 species, descriptions and illustrations)
Onions, A. H. S., and Barron, G. L. 1967. Monophialidic species of *Paecilomyces*. Mycol. Pap. 107:1-25. (reviews related genera, key to 11 species, descriptions and illustrations)
Pitt, J. I., and Hocking, A. D. 1985. Interfaces among genera related to *Aspergillus* and *Penicillium*. Mycologia 77:810-824. (discusses characteristics of several related genera including *Paecilomyces*, describes *Paecilomyces pascua*, illustrations)
Samson, R. A. 1974. *Paecilomyces* and some allied Hyphomycetes. Stud. Mycol. 6:1-119. (discusses related genera, key to genera, key to 31 species, descriptions and illustrations)
Several good references are available for identification of *Paecilomyces* and its teleomorphs in *Byssochlamys* and *Talaromyces*, including Domsch et al. (1980). Some species cause food spoilage

PAECILOMYCES - cont.
especially of canned products; others occur on insects or are
isolated from soil. Species include: *Paecilomyces breviramosus*
Bissett, on larvae of Lepidoptera (F. Can. 159); *P. carneus* (Duche &
R. Heim) A. H. S. Brown & G. Smith, on insects (F. Can. 152); *P.
farinosus* (Holm:Fr.) A. H. S. Brown & G. Smith, on insects (CMI 613,
F. Can. 153); *P. fumosoroseus* (Wize) A. H. S. Brown & G. Smith, on
insects (CMI 614, F. Can. 154); *P. inflatus* (Burnside) J. W.
Carmichael, from soil (F. Can. 155); *P. lilacinus* (Thom) Samson,
from soil and a nematode parasite (F. Can. 156); *P. marquandii*
(Massee) Hughes, from soil (F. Can. 157); *P. tenuipes* (Peck) Samson,
on larvae of Lepidoptera and pupae and other insects (CMI 615, F.
Can. 158); and *P. variotii* Bainier, on soil, seeds, and wood (F.
Can. 151).

PAPULASPORA G. Preuss
    Hyphomycetes   10 spp.
Weresub, L. K., and LeClair, P. M. 1971. On *Papulaspora* and
  bulbilliferous basidiomycetes *Burgoa* and *Minimedusa*. Can. J.
  Bot. 49:2203-2213. (reviews definition of papulospores, as
  papulaspores, and redisposes several species)
This genus is characterized by having papulospores as defined by
Weresub and LeClair (1971). No comprehensive treatment exists.
Papulaspora sepedonioides G. Preuss occurs on corms of *Crocus* and
Gladiolus as well as *Helianthus* tubers (F. Can. 27).

PARACERCOSPORA   See *Cercospora*.

PARAPHAEOSPHAERIA O. Eriksson
    Pleosporales   6 spp.
Koponen, H., and Makela, K. 1975. *Leptosphaeria* s. lat.
  (*Keissleriella, Paraphaeosphaeria, Phaeosphaeria*) on Gramineae in
  Finland. Ann. Bot. Fenn. 12:141-160. (descriptions and
  illustrations of *Paraphaeosphaeria michotii*)
Shoemaker, R. A., and Babcock, C. E. 1985. Canadian and some
  extralimital *Paraphaeosphaeria*. Can. J. Bot. 63:1284-1291. (key
  to six species, descriptions and illustrations)
This genus is a recent segregate of *Leptosphaeria* and *Phaeosphaeria*.
Sivanesan (1984) provides a key to four species with *Coniothyrium*
anamorphs. *Paraphaeosphaeria michotii* (Westendorp) O. Eriksson
causes a disease of sugarcane (CMI 144 as *Leptosphaeria michotii*).
The anamorph of *P. michotii* is *Coniothyrium scirpi* Trail.

PASSALORA Fr.
    Hyphomycetes   3 spp.
Deighton, F. C. 1967. Studies on *Cercospora* and allied genera II.
  Passalora, Cercosporidium and some species of *Fusicladium* on
  Euphorbia. Mycol. Pap. 112:1-80. (key to three *Passalora* spp.,

PASSALORA - cont.
descriptions and illustrations)
See Ellis (1971, 1976) and *Cercospora*.

PELLICULARIA   See *Ceratobasidium* and *Thanatephorus*.

PENICILLIUM Link:Fr.
Hyphomycetes   96 spp.
Bridge, P. D., and Hawksworth, D. L. 1984. The API ZYM enzyme testing system as an aid to the rapid identification of *Penicillium* isolates. Microbiol. Sci. 1:232-234. (useful for identification to sections and series)
Kulik, M. M. 1968. A compilation of descriptions of new *Penicillium* species. U.S. Dep. Agric., Agric. Handb. 351:1-80. (key to 113 species described since Raper and Thom, 1949, descriptions)
Onions, A. H. S., Bridge, P. D., and Paterson, R. R. 1984. Problems and prospects for the taxonomy of *Penicillium*. Microbiol. Sci. 1:185-189. (summarizes problems in identification despite several modern monographic treatments)
Pitt, J. I. 1974. A synoptic key to the genus *Eupenicillium* and to sclerotigenic *Penicillum* spp. Can. J. Bot. 52:2231-2236. (synoptic key to 36 European *Eupenicillium* spp. and 22 sclerotigenic *Penicillium* spp.)
Pitt, J. I. 1979. The Genus *Penicillium* and its Teleomorphic States *Eupenicillium* and *Talaromyces*. Academic Press, New York. 634 pp. (a comprehensive account with keys, descriptions and illustrations)
Pitt, J. I. 1985. A Laboratory Guide to the Common *Penicillium* Species. Commonwealth Scientific and Industrial Research Organization, North Ryde, New South Wales, Australia. 182 pp. (key to subgenera, synoptic key to 30 species, descriptions and illustrations)
Pitt, J. I., and Hocking, A. D. 1985. Interfaces among genera related to *Aspergillus* and *Penicillium*. Mycologia 77:810-824. (discusses characteristics of similar genera, describes two *Penicillium* spp., illustrations)
Ramirez, C. 1982. Manual and Atlas of the Penicillia. Elsevier Biomedical, New York. 974 pp. (keys to sections and species, descriptions and illustrations)
Raper, K. B., and Thom, C. 1949. A Manual of the Penicillia. The Williams and Wilkins Co., Baltimore, MD. 875 pp. (the classic work, now outdated, keys, descriptions and illustrations)
Samson, R. A., and Pitt, J. I., eds. 1985. Advances in *Penicillium* and *Aspergillus* Systematics. Plenum Press, New York. 483 pp. (keys provided in some chapters, descriptions and illustrations of groups)
Samson, R. A., Stolk, A. C., and Hadlok, R. 1976. Revision of the subsection *Fasciculata* of *Penicillium* and some allied species.

PENICILLIUM - cont.
Stud. Mycol. 11:1-47. (synoptic key, descriptions and
illustrations)
Stolk, A. C., and Samson, R. A. 1983. The ascomycete genus
Eupenicillium and *Penicillium* anamorphs. Stud. Mycol. 23:1-149.
(monograph of *Eupenicillium* spp. and sclerotial *Penicillium*
anamorphs, synoptic key using teleomorph and anamorph characters,
keys to 33 *Eupenicillium* taxa, descriptions and illustrations)
Penicillium and its teleomorphs *Eupenicillium* and *Talaromyces* are
well-documented, although it is still difficult to identify species.
Several diseases are caused by *Penicillium* species including:
Penicillium digitatum Sacc., green mold of citrus fruit (CMI 96); *P.
expansum* Link, blue mold of apple (CMI 97); *P. gladioli* McCulloch &
Thom, storage rot of gladiolus corms (CMI 98); and *P. italicum*
Wehmer, blue mold of citrus (CMI 99).

PERICONIA Tode:Fr.
    Hyphomycetes   30 spp.
Ellis (1971, 1976) provides the most recent, comprehensive treatment
of these species causing leaf spots on a variety of hosts. *Periconia
circinata* (Mangin) Sacc. causes milo disease of *Sorghum* (CMI 167)
and *P. macrospinosa* Lefebvre & A. G. Johnson is isolated from
various plants (CMI 168).

PERICONIELLA Sacc.
    Hyphomycetes   23 spp.
Ellis, M. B. 1967. Dematiaceous Hyphomycetes. VIII. *Periconiella,
Trichodochium*, etc. Mycol. Pap. 111:1-46. (key to 24
Periconiella spp., descriptions and illustrations)
Ellis (1971, 1976) provides the most recent account of these species
causing leaf spots on a variety of hosts.

PERIDERMIUM (Link) Schmidt & Kunze
    Uredinales   20 spp.
Saho, H. 1981. Notes on Japanese rust fungi VII. *Peridermium
yamabense* sp. nov., a pine-to-pine stem rust of white pines.
Trans. Mycol. Soc. Jpn. 22:27-36. (describes and illustrates
three species)
Anamorphs of Uredinales placed in *Peridermium* are restricted to the
peridiate aecial states on gymnosperms of the genera *Coleosporium,
Cronartium, Endocronartium, Milesina*, and *Pucciniastrum*. See
Aecidium and Uredinales.

PERONOPLASMOPARA   See *Pseudoperonospora*.

PERONOSCLEROSPORA (Ito) Shirai & K. Hara
    Peronosporales   9 spp.
Shaw, C. G. 1978. *Peronosclerospora* species and other downy mildews

PERONOSCLEROSPORA - cont.
of the Gramineae. Mycologia 70:594-604. (describes genus, lists
nine *Peronosclerospora* spp. and synonyms)
Shaw, C. G., and Waterhouse, G. M. 1980. *Peronosclerospora* (Ito)
Shirai & K. Hara antedates *Peronosclerospora* (Ito) C. G. Shaw.
Mycologia 72:425-426. (nomenclatural note)
The genus *Peronosclerospora* was established for those *Sclerospora*
species producing true conidia that germinate by germ tubes. These
species are obligate parasites of members of the Poaceae. No recent
monograph of the genus exists. See Peronosporales for additional
references. Diseases include: *Peronosclerospora philippinensis*
(Weston) C. G. Shaw, downy mildew of corn (CMI 454 as *Sclerospora
philippinensis*); *P. sacchari* (Miyake) Shirai & K. Hara, downy mildew
of sugarcane and corn (CMI 453 as *Sclerospora sacchari*); and *P.
sorghi* (Western & Uppal) C. G. Shaw, sorghum downy mildew (CMI 451
as *Sclerospora sorghi*, CMI 761).

PERONOSPORA Corda
   Peronosporales   75 spp.
Chereponova, P. 1982. [Species of the genus *Peronospora* in the flora
   of Leningrad Oblast, Russian SFSR, USSR.] Vestn. Leningr. Univ.
   Biol. 1:118-122. (in Russian with English summary, host-pathogen
   list)
Rao, V. G. 1968. The genus *Peronospora* Corda in India. Nova Hedw.
   16:269-282. (includes 28 species with brief descriptions)
Solheim, W. G., and Gilbertson, R. L. 1973. *Peronospora* species in
   Arizona. Mycopathol. Mycol. Appl. 49:153-159. (synoptic table
   comparing eight species on Polygonaceae, briefly describes and
   illustrates eight additional species)
Yerkes, W. D., Jr., and Shaw, C. G. 1959. Taxonomy of the *Peronospora*
   species on Cruciferae and Chenopodiaceae. Phytopathology
   49:499-507. (a single species is recognized on each family, *P.
   parasitica* on Cruciferae (Brassicaceae) and *P. farinosa* on
   Chenopodiaceae.)
This genus includes many important downy mildews. The Peronosporales
are often treated based on host family or on a specific geographic
area. Much of the literature on the Peronosporales has been reviewed
in Spencer (1981). See Peronosporales. Important diseases include:
Peronospora alta Fuckel, disease of psyllium (CMI 683); *P. anemones*
Tramier, downy mildew of anemones (CMI 684); *P. antirrhini* J.
Schroet., downy mildew of antirrhinum (CMI 685); *P. arborescens*
(Berk.) Casp., downy mildew of opium poppy (CMI 686); *P. cytisi*
Rostr., downy mildew of laburnum (CMI 762); *P. destructor* (Berk.)
Casp. ex Berk., downy mildew of onion (CMI 456); *P. dianthi* de Bary
sensu lato, downy mildew of annual *Dianthus* spp. (CMI 763); *P.
dianthicola* Barthelet, downy mildew of carnation (CMI 764); *P.
farinosa* (Fr.) Fr. f. sp. *betae* Byford, downy mildew of beet (CMI
765); *P. grisea* (Unger) Unger, downy mildew of *Hebe* spp. (CMI 766);

PERONOSPORA - cont.
P. hariotii Gaeumann, downy mildew of *Buddleja* (CMI 767); *P. jaapiana* Magnus, downy mildew of rhubarb (CMI 687); *P. lamii* A. Braun in Rabenh., downy mildew of *Lamium* spp. and other hosts (CMI 688); *P. manshurica* (Naoum.) Sydow, downy mildew of soybean (CMI 689); *P. sparsa* Berk., downy mildew of rose (CMI 690); *P. trifoliorum* de Bary *sensu lato*, downy mildew of alfalfa (CMI 768); and *P. viciae* (Berk.) Casp., downy mildew of peas (CMI 455).

PESOTUM Crane & Schoknecht
    Hyphomycetes   2 spp.
Crane, J. L., and Schoknecht, J. D. 1973. Conidiogenesis in *Ceratocystis ulmi*, *Ceratocystis piceae*, and *Graphium penicillioides*. Amer. J. Bot. 60:346-354. (places anamorphs of two *Ceratocystis* spp. in *Pesotum* based on conidiogenesis, descriptions and illustrations)
Pesotum ulmi (Schwarz) Crane & Schoknecht (syn. *Graphium ulmi* Schwarz) is the anamorph of *Ophiostoma ulmi* (Buisman) Nannf., cause of Dutch elm disease (CMI 361 as *Ceratocystis ulmi*).

PESTALOSPHAERIA Barr
    Sphaeriales   5 spp.
Barr, M. E. 1975. *Pestalosphaeria*, a new genus in the Amphisphaeriaceae. Mycologia 67:187-194. (describes and illustrates the new genus with one species)
Nag Raj, T. R. 1985. Redisposal and redescriptions in the *Monochaetia-Seiridium, Pestalotia-Pestalotiopsis* complexes. II. *Pestalotiopsis besseyii* (Guba) comb. nov. and *Pestalosphaeria varia* sp. nov. Mycotaxon 22:52-63. (key to five *Pestalosphaeria* spp.)
Shoemaker, R. A., and Simpson, J. A. 1981. A new species of *Pestalosphaeria* on pine with comments on the generic placement of the anamorph. Can. J. Bot. 59:986-991. (key to four species, discussion and illustrations)
Pestalosphaeria species have *Pestalotiopsis* anamorphs.

PESTALOTIA De Not.
    Coelomycetes   222 spp.
Griffiths, D. A., and Swart, H. J. 1974. Conidial structure in *Pestalotia pezizoides*. Trans. Br. Mycol. Soc. 63:169-173. (description and illustrations of conidial wall of *P. pezizoides* using transmission electron microscopy, discussion of implications for distinguishing *Pestalotia* and *Pestalotiopsis*)
Guba, E. F. 1961. Monograph of *Monochaetia* and *Pestalotia*. Harvard University Press, Cambridge, MA. 342 pp. (keys to over 300 *Pestalotia* spp., many of which belong in *Pestalotiopsis*, descriptions and illustrations, based primarily on host)
Nag Raj, T. R. 1985. Redisposals and redescriptions in the

PESTALOTIA - cont.
Monochaetia-Seiridium, Pestalotia-Pestalotiopsis complexes. I.
The correct name for the type species of *Pestalotiopsis*.
Mycotaxon 22:43-51. (key to seven genera)
Steyaert, R. L. 1949. Contribution a l'etude monographique de
Pestalotia de Not. et *Monochaetia* Sacc. (*Truncatella* gen. nov. et
Pestalotiopsis gen. nov.). Bull. Jard. Bot. Brux. 19:285-354.
(in French, keys to sections and 46 species, descriptions and
illustrations)
Sutton, B. C. 1969. Forest microfungi. III. The heterogeneity of
Pestalotia deNot. section sexloculatae Klebahn sensu Guba. Can.
J. Bot. 47:2083-2094. (discussion of generic concept of
Pestalotia, retains only *P. pezizoides*)
Sutton (1969, 1980) regards this genus as monotypic and provides a
description of *Pestalotia pezizoides* De Not., the type species. Many
species have been transferred to *Pestalotiopsis*. Guba (1961)
provides the only monographic treatment of *Pestalotia* in the broad
sense. See *Pestalotiopsis*.

PESTALOTIOPSIS Steyaert
    Coelomycetes   200 spp.
Guba, E. F. 1961. Monograph of *Monochaetia* and *Pestalotia*. Harvard
    University Press, Cambridge, MA. 342 pp. (keys to over 300
    Pestalotia spp., many of which belong in *Pestalotiopsis*,
    descriptions and illustrations, based primarily on host)
Nag Raj, T. R. 1985a. Redisposals and redescriptions in the
    Monochaetia-Seiridium, Pestalotia-Pestalotiopsis complexes. I.
    The correct name for the type species of *Pestalotiopsis*.
    Mycotaxon 22:43-51. (key to seven genera, describes and
    illustrates *P. maculans* of which *P. guepini* is a synonym)
Nag Raj, T. R. 1985b. Redisposal and redescriptions in the
    Monochaetia-Seiridium, Pestalotia-Pestalotiopsis complexes. II.
    Pestalotiopsis besseyii (Guba) comb. nov. and *Pestalosphaeria
    varia* sp. nov. Mycotaxon 22:52-63. (describes and illustrates
    anamorphs and teleomorphs)
Nag Raj, T. R. 1985c. Redisposals and redescriptions in the
    Monochaetia-Seiridium, Pestalotia-Pestalotiopsis complexes. IV.
    On *Monochaetia miersi*. Mycotaxon 22:71-75. (places *M. miersi* in
    Pestalotiopsis)
Nag Raj, T. R. 1986. Redisposals and redescriptions in the
    Monochaetia-Seiridium, Pestalotia-Pestalotiopsis complexes. VI.
    Pestalotia decolorata and *Pestalotia heteromorpha*. Mycotaxon
    26:199-210. (descriptions and illustrations of two *Pestalotiopsis*
    spp., now placed in *Pestalozzina*, provides synoptic plate
    comparing *Pestalozzina, Pestalotiopsis*, and *Bartilinia*)
Nag Raj, T. R. 1986. Redisposals and redescriptions in the
    Monochaetia-Seiridium, Pestalotia-Pestalotiopsis complexes. VII.
    Pestalotia citrina, P. maura and *Pestalotiopsis uvicola*.

PESTALOTIOPSIS - cont.
Mycotaxon 26:211-222. (descriptions and illustrations of three *Pestalotiopsis* spp.)
Steyaert, R. L. 1961. Type specimens of Spegazzini's collections in the *Pestalotiopsis* and related genera. (Fungi Imperfecti: Melanconiales). Darwiniana 12:157-175. (type studies of 29 taxa, indicates taxonomic disposition, descriptions and illustrations)
Sutton (1980) treats the type species, *Pestalotiopsis guepini* (Desmaz.) Steyaert, now *P. maculans* (Corda) Nag Raj (Nag Raj, 1985a), and discusses the generic confusion and taxonomic problems of *Pestalotiopsis* and *Pestalotia*. Guba (1961) remains the only comprehensive treatment of *Pestalotiopsis* species, mostly as *Pestalotia*. Important diseases caused by *Pestalotiopsis* species include: *Pestalotiopsis dichaeta* (Speg.) Steyaert, diseases of *Araucaria* spp. and other hosts (CMI 675); *P. funerea* (Desmaz.) Steyaert, diseases of conifers (CMI 514); *P. maculans* (Corda) Nag Raj, diseases of *Camellia* and other hosts (CMI 320 as *P. guepini*); *P. mangiferae* (Henn.) Steyaert, grey leaf spot of mango (CMI 676); *P. palmarum* (Cooke) Steyaert, leaf spot of Palmaceae (CMI 319); *P. psidii* (Pat.) Mordue, grey leaf spot and post-harvest fruit canker of guava (CMI 515); and *P. theae* (Sawada) Steyaert, grey blight of tea (CMI 318). See the teleomorph genus *Pestalosphaeria*.

PESTALOZZINA See *Pestalotiopsis*.

PEZICULA Tul. & C. Tul.
    Helotiales   30 spp.
Dennis, R. W. G. 1974. New or interesting British microfungi, II. Kew Bull. 29:157-179. (key to 16 *Pezicula* spp., comments and drawings of seven species)
Groves, J. W. 1939. Some *Pezicula* species and their conidial stages. Can. J. Res. Sect. C. 17:125-143. (establishes genetic connection for five *Pezicula* spp. and their anamorphs, descriptions and illustrations)
Groves, J. W. 1940. Three *Pezicula* species occurring on *Alnus*. Mycologia 32:112-123. (descriptions of three species on host and in culture, illustrations)
Kennel, W., and Weiler, R. 1984. Zur Ursaeche der Lentizellenroete beim Apfel. Z. Pflanzenkr. Pflanzensch. 91:552-555. (in German with English summary, the cause of the red lenticel diseases of apples caused by *Pezicula alba*, not *P. malicorticis*)
Members of this genus cause cankers and other diseases on both conifers and hardwood trees. Funk (1981) includes two species. No monographic treatment exists.

PEZIZELLA Fuckel
    Helotiales   40 spp.
Dennis, R. W. G. 1956. A revision of the British Helotiaceae in the

PEZIZELLA - cont.
herbarium of the Royal Botanic Gardens, Kew, with notes on
related European species. Mycol. Pap. 62:1-216. (key to 12
Pezizella spp., descriptions and illustrations)
Funk (1981) includes *Pezizella chapmannii* Whitney & Funk occurring
inside bark beetle galleries of conifers. *Pezizella oenotherae*
(Cooke & Ellis) Sacc. causes a disease of strawberry (CMI 535).

PHACIDIOPYCNIS Potebnia
    Coelomycetes   3 spp.
DiCosmo, F., Nag Raj, T. R., and Kendrick, W. B. 1984. A revision of
the Phacidiaceae and related anamorphs. Mycotaxon 21:1-234.
(descriptions and illustrations of *P. balsamicola* var.
balsamicola and *P. pseudotsugae*, both as *Apostrasseria* spp., and
P. piri)
DiCosmo et al. (1984) transfer *Phacidiopycnis pseudotsugae* (M.
Wilson) G. Hahn and *P. balsamicola* Funk var. *balsamicola* to
Apostrasseria. They retain *P. piri* (Fuckel) Weindlmayr, anamorph of
Potebniamyces pyri (Berk. & Broome) Dennis. Sutton (1980) includes
descriptions and illustrations of two additional *Phacidiopycnis*
species. *Phacidiopycnis tuberivora* (H. T. Guessow & W. R. Foster)
Sutton (syn. *Phomopsis tuberivora* H. T. Guessow & W. R. Foster)
causes stem-end hard rot or dry rot of potatoes (CMI 823).

PHACIDIUM Fr.
    Helotiales   26 spp.
DiCosmo, F., Nag Raj, T. R., and Kendrick, W. B. 1984. A revision of
the Phacidiaceae and related anamorphs. Mycotaxon 21:1-234. (a
complete monographic treatment, synoptic key to 26 *Phacidium*
spp., descriptions and illustrations, disposition of an
additional 153 names)
Reid, J., and Cain, R. F. 1962. Studies on the organisms associated
with "snow blight" of conifers in North America. II. Some
species of the genera *Phacidium, Lophophacidium, Sarcotrochila,*
and *Hemiphacidium*. Mycologia 54:481-497. (descriptions and
illustrations)
DiCosmo et al. (1984) present an excellent monograph and update the
taxonomy of this group. Anamorphs are placed in *Apostrasseria* and
Ceuthospora. Funk (1985) compares four species of *Phacidium* on
conifers. Funk (1981) treats two varieties of *Phacidium balsamicola*
(Smerlis) DiCosmo, Nag Raj & Kendrick and *P. coniferarum* (Hahn)
DiCosmo, Nag Raj & Kendrick causing twig dieback and basal canker
(CMI 517 as *Potebniamyces coniferarum*), both in *Potebniamyces*.
Additional diseases caused by *Phacidium* spp. include: *Phacidium
infestans* Karst., snow blight of pine (CMI 652) and *P. pini-cembrae*
(Rehm) Terrier, snow blight of pine (CMI 653).

PHAEOCRYPTOPUS Naumov
  Pleosporales   5 spp.
Stone, J., and Carroll, G. 1985. Observations of the development of ascocarps in *Phaeocryptopus gaeumanni* and on the possible existence of an anamorphic state. Sydowia 38:317-323. (description of ascomata development and of possible phialidic anamorph)
Funk (1985) includes descriptions and illustrations of two species. *Phaeocryptopus gaeumannii* (Rohde) Petr. causes Swiss needle cast on *Pseudotsuga menziesii*.

PHAEOCYTOSTROMA Petr.
  Coelomycetes   4 spp.
Sutton, B. C. 1964. Coelomycetes III. *Annellolacinia* gen. nov., *Aristostoma*, *Phaeocytostroma*, *Seimatosporium*, etc. Mycol. Pap. 97:1-42. (key to four *Phaeocytostroma* taxa, descriptions and illustrations)
Sutton (1980) presents a key to six taxa with descriptions and illustrations. *Phaeocytostroma sacchari* (Ellis & Everh.) Sutton (syn. *Melanconium sacchari* Massee) causes rind disease and sour rot of sugarcane (CMI 87).

PHAEOISARIOPSIS Ferraris
  Hyphomycetes   15 spp.
Brown, L. G., and Morgan-Jones, G. 1976. Notes on Hyphomycetes. XI. Additions to the genera *Cercosporidium*, *Passalora*, and *Phaeoisariopsis*. Mycotaxon 4:299-306. (includes *Phaeoisariopsis sphaeroidea*, description and illustrations)
Jong, S. C., and Morris, E. F. 1968. Studies on the synnematous Fungi Imperfecti III. *Phaeoisariopsis*. Mycopathol. Mycol. Appl. 34:263-272. (descriptions of 11 species with illustrations)
Morgan-Jones, G. 1978. Notes on Hyphomycetes. XXII. *Phaeoisariopsis bambusicola* sp. nov. Mycotaxon 7:130-132. (description and illustration)
Morgan-Jones, G., and Brown, L. G. 1976. Notes on Hyphomycetes. XIII. Concerning two species of *Phaeoisariopsis*. Mycotaxon 4:493-497. (includes two *Phaeoisariopsis* spp. on magnoliaceous hosts, descriptions and illustrations)
Ellis (1971, 1976) includes descriptions and illustrations of eleven species according to host family. Species include: *Phaeoisariopsis bonducellae* (Henn.) Deighton, leaf spot of *Caesalpinia bonducella* (CMI 844); *P. chonemorphae* Rajak & Pandey, leaf spot of *Chonemorpha macrophylla* (CMI 845); *P. glochidii* (Petch) M. B. Ellis, leaf spot of *Glochidion* spp. (CMI 846); *P. griseola* (Sacc.) Ferraris, angular leaf spot of beans (CMI 847); *P. melanochaeta* (Ellis & Everh.) Deighton, leaf spot of *Celastrus* spp. (CMI 848); *P. robiniae* (Shear) Deighton, leaf spot of *Robinia pseudoacacia* (CMI 849); and *P. simulata* (Ellis & Everh.) L. G. Brown & Morgan-Jones, leaf spot of

PHAEOISARIOPSIS - cont.
Cassia spp. (CMI 850).

PHAEOLUS Pat.
    Aphyllophorales   2 spp.
Phaeolus schweinitzii (Fr.:Fr.) Pat. (syn *Polyporus schweinitzii*
Fr.:Fr.) causes red-brown butt rot of conifers. See Aphyllophorales.

PHAEORAMULARIA Muntanola
    Hyphomycetes   22 spp.
Deighton, F. C. 1979. Studies on *Cercospora* and allied genera. VII.
New species and redispositions. Mycol. Pap. 144:1-56. (describes
and illustrates 12 *Phaeoramularia* spp., mostly transferred from
Cercospora)
Ellis (1976) lists numerous species according to host family with
brief descriptions and illustrations. *Phaeoramularia angolensis* (de
Carvalho & O. Mendes) P. M. Kirk causes leaf and fruit spot on
Citrus spp. (CMI 843). See *Cercospora*.

PHAEOSEPTORIA Speg.
    Coelomycetes   17 spp.
Punithalingam, E. 1980. New microfungi from cereals. Nova Hedw.
32:585-606. (key to 13 graminicolous *Phaeoseptoria* spp., brief
descriptions)
Punithalingam, E. 1981. New microfungi from cereals and grasses. II.
Nova Hedw. 34:67-95. (describes and illustrates *P. setariae* on
Setaria)
No comprehensive account exists for species occurring on
non-graminicolous hosts. Species causing diseases include:
Phaeoseptoria musae Punithalingam, leaf spot of *Musa* spp. (CMI 772);
P. oryzae Miyake, a minor leaf disease of rice (CMI 664); and *P.
vermiformis* Punithalingam & J. M. Waller, leaf spot of *Triticale*,
wheat, and oats (CMI 824). See the teleomorph *Leptosphaeria*.

PHAEOSPHAERIA Miyake
    Pleosporales   20 spp.
Eriksson, O. 1967. On graminicolous pyrenomycetes from Fennoscandia
2. Phragmosporous and scolecosporous species. Ark. Bot.
6:381-440. (key to 16 *Phaeosphaeria* spp., descriptions and
illustrations)
Hedjaroude, G. A. 1968. Etudes taxonomiques sur les *Phaeosphaeria*
Miyake et leurs formes voisines (Ascomycetes). Sydowia
22:57-107. (in French, comprehensive, later authors disagree with
species concepts)
Holm, L., and Holm, K. 1981. Nordic equiseticolous pyrenomycetes.
Nord. J. Bot. 1:109-119. (key to five *Phaeosphaeria* taxa,
descriptions and illustrations)

PHAEOSPHAERIA - cont.
Koponen, H., and Makela, K. 1975. *Leptosphaeria* s. lat.
  (*Keissleriella, Paraphaeosphaeria, Phaeosphaeria*) on Gramineae in
  Finland. Ann. Bot. Fenn. 12:141-160. (includes descriptions and
  illustrations of ten *Phaeosphaeria* spp.)
Leuchtmann, A. 1984. Ueber *Phaeosphaeria* Miyake und andere bitunicate
  Ascomyceten mit mehrfach querseptierten Ascosporen. Sydowia
  37:75-194. (in German with English summary, key to related genera
  and 67 species, descriptions, illustrations, host index)
Otani, Y. 1976. Graminicolous fungi of the genus *Phaeosphaeria* and
  its allied genera in Japan. 1. Bull. Nat. Sci. Mus., Ser. B.
  (Tokyo) 2:87-98. (describes and illustrates six *Phaeosphaeria*
  spp., one *Paraphaeosphaeria* sp.)
Phaeosphaeria is a segregate of *Leptosphaeria* and includes mostly
graminicolous species. See the related genus *Paraphaeosphaeria*. The
anamorphs are placed in *Hendersonia* and *Stagonospora*. Diseases
include *Phaeosphaeria avenaria* (G. F. Weber) O. Erikss. f. sp.
avenaria, speckle blotch of oats (CMI 312 as *Leptosphaeria avenaria*
f. sp. *avenaria*), and *P. nodorum* (E. Mueller) Hedjaroude, glume
blotch of wheat (CMI 86 as *Leptosphaeria nodorum*, F. Can. 240 under
the anamorph *Stagonospora nodorum*).

PHAKOPSORA Dietel
  Uredinales   50 spp.
Bromfield, K. R. 1984. Soybean rust. American Phytopathological
  Society Monograph 11:1-65. (all aspects of the disease)
Sinclair, J. B. 1982. Compendium of Soybean Diseases. Second
  Edition. American Phytopathological Society, St. Paul, MN. 104
  pp. (pathology, description and illustrations)
No comprehensive account of this genus exists. See Cummins and
Hiratsuka (1983). *Phakopsora pachyrhizi* Sydow, the cause of brown
rust of sorghum and other legumes (CMI 589), is discussed in
Bromfield (1984) and Sinclair (1982). *Phakopsora gossypii* (Arth.)
Hirat. causes cotton rust (CMI 172).

PHELLINUS Quel.
  Aphyllophorales   80 spp.
Fiasson, J.-L., and Niemela, T. 1984. The Hymenochaetales: a
  revision of the European poroid taxa. Karstenia 24:14-28. (deals
  primarily with taxa above the species level)
Gilbertson, R. L. 1979. The genus *Phellinus* (Aphyllophorales:
  Hymenochaetaceae) in western North America. Mycotaxon 9:51-89.
  (key to 29 species, descriptions and illustrations)
Jahn, H. 1981. Die resupinaten *Phellinus*-Arten in Mitteleuropa.
  Bibl. Mycol. 81:37-151. (in German with English summary, key to
  22 species, descriptions and illustrations, originally published
  in 1966, extensive supplement in 1981 version)
Roy, A. 1979. Taxonomy of *Fomes durissimus*. Mycologia 71:1005-1009.

PHELLINUS - cont.
(transfers this pathogenic species to *Phellinus*)
This genus contains a number of wood-rotting pathogens. Monographic studies of limited geographical areas are available. Diseases include: *Phellinus igniarius* (L.:Fr.) Quelet, white heart rot of many hardwoods (CMI 194); *P. noxius* (Corner) G. Cunn., brown root rot especially of rubber, oil palm, and tea (CMI 195); *P. pomaceus* (Pers.) Maire, heart rot of plum (CMI 196); *P. robustus* (P. Karst.) Bourd., yellow trunk rot of oak (CMI 197); and *P. weirii* (Murrill) R. Gilb. (syn. *Poria weirii* (Murrill) Murrill), laminated butt rot and yellow ring rot of conifers (CMI 323 as *Inonotus weirii*). See Aphyllophorales.

PHIALOPHORA Medlar
  Hyphomycetes   12 spp.
Cole, G. T., and Kendrick, W. B. 1973. Taxonomic studies of *Phialophora*. Mycologia 65:661-688. (key to six *Phialophora* spp. causing blue stain in wood, descriptions and illustrations)
Gams, W., and Holubova-Jechova, V. 1976. *Chloridium* and some other dematiaceous hyphomycetes growing on decaying wood. Stud. Mycol. 13:1-99. (includes *Phialophora* sect. *Catenulatae* with key to nine *Phialophora* spp., descriptions and illustrations)
Schol-Schwarz, M. B. 1970. Revision of the genus *Phialophora* (Moniliales). Persoonia 6:59-94. (key to 12 *Phialophora* spp., descriptions and illustrations)
Sivasithamparam, K. 1975. *Phialophora* and *Phialophora*-like fungi occurring in the root region of wheat. Aust. J. Bot. 23:193-212. (descriptions and illustrations of eight *Phialophora* spp.)
This heterogeneous genus has been well-studied. Several references with keys, illustrations, and descriptions are available including Ellis (1971, 1976) and Domsch et al. (1980). Teleomorph states include *Ascocoryne, Coniochaeta, Lasiosphaeria, Mollisia, Pseudopezicula*, and *Pyrenopeziza*. Some species of *Phialophora* are saprophytic causing blue stain in wood. Others are parasitic on vascular plants including: *Phialophora asteris* (Dowson) Burge & Isaac, aster wilt, vascular wilt of michaelmas daisies (CMI 505); *P. cinerescens* (Wollenw.) Beyma, carnation wilt (CMI 503); and *P. parasitica* Ajello, Georg & Wang, diseases of various hosts (CMI 504). See also *Gaeumannomyces* for references dealing with *Phialophora* species associated with roots of Poaceae.

PHLOEOSPORA Wallr.
  Coelomycetes   160 spp.
Sutton, B. C., and Pollack, F. G. 1974. Microfungi on *Cercocarpus*. Mycopathol. Mycol. Appl. 52:331-351. (discusses generic typification, describes and illustrates *Phloeospora cercocarpi* and *P. ulmi*)
This genus is related to *Cercospora, Cylindrosporium, Septoria*, and

PHLOEOSPORA - cont.
Septogloeum. Sutton (1980) provides an account of six species; otherwise little taxonomic literature exists. Teleomorphs are placed in *Mycosphaerella*.

PHLOEOSPORELLA Hoehn.
    Coelomycetes   5 spp.
Constantinescu, O. 1984. Taxonomic revision of *Septoria*-like fungi parasitic on Betulaceae. Trans. Br. Mycol. Soc. 83:383-398. (describes and illustrates *Phloeosporella borealis*)
Sutton (1980) includes descriptions and illustrations of five species causing leaf lesions and shot-hole of woody plants. Teleomorphs are placed in *Blumeriella*.

PHOMA Sacc.
    Coelomycetes   40 spp.
Aa, H. A. van der, and Kesteren, H. A. van 1979. Some pycnidial fungi occurring on *Atriplex* and *Chenopodium*. Persoonia 10:267-276. (describes and illustrates *Phoma variospora* and *P. dimorphospora*)
Aa, H. A. van der, and Kesteren, H. A. van 1980. *Phoma heteromorphospora* nom. nov. Persoonia 10:542. (*P. heteromorphospora* replaces *P. variospora*, a later homonym)
Baker, R. F., Davis, L. H., Wilhelm, S., and Snyder, W. C. 1985. An agressive vascular-inhabiting *Phoma* (*Phoma tracheiphila* f. sp. *chrysanthemi* nov. f. sp.) weakly pathogenic to chrysanthemum. Can. J. Bot. 63:1730-1735. (morphologically indistinguishable from *P. tracheiphila* f. sp. *tracheiphila*, the cause of mal secco disease of citrus)
Boerema, G. H. 1976. The *Phoma* species studied in culture by Dr. R. W. G. Dennis. Trans. Br. Mycol. Soc. 67:289-319. (includes 15 *Phoma* spp. and five varieties of *P. exigua*)
Boerema, G. H., and Bollen, G. J. 1975. Conidiogenesis and conidial septation as differentiating criteria between *Phoma* and *Ascochyta*. Persoonia 8:111-144. (defines *Ascochyta* with annellidic ontogeny and *Phoma* with phialidic ontogeny)
Boerema, G. H., and Dorenbosch, M. M. J. 1973. The *Phoma* and *Ascochyta* species described by Wollenweber and Hochapfel in their study on fruit-rotting. Stud. Mycol. 3:1-50. (descriptions and illustrations of 13 *Phoma* spp.)
Boerema, G. H., and Kesteren, H. A. van 1981. Nomenclatural notes on some species of *Phoma* sect. *Plenodomus*. Persoonia 11:317-331. (reviews synonymy and characteristics of *Phoma* spp. with teleomorphs in *Leptosphaeria*, host records)
Boerema, G. H., Kesteren, H. A. van, and Loerakker, W. M. 1981. Notes on *Phoma*. Trans. Br. Mycol. Soc. 77:61-74. (describes and illustrates six species in *Phoma* sect. *Plenodomus*)
Byford, W. J., and Gambogi, P. 1985. *Phoma* and other fungi on beet seed. Trans. Br. Mycol. Soc. 84:21-28. (descriptions and

PHOMA - cont.
illustrations of four *Phoma* spp.)
Dorenbosch, M. M. J. 1970. Key to nine ubiquitous soil-borne *Phoma*-like fungi. Persoonia 6:1-14. (descriptions and illustrations)
Johnston, P. R. 1981. *Phoma* on New Zealand grasses and pasture legumes. N. Z. J. Bot. 19:173-186. (presents table comparing 15 *Phoma* spp., descriptions and illustrations)
Kliejunas, J. T., Allison, J. R., McCain, A. H., and Smith, R. S., Jr. 1985. *Phoma* blight of fir and douglas-fir seedlings in a California nursery. Plant Dis. 69:773-775. (describes and illustrates *P. eupyrena*)
Morgan-Jones, G., and White, J. F. 1983. Studies in the genus *Phoma*. I. *Phoma americana* sp. nov. Mycotaxon 16:403-413. (descriptions and illustrations)
Morgan-Jones, G., and White, J. F. 1983. Studies in the genus *Phoma*. III. *Paraphoma*, a new genus to accommodate *Phoma radicina*. Mycotaxon 18:57-65. (description and illustrations)
Rajak, R. C., and Rai, M. K. 1982. Species of *Phoma* from legumes. Indian Phytopathol. 35:609-612. (key to six *Phoma* spp., descriptions and illustrations)
Rajak, R. C., and Rai, M. K. 1983. Effect of different factors on the morphology and cultural characters of 18 species and five varieties of *Phoma*. I. Effect of different media. Bibl. Mycol. 91:301-317. (brief descriptions)
Rajak, R. C., and Rai, M. K. 1984. Effect of different factors on the morphology and cultural characters of *Phoma*, II. Effect of different hydrogen-ion-concentrations. Nova Hedw. 40:299-311. (compares 20 taxa)
Sutton, B. C. 1964. *Phoma* and related genera. Trans. Br. Mycol. Soc. 47:497-509. (characterizes *Phoma* and related genera, describes and illustrates two taxa, discusses conidial formation)
White, J. F., Jr., and Morgan-Jones, G. 1987. Studies in the genus *Phoma*. VI. Concerning *Phoma medicaginis* var. *pinodella*. Mycotaxon 28:241-248. (describes and illustrates cultural characteristics and pycnidial wall anatomy)
White, J. F., and Morgan-Jones, G. 1983. Studies in the genus *Phoma*. II. Concerning *Phoma sorghina*. Mycotaxon 18:5-13. (descriptions and illustrations)
White, J. F., and Morgan-Jones, G. 1984. Studies in the genus *Phoma*. IV. Concerning *Phoma macrostoma*. Mycotaxon 20:197-204. (descriptions and illustrations)
White, J. F., and Morgan-Jones, G. 1986. Studies in the genus *Phoma*. V. Concerning *Phoma pomorum*. Mycotaxon 25:461-466. (descriptions and illustrations)
Most of the works cited above exemplify the modern approach to the taxonomy of *Phoma*; only a fraction of the described species have been considered. No useable system for the identification of these

PHOMA - cont.
heterogeneous fungi exists and even regional treatments are lacking.
Sutton (1980) provides a key to 27 species. Teleomorphs have been
placed in *Didymella, Leptosphaeria, Mycosphaerella*, and *Pleospora*
(Sivanesan, 1984). *Deuterophoma* is now considered a synonym of
Phoma. See also the similar genus *Phyllosticta*. Diseases caused by
Phoma species include: *Phoma caricae* (Pat.) Punithalingam, diseases
of papaya (CMI 634); *P. epicoccina* Punithalingam, Tulloch & Leach,
on various hosts (CMI 738); *P. exigua* Desmaz., leaf spot of bean
(CMI 81 as *Ascochyta phaseolorum*); *P. glomerata* (Corda) Wollenw. &
Hochapfel, on various hosts (CMI 134); *P. lingam* (Tode:Fr.) Desmaz.,
blackleg of cabbage and diseases of other Brassicaceae (CMI 331
under the teleomorph *Leptosphaeria maculans*); *P. medicaginis* Malbr.
& Roum. var. *pinodella* (Jones) Boerema, clover black stem, summer
black stem, and leaf spot of clover (CMI 518); *P. prunicola* (Opiz)
Wollenw. & Hochapfel, on apple, pear, and *Prunus* spp. (CMI 135); *P.
sorghina* (Sacc.) Boerema, Dorenbosch & van Kesteren, diseases of
various Poaceae (CMI 333 as *P. insidiosa*, CMI 584 under
Mycosphaerella holci, CMI 825); and *Phoma tracheiphila* (Petri)
Kantachveli & Gikachvili, mal secco disease of *Citrus* spp.,
especially lemon (CMI 399 as *Deuterophoma tracheiphila*).

PHOMATOSPORA Sacc.
   Sphaeriales   20 spp.
Kobayashi, T., and Sasaki, K. 1982. Two new species of Diaporthaceae
   from Japan causing leaf spot disease in the evergreen oak.
   Trans. Mycol. Soc. Jpn. 23:251-258. (describes and illustrates *P.
   albomaculans*)
Kobayashi, T., and Suto, Y. 1983. Material for the fungus flora of
   Japan (34). Trans. Mycol. Soc. Jpn. 24:277-282. (describes and
   illustrates *P. aucubae*)
Webster, J. 1955. Graminicolous Pyrenomycetes. V. Conidial states
   of *Leptosphaeria michotii, L. microscopica, Pleospora vagans* and
   the perfect state of *Dinemasporium graminum*. Trans. Br. Mycol.
   Soc. 38:347-365. (describes and illustrates *Phomatospora
   dinemasporium* and its anamorph)
Arx and Mueller (1954) provide an account of this genus and include
descriptions of two species. Anamorphs are placed in *Dinemasporium*
and *Phomatosporella*.

PHOMOPSIS (Sacc.) Bubak
   Coelomycetes   100 spp.
Hobbs, T. W., Schmitthenner, A. F., and Kuter, G. A. 1985. A new
   Phomopsis species from soybean. Mycologia 77:535-544. (compares
   P. longicolla with *P. sojae*)
Kulik, M. M. 1984. Symptomless infection, persistence, and production
   of pycnidia in host and non-host plants by *Phomopsis batatae,
   Phomopsis phaseoli*, and *Phomopsis sojae*, and the taxonomic

PHOMOPSIS - cont.
implications. Mycologia 76:274-291. (compares morphological data from anamorphs and teleomorphs)

Kulik, M. M. 1985. Some observations on the ascospore and cultural morphology of *Diaporthe phaseolorum* f. sp. *caulivora*, cause of soybean stem canker. Can. J. Plant Pathol. 7:387-388. (description and illustration)

Maffee, H. M., and Morton, H. L. 1983. *Phomopsis* canker of Russian olive in southeastern Michigan. Plant Dis. 67:964-965. (describes disease caused by *P. elaeagni*)

McPartland, J. M. 1983. *Phomopsis ganjae* sp. nov. on *Cannabis sativa*. Mycotaxon 18:527-530. (description and illustration)

Milholland, R. D., and Daykin, M. E. 1983. Blueberry fruit rot caused by *Phomopsis vaccinii*. Plant Dis. 67:325-326. (describes pathogenicity)

Moller, W. J., and Kasimatis, A. N. 1981. Further evidence that *Eutypa armeniacae*-not *Phomopsis viticola*-incites dead arm symptoms on grape. Plant Dis. 65:429-431. (based on inoculation study)

Muntanola-Cvetkovic, M., Bojovic-Cvetic, D., and Vukojevic, J. 1985. An ultrastructural study of alpha- and beta-conidia in the fungal genus *Phomopsis*. Crypt. Mycol. 6:171-184. (detailed observations of conidial germination in *P. helianthi*)

Muntanola-Cvetkovic, M., Mihaljcevic, M., and Petrov, M. 1981. On the identity of the causative agent of a serious *Phomopsis-Diaporthe* disease in sunflower plants. Nova Hedw. 34:417-435. (describes and illustrates *Diaporthe helianthi-Phomopsis helianthi* and the disease it causes)

Muntanola-Cvetkovic, M., Mihaljcevic, M., Vukojevic, J., and Petrov, M. 1985. Comparisons of *Phomopsis* isolates obtained from sunflower plants and debris in Yugoslavia. Trans. Br. Mycol. Soc. 85:477-483. (compares *P. helianthi* with *P. sojae* and *Phomopsis* sp. in culture)

Ponnappa, K. M., and Nag Raj, T. R. 1974. Some interesting additions to Indian mycoflora. Nova Hedw. Beih. 47:571-578. (describes and illustrates six *Phomopsis* spp.)

Punithalingam, E. 1975. Some new species and combinations in *Phomopsis*. Trans. Br. Mycol. Soc. 64:427-435. (six new species with descriptions and illustrations, two new combinations)

Punithalingam, E. 1980. New microfungi from cereals. Nova Hedw. 32:585-606. (key to nine *Phomopsis* spp., describes and illustrates *P. tritici*)

Punithalingam, E. 1981. Studies on Sphaeropsidales in culture. III. Mycol. Pap. 149:1-42. (describes and illustrates *Diaporthe capsici-Phomopsis capsici*)

Punithalingam, E., and Sharma, N. D. 1979. New microfungi from South East Asia. Nova Hedw. 31:881-897. (key to two *Phomopsis* spp. on *Oryza*, descriptions and illustrations)

PHOMOPSIS - cont.

Rosenberger, D. A., and Burr, T. J. 1982. Fruit decays of peach and apple caused by *Phomopsis mali*. Plant Dis. 66:1073-1075. (describes diseases and cultural characteristics of the fungus)

Webber, J. F., and Gibbs, J. N. 1984. Colonization of elm bark by *Phomopsis oblonga*. Trans. Br. Mycol. Soc. 82:348-352. (biocontrol agent to prevent breeding of scolytid bark beetles)

Numerous *Phomopsis* species have been described but no adequate treatment of the genus exists and keys are almost entirely lacking. Sutton (1980) includes descriptions and illustrations of nine species. Where known, *Diaporthe* is the teleomorph. See *Dendrophoma* and *Phacidiopycnis*. Diseases include: *Phomopsis anacardii* Early & Punithalingam, drying of shoots and leaf blotch of cashew (CMI 826); *P. capsici* (Magnus) Sacc., dieback and fruit rot of *Capsicum* spp. (CMI 733 under *Diaporthe capsici*); *P. caricae-papayae* Petr. & Cif., stem rot and fruit rot of papaya (CMI 827); *P. cinerescens* (Sacc.) Trav., fig canker; *P. citri* H. Fawc., melanose of *Citrus* spp. (CMI 396 under *Diaporthe citri*); *P. cocoina* (Cooke) Punithalingam, leaf spot of coconut (CMI 828); *P. cucurbitae* McKeen, black rot and other diseases of cucumber (CMI 469); *P. ipomoeae-batatas* Punithalingam, leaf blight or leaf spot of sweet potato (CMI 739); *P. juniperivora* Hahn, blight and dieback of junipers and other conifers (CMI 370); *P. leptostromiformis* (Kuehn) Bubak, *Phomopsis* stem blight of lupine (CMI 476 under *Diaporthe woodii*); *P. manihotis* Swarup, Chauhan & Tripathi, leaf spot of cassava (CMI 734 under *Diaporthe manihotis*); *P. obscurans* (Ellis & Everh.) Sutton (syn. *Dendrophoma obscurans* (Ellis & Everh.) H. W. Anderson), leaf blight of strawberry (CMI 227); *P. oryzae-sativae* Punithalingam, collar rot of rice (CMI 665); *P. phaseoli* (Desmaz.) Grove, various diseases of soybean, dry rot of sweet potato, and pod blight of lima bean (CMI 336 under *Diaporthe phaseolorum*); *P. sclerotioides* van Kesteren, black rot of cucumber (CMI 470); *P. theae* Petch, collar and branch canker of tea (CMI 330); *P. vexans* (Sacc. & Sydow) Harter, diseases of eggplant (CMI 338); and *P. viticola* (Sacc.) Sacc., associated with dead arm of grapevine (CMI 635) but see Moller and Kasimatis (1981).

PHRAGMIDIUM Link
Uredinales   60 spp.

Bedlan, G. 1984. Die Gattung *Phragmidium* Link. mit besonderer Beruecksichtigung des Formenkreises um *Phragmidium mucronatum* und *Phragmidium potentillae* in Mitteleuropa. Pflanzenschutz Berichte 46(6/12):33-60. (in German with English summary, includes six species)

Hiratsuka, N., Kaneko, S., and Nishigaki, H. 1980. A taxonomic revision of the species of *Phragmidium* in the Japanese Archipelago. Contributions to the rust flora of Eastern Asia XIV. Rep. Tottori Mycol. Inst. 18:53-88. (key to 28 species, descriptions, illustrations, host index)

PHRAGMIDIUM - cont.
Peterson, R. S. 1962. Notes on western rust fungi. II.
Pucciniaceae. Mycologia 54:389-394. (key to *Phragmidium* subgenus
Earlea on Potentilleae using teliospores)
All species occur on Rosaceae, predominantly in the northern
hemisphere. No world monograph exists. See Uredinales, especially
Cummins and Hiratsuka (1983). Diseases include: *Phragmidium arctium*
Lagerh. ex Liro, on *Rubus* subg. *Cylactis* (F. Can. 79); *P. bulbosum*
(Strauss) Schlectend., blackberry leaf rust (CMI 203); *P. fusiforme*
J. Schroet. var. *novi-boreale* Savile, rust of *Rosa* spp. (F. Can.
54); *P. mucronatum* (Pers.) Schlechtend., rust of *Rosa* spp. (CMI
204); *P. occidentale* Arth., on *Rubus* subg. *Anaplobatus* (F. Can. 80);
P. potentillae (Pers.) P. Karst., rust of *Potentilla* spp. (F. Can.
41); *P. rosae-pimpinellifoliae* Dietel, rust of burnet rose and its
hybrids (CMI 205); *P. rosae-setigerae* Dietel, rust of setigera rose
and its hybrids (CMI 206); *P. rubi-idaei* (DC.) P. Karst., cane rust
of raspberry (CMI 207); *P. tuberculatum* Mueller, rust of *Rosa* spp.
(CMI 208); and *P. violaceum* (C. F. Schultz) Wint., blackberry leaf
rust (CMI 209).

PHYLLACHORA Nitschke ex Fuckel
    Polystigmatales   101 spp.
Hosagoudar, V. B. 1985. New and noteworthy species of *Phyllachora*
from South India. Indian Phytopathol. 38:447-451. (lists 12
species, descriptions and illustrations, compares species on
Theaceae)
Kamat, M. N., Seshadri, V. S., and Pande, A. H. 1978. A monographic
study of Indian species of *Phyllachora*. Univ. Agric. Sci.
Bangalore, Curr. Res. 4:1-100. (key to 88 spp., descriptions,
illustrations, host index)
Orton, C. R. 1944. Graminicolous species of *Phyllachora* in North
America. Mycologia 36:18-53. (key to 46 species, descriptions)
Parberry, D. G. 1967. Studies on graminicolous species of *Phyllachora*
Nke. in Fckl. V. A taxonomic monograph. Aust. J. Bot.
15:271-375. (key to 96 species, descriptions, illustrations, host
index)
Parberry, D. G. 1971. Studies on graminicolous species of *Phyllachora*
Nke. in Fckl. VI. Additions and corrections to part V. Aust. J.
Bot. 19:207-235. (extends geographic and host range)
The species occurring on grasses have been well-studied. Arx and
Mueller (1954) include ten mostly non-graminicolous species.
Phyllachora sacchari Henn. causes tar spot of sorghum (CMI 588). See
Atopospora.

PHYLLACTINIA Lev.
    Erysiphales   5 spp.
Braun, U. 1985. Taxonomic notes on some powdery mildews. (V).
Mycotaxon 22:87-96. (key to five *Phyllactinia* spp. in North

PHYLLACTINIA - cont.
America with brief descriptions and illustrations)
Phyllactinia dalbergiae Pirozynski causes mildew of *Dalbergia* spp. and *P. guttata* (Wallr.:Fr.) Lev. (syn. *P. corylea* (Pers.) P. Karst.) causes mildew of hazel, birch, ash, and other trees (CMI 157). See *Erysiphe, Oidium,* and Erysiphales.

PHYLLOSTICTA Pers.
    Coelomycetes   46 spp.
Aa, H. A. van der 1973. Studies in *Phyllosticta* I. Stud. Mycol. 5:1-110. (key to 46 species previously placed in *Phyllostictina*, descriptions and illustrations)
Aa, H. A. van der, and Kesteren, H. A. van 1971. The identity of *Phyllosticta destructiva* Desm. and similar *Phoma*-like fungi described from Malvaceae and *Lycium halimifolium*. Acta Bot. Neerl. 20:552-563. (a synonym of *Phoma exigua*, lists other *Phyllosticta* and *Phoma* species on these hosts)
Batista, A. C., and Vital, A. F. 1952. Monografia das especies de *Phyllosticta* em Pernambuco. Bol. Sec. Agric., Ind., Comer. 19:1-80. (in Portugese, describes 115 species, host index)
Bissett, J. 1979. Coelomycetes on Liliales: the genus *Phyllosticta*. Can. J. Bot. 57:2082-2095. (describes and illustrates seven species)
Chuang, T. Y. 1981. Isolation of *Phyllosticta musarum*, causal organism of banana freckle. Trans. Br. Mycol. Soc. 77:670-671. (describes variability)
Mukunya, D. M., and Boothroyd, C. W. 1973. *Mycosphaerella zeae-maydis* sp. n., the sexual stage of *Phyllosticta maydis*. Phytopathology 63:529-532. (description and illustrations)
Punithalingam, E., and Woodhams, J. E. 1982. The conidial appendage in *Phyllosticta* spp. Nova Hedw. 36:151-198. (describes morphological details for ten species)
Rao, V. G. 1964. The genus *Phyllosticta* in Bombay-Maharashtra IV. Mycopathol. Mycol. Appl. 22:157-166. (brief descriptions and illustrations of 17 species)
Seaver, F. J. 1922. Phyllostictales, Phyllostictaceae. North Am. Flora 6:1-84. (key to 300 species with brief descriptions, outdated but useful as a last resort)
Weidemann, G. J., Boone, D. M., and Burdsall, H. H., Jr. 1982. Taxonomy of *Phyllosticta vaccinii* (Coelomycetes) and a new name for the true anamorph of *Botryosphaeria vaccinii* (Dothideales, Dothioraceae). Mycologia 74:59-65. (describes *P. elongata* and compares with *P. vaccinii*)
Westhuizen, G. C. A. van der 1980. *Phyllosticta maydis* on maize in South Africa. Phytophylactica 12:27-29. (complete description and illustration)
Phoma-like fungi occurring on living leaves have been placed in *Phyllosticta*. Van der Aa (1973) states that for nomenclatural

PHYLLOSTICTA - cont.
reasons species formerly placed in *Phyllostictina* must be included in *Phyllosticta*. Thus *Phyllosticta* is a large, heterogeneous genus whose species remain difficult to identify. Sivanesan (1984) and van der Aa (1973) treat species formerly placed in *Phyllostictina* and their *Guignardia* anamorphs. More work on the genus is needed, especially on the proper generic placement of *Phoma*-like species. Teleomorphs belong to *Botryosphaeria, Guignardia*, and *Mycosphaerella*. See also *Phoma*. Diseases include: *Phyllosticta ampelicida* (Engleman) van der Aa, black rot of grapevine (CMI 710 under *Guignardia bidwellii*); *P. citricarpa* (McAlpine) van der Aa, *Citrus* black spot (CMI 85 under *Guignardia citricarpa*); *P. musarum* (Cooke) van der Aa, freckle of banana (CMI 467 under *Guignardia musae*); *P. pervincae* on *Vinca minor* (F. Can. 282); and *P. toxica* on *Rhus radicans* (F. Can. 281).

PHYLLOSTICTINA   See *Guignardia* and *Phyllosticta*.

PHYMATOTRICHOPSIS Hennebert
   Hyphomycetes   1 sp.
Baniecki, J. F., and Bloss, H. E. 1969. The basidial stage of *Phymatotrichum omnivorum*. Mycologia 61:1054-1059. (describes the teleomorph)
Hennebert, G. L. 1973. *Botrytis* and *Botrytis*-like genera. Persoonia 7:183-204. (synoptic key to related genera, generic descriptions)
Lyda, S. D. 1978. Ecology of *Phymatotrichum omnivorum*. Ann. Rev. Phytopathol. 16:193-209. (reviews all aspects of the fungus)
Mouton, A. 1953. *Phymatotrichum omnivorum* (Shear) Dug., pourridie du Cotonnier. Rev. Mycol. Suppl. coloniae no. 2 18:69-87. (in French, presents a detailed morphological account)
Streets, R. B., and Bloss, H. E. 1973. *Phymatotrichum* root rot. American Phytopathological Society Monograph 8:1-38. (reviews all aspects of the disease including a description and illustrations of the fungus)
Hennebert (1973) revealed that the generic name *Phymatotrichum* Bonorden was a synonym of *Botrytis* and erected the genus *Phymatotrichopsis* to accommodate the ubiquitous plant pathogen, *Phymatotrichopsis omnivorum* (Duggar) Hennebert. This species is the anamorph of the basidiomycete *Trechispora brinkmanii* (Bres.) D. P. Rogers, the cause of root rot of cotton and many other crop plants.

PHYMATOTRICHUM   See *Phymatotrichopsis*.

PHYSALOSPORA Niessl
   Sphaeriales   30 spp.
Barr, M. E. 1970. Some amerosporous Ascomycetes on Ericaceae and Empetraceae. Mycologia 62:377-394. (key to 12 species in *Physalospora* and related genera, descriptions and illustrations)

PHYSALOSPORA - cont.
Narendra, D. V., and Rao, V. G. 1977. A new species of *Physalospora* from India. Mycologia 69:1191-1193. (describes and illustrates *P. eucalypti*, compares seven species from India)
Arx and Mueller (1954) include descriptions of nine species and account for numerous *Physalospora* names. *Physalospora miyabeana* Fukushi causes black canker of willow and *P. obtusa* (Schwein.) Cooke causes apple black rot and leaf spot.

PHYSODERMA Wallr.
  Blastocladiales   80 spp.
Karling, J. S. 1950. The genus *Physoderma* (Chytridiales). Lloydia 13:29-71. (monograph, provides descriptions for approximately 50 species)
Karling, J. S. 1977. Chytridiomycetarum Iconographia. J. Cramer, Vaduz. 414 pp. (includes a chapter on the Physodermataceae, illustrations, references)
Diseases caused by *Physoderma* (syn. *Urophlyctis*) include: *Physoderma alfalfae* (Pat. & Lagerh.) Karling, crownwart of alfalfa (CMI 751); *P. leproides* (Trabut) Karling, galls on sugarbeet (CMI 752); *P. maydis* (Miyabe) Miyabe, brown spot of maize (CMI 753); and *P. pulposum* Wallr., galls on Apiaceae, Asteraceae, and Chenopodiaceae (CMI 754).

PHYSOPELLA Arth.
  Uredinales   18 spp.
Cummins, G. B., and Ramachar, P. 1958. The genus *Physopella* (Uredinales) replaces *Angiopsora*. Mycologia 50:741-744. (lists 11 species)
Arthur (1935) provides a key to species in the United States. Additional references are listed in Cummins and Hiratsuka (1983). *Physopella ampelopsidis* (Dietel & P. Sydow) Cummins & Ramachar causes leaf rust of grapes and *Ampelopsis* (CMI 173) and *P. zeae* (Mains) Cummins & Ramachar occurs on *Zea mays* (CMI 5).

PHYTOPHTHORA de Bary
  Peronosporales   39 spp.
Ann, P. J., and Ko, W. H. 1985. Variants of *Phytophthora cinnamomi* extend the known limits of the species. Mycologia 77:946-950. (isolates from a broad host and geographical range have identical protein patterns)
Cristinzio, G., Scala, F., and Noviello, C. 1983. Differenziazione di alcune specie di *Phytophthora* mediante l'uso dell'-immunoelectroforesi in due dimensioni. Ann. Fac. Sci. Agrar. Univ. Studi Napoli, Portici 17:77-89. (in Italian with English summary, uses immunoelectrophoresis to differentiate *Phytophthora* species)
Dantanarayana, D. M., Peries, O. S., and Liyanage, A. de S. 1984.

PHYTOPHTHORA - cont.

Taxonomy of *Phytophthora* species isolated from rubber in Sri Lanka. Trans. Br. Mycol. Soc. 82:113-126. (compares eight species using morphological characters)

Erselius, L. J., and Vallavieille, C. de 1984. Variation in protein profiles of *Phytophthora*: comparison of six species. Trans. Br. Mycol. Soc. 83:463-472. (finds protein profiles determined by isoelectric focusing useful in taxonomy)

Erwin, D. C., Bartnicki-Garcia, S., and Tsao, P. H., eds. 1983. *Phytophthora*: Its Biology, Taxonomy, Ecology, and Pathology. American Phytopathological Society, St. Paul, MN. 392 pp. (includes four chapters on taxonomy)

Gerrettson-Cornell, L. 1985. A working key to the species of *Phytophthora* de Bary. Acta Bot. Hung. 31:89-97. (key to 39 *Phytophthora* spp.)

Gregory, P. H. 1974. *Phytophthora* Disease of Cocoa. Longman, London. 348 pp. (*P. palmivora*, 29 chapters on various aspects)

Ho, H. H. 1981. Synoptic keys to the species of *Phytophthora*. Mycologia 73:705-714. (keys to plant pathogenic species in culture, to those known only on hosts, and to aquatic species)

Ho, H. H., Liang, Z. R., Zhuang, W. Y., and Yu, Y. N. 1984. *Phytophthora* spp. from rubbertree plantations in Yunnan Province of China. Mycopathologia 86:121-124. (compares three species)

Kaosiri, T., and Zentmeyer, G. A. 1980. Protein, esterase, and peroxidase patterns in the *Phytophthora palmivora* complex from cacao. Mycologia 72:988-1000. (distinguishes *P. capsici*, *P. cinnamomi*, and three morphological forms of *P. palmivora*)

Newhook, F. J., and Podger, F. D. 1972. The role of *Phytophthora cinnamomi* in Australian and New Zealand forests. Ann. Rev. Phytopathol. 10:299-326. (presents an ecological appraisal)

Newhook, F. J., Waterhouse, G. M., and Stamps, D. J. 1978. Tabular key to the species of *Phytophthora* De Bary. Mycol. Pap. 143:1-20. (tabulates 47 taxa, includes illustrations of characters)

Old, K. M., Moran, G. F., and Bell, J. C. 1984. Isozyme variability among isolates of *Phytophthora cinnamomi* from Australia and Papua New Guinea. Can. J. Bot. 62:2016-2022. (assesses genetic variability)

Ribeiro, O. K. 1978. A Source Book on the Genus *Phytophthora*. J. Cramer, Vaduz. 417 pp. (contains a chapter on identification)

Waterhouse, G. M. 1963. Key to the species of *Phytophthora* De Bary. Mycol. Pap. 92:1-22. (key to 42 species, descriptions)

Waterhouse, G. M. 1970. The genus *Phytophthora* De Bary. Mycol. Pap. 122:1-59. (a compilation of names, original descriptions, and comments)

Zentmeyer, G. A. 1980. *Phytophthora cinnamomi* and the diseases it causes. American Phytopathological Society Monograph 10:1-96. (comprehensive, includes host list)

PHYTOPHTHORA - cont.
Despite the recent publications, identification of *Phytophthora* species remains difficult. Domsch et al. (1980) include descriptions and illustrations to two species as well as numerous references. Important species include: *Phytophthora arecae* (L. Coleman) Pethybr., mahli disease of *Areca catechu*, nut fall of coconut (CMI 833); *P. boehmeriae* Sawada, various diseases on several hosts (CMI 591); *P. botryosa* Chee, leaf fall and blackstripe of *Hevea* rubber (CMI 835); *P. cactorum* (Lebert & Cohn) J. Schroet., damping off, rot, root rots, and other diseases of various hosts (CMI 111); *P. cambivora* (Petri) Buisman, root rot of various hosts (CMI 112); *P. capsici* Leonian, various rots of *Capsicum*, tomato, and members of Cucurbitaceae (CMI 836); *P. cinnamomi* Rands, root rot of many hosts (CMI 113); *P. citricola* Sawada, various diseases of numerous hosts (CMI 114); *P. citrophthora* (R. E. Sm. & E. H. Sm.) Leonian, diseases of *Citrus* spp. and many other hosts (CMI 33); *P. cryptogea* Pethybr. & Lafferty, diseases of a wide range of glasshouse and field crops (CMI 592); *P. drechsleri* Tucker, tuber rot of potato, various root rots, damping off, and soft rots (CMI 840); *P. erythroseptica* Pethybr., on various hosts (CMI 593); *P. heveae* Thompson, pod rot and black stripe of rubber (CMI 594); *P. hibernalis* Carne, brown rot of *Citrus* fruit and leaf and twig blight (CMI 31); *P. infestans* (Mont.) de Bary, late blight of potato and tomato (CMI 838); *P. katsurae* Ko & Chang, trunk rot of chestnut (CMI 837); *P. meadii* McRae, abnormal leaf fall of rubber (CMI 834); *P. megakarya* Brasier & Griffin, black pod of cacao (CMI 832); *P. megasperma* Drechs., rots of various hosts (CMI 115); *P. nicotianae* Breda de Haan var. *nicotianae*, black shank of tobacco (CMI 34); *P. nicotianae* var. *parasitica* (Dastur) G. M. Waterhouse, damping off of seedlings and root rots (CMI 35); *P. palmivora* (E. J. Butler) E. J. Butler, black pod and canker of cacao, diseases of *Hevea*, palms, pawpaw, and root rots and damping-off of seedlings (CMI 831): *P. phaseoli* Thaxter, lima bean downy mildew; *P. porri* Foister, diseases of *Allium* spp. and others (CMI 595); *P. primulae* Tomlinson, brown core root rot of *Primula* (CMI 839); and *P. syringae* (Kleb.) Kleb., twig blight and wilt of lilac and many other hosts (CMI 32).

PILEOLARIA Castagne
    Uredinales   20 spp.
Kakishima, M., Sato, T., and Sato, S. 1984. Notes on two rust fungi, *Pileolaria klugkistiana* and *Nyssopsora cedrelae*. Trans. Mycol. Soc. Jpn. 25:355-359. (clarifies life cycle of *P. klugkistiana* on *Rhus javanica*)
Katsuya, K., Kakishima, M., and Sato, S. 1980. Spore surface structure of three *Pileolaria* species in Japan. Rep. Tottori Mycol. Inst. 18:163-167. (useful in distinguishing species)
Species occur on *Rhus* and *Pistacia* in the Anacardiaceae. No monograph exists. See Arthur (1935) for species that occur in the United

PILEOLARIA - cont.
States, and Cummins and Hiratsuka (1983) for additional references.
Pileolaria brevipes Berk. & Rav. causes a rust on *Rhus radicans* (F. Can. 50).

PITHOMYCES Berk. & Broome
    Hyphomycetes   24 spp.
Rao, V., and Hoog, G. S. de 1986. New or critical hyphomycetes from India. Stud. Mycol. 28:1-83. (key to 24 *Pithomyces* spp., description and illustrations of *P. ellisii*)
Roux, C. 1986. *Leptosphaerulina chartarum* sp. nov., the teleomorph of *Pithomyces chartarum*. Trans. Br. Mycol. Soc. 86:319-323. (descriptions and illustrations)
Ellis (1971, 1976) and Rao and de Hoog (1986) provide the most comprehensive treatments of this genus. *Pithomyces chartarum* (Berk. & M. A. Curtis) M. B. Ellis causes glume blotch of rice and sorghum and facial eczema of sheep (CMI 540). *Leptosphaerulina chartarum* Roux was recently identified as the teleomorph of *P. chartarum* (Roux, 1986).

PLAGIOSPHAERA Petr.
    Diaporthales   7 spp.
Kobayashi, T., and Sasaki, K. 1982. Two new species of Diaporthaceae from Japan causing leaf spot disease in the evergreen oak. Trans. Mycol. Soc. Jpn. 23:251-258. (compares two *Plagiosphaera* spp., descriptions and illustrations)
Walker, J. 1980. *Gaeumannomyces, Linocarpon, Ophiobolus*, and several other genera of scolecospored Ascomycetes and *Phialophora* conidial states, with a note on hyphopodia. Mycotaxon 11:1-129. (treats seven *Plagiosphaera* spp.)
Most species are saprophytic. Barr (1978) includes two species.

PLASMODIOPHORA Woron.
    Plasmodiophorales   6 spp.
Buczacki, S. T., and Humphrey, J. G. 1973. Problems associated with physiologic specialization in *Plasmodiophora brassicae*, with a description of two new races infecting *Brassica napus*. Trans. Br. Mycol. Soc. 60:588-590. (discusses problems of race differentiation)
Karling, J. S. 1968. The Plasmodiophorales. Hafner Publishing, New York. 256 pp. (a comprehensive monograph covering all aspects of the fungi, their diseases and control, key to six *Plasmodiophora* spp., descriptions and illustrations)
The work by Karling (1968) is excellent. *Plasmodiophora brassicae* Woron. causes clubroot of Brassicaceae (CMI 621).

PLASMOPARA J. Schroet.
  Peronosporales   20 spp.
References to these downy mildews as well as other members of the Peronosporales are generally presented according to host or geographical area. See Peronosporales. *Plasmopara viticola* (Berk. & M. A. Curtis ex de Bary) Berl. & de Toni causes vine downy mildew.

PLECTOPHOMELLA Moesz
  Coelomycetes   3 spp.
Redfern, D. B., and Sutton, B. C. 1981. Canker and dieback of *Ulmus glabra* caused by *Plectophomella concentrica*, and its relationship to *P. ulmi*. Trans. Br. Mycol. Soc. 77:381-390. (descriptions and illustrations)
The type species, *Plectophomella visci* Moesz, is included in Sutton (1980).

PLEIOCHAETA (Sacc.) S. J. Hughes
  Hyphomycetes   4 spp.
Bhat, D. J. 1983. An undescribed species of *Pleiochaeta* from Asmara, Ethiopia. Trans. Br. Mycol. Soc. 81:405-406. (describes and illustrates *P. ghindensis* from leaves of a leguminous tree)
Hughes, S. J. 1951. Studies on micro-fungi. III. *Mastigosporium, Camposporium*, and *Ceratophorum*. Mycol. Pap. 36:1-43. (places two species in *Pleiochaeta*, illustrations)
Rambelli, A., Onofri, S., and Lunghini, D. 1981. New dematiaceous hyphomycetes from Ivory Coast forest litter. Trans. Br. Mycol. Soc. 76:53-58. (description and illustration of *P. stellaris*)
Ellis (1971) describes and illustrates *Pleiochaeta setosa* (Kirchn.) S. J. Hughes, the cause of leaf, stem, and pod lesions on *Cytisus* and *Lupinus* (CMI 495, F. Can. 12).

PLEOSPORA Rabenh. ex Ces. & De Not.
  Pleosporales   200 spp.
Eriksson, O. 1967. On graminicolous pyrenomycetes from Fennoscandia 1. Dictyosporous species. Ark. Bot. 6:339-380. (key to genera and 12 *Pleospora* spp.)
Lamprecht, S. C., Baxter, A. P., and Thompson, A. H. 1984. *Stemphylium vesicarium* on *Medicago* spp. in South Africa. Phytophylactica 16:73-75. (describes and illustrates anamorph and its teleomorph *P. allii*)
Mueller, E. 1951. Die schweizerischen Arten der Gattungen *Clathrospora, Pleospora, Pseudoplea* und *Pyrenophora*. Sydowia 5:248-310. (in German, key to 53 *Pleospora* spp., descriptions and illustrations)
Simmons, E. G. 1969. Perfect states of *Stemphylium*. Mycologia 61:1-26. (describes five *Pleospora-Stemphylium* holomorphs)
Simmons, E. G. 1985. Perfect states of *Stemphylium*. II. Sydowia 38:284-293. (describes and illustrates three *Pleospora-*

PLEOSPORA - cont.
Stemphylium holomorphs from *Medicago sativa*, determines that *Pleospora herbarum* and *Stemphylium botryosum* are not a holomorph, elucidates their respective genetic connections)
Simmons, E. G. 1986. *Alternaria* themes and variations (22-26). Mycotaxon 25:287-308. (describes new teleomorph genus *Lewia* for *Pleospora*-like species with *Alternaria* anamorphs)
Wehmeyer, L. E. 1961. A World Monograph of the Genus *Pleospora* and its segregates. University of Michigan, Ann Arbor. 451 pp. (key to related genera and species, treats over 100 *Pleospora* spp. and seven *Pyrenophora* spp., descriptions and illustrations)
Several major references to *Pleospora* and related genera are available including Wehmeyer (1961) and Barr (1972). The anamorphs are placed in *Dendryphion* and *Stemphylium*. See also the related genus *Pyrenophora*. Diseases include: *Pleospora bjorlingii* Byford, black leg of sugarbeet, marigold, and various other diseases (CMI 149); *P. herbarum* (Fr.) Rabenh., various leaf spots and net blotch of field and broad bean (CMI 150, F. Can. 232); and *P. papaveracea* (De Not.) Sacc., leaf blight of opium poppy (CMI 730).

PLOIODERMA Darker
    Rhytismatales   3 spp.
Hunt, R. S., and Ziller, W. G. 1978. Host-genus keys to the Hypodermataceae of conifer leaves. Mycotaxon 6:481-496. (key to three *Ploioderma* spp. on *Pinus*)
Ploioderma hedgecockii (Dearn.) Darker, anamorph *Leptostroma hedgecockii* Dearn., causes needle blight of pines (CMI 799) and *P. lethale* (Dearn.) Darker causes *Hypoderma* needle blight of southern pines (CMI 570). See *Lophodermium* and the anamorph *Leptostroma*.

PODOSPHAERA Kunze
    Erysiphales   9 spp.
Boesewinkel, H. J. 1979. Differences between the conidial states of *Podosphaera tridactyla* and *Sphaerotheca pannosa*. Ann. Phytopathol. 11:525-527. (descriptions and illustrations)
Braun, U. 1984. Taxonomic notes on some powdery mildews. (III). Mycotaxon 19:369-374. (key to nine *Podosphaera* spp., descriptions and illustrations)
The species are treated by Braun (1984). *Podosphaera clandestina* (Wallr.:Fr.) Lev. causes hawthorn mildew (CMI 478); *P. leucotricha* (Ellis & Everh.) Salmon causes powdery mildew of apple (CMI 158); and *P. tridactyla* (Wallr.) de Bary causes plum mildew (CMI 187). See *Erysiphe* and Erysiphales.

POLLACCIA Baldacci & Cif.
    Hyphomycetes   3 spp.
Ondrej, M. 1984. *Pollaccia spiraeae* (Karakulin) Ondrej. Ceska Mykol. 38:46-48. (in Czechoslovakian with English summary, on *Spiraea*)

POLLACCIA - cont.
Three species of *Pollaccia* are included as anamorphs of *Venturia* species in Sivanesan (1977 under *Venturia*). Keys, descriptions, and illustrations are included in Ellis (1971, 1976). Diseases include: *Pollaccia elegans* Servazzi, blight or dieback of poplars (CMI 483 under *Venturia populina*); *P. radiosa* (Lib.) Baldacci & Cif., leaf and shoot blight of poplar or poplar scab (CMI 403 under *Venturia macularis*); and *P. saliciperda* (Allesch. & Tubeuf) Arx, willow scab (CMI 482, F. Can. 247 under *Venturia saliciperda*).

POLYMYXA Ledingham
    Plasmodiophorales   2 spp.
Barr, D. J. S. 1979. Morphology and host range of *Polymyxa graminis*, *Polymyxa betae*, and *Ligniera pilorum* from Ontario and some other areas. Can. J. Plant Pathol. 1:85-94. (descriptions and illustrations)
Karling, J. S. 1968. The Plasmodiophorales. Hafner Publishing, New York. 256 pp. (a comprehensive monograph covering all aspects of these fungi, their diseases and control, describes and illustrates two *Polymyxa* spp.)
Karling (1968) and Barr (1979) provide an excellent account of both *Polymyxa graminis* Ledingham in root cells of various Poaceae (F. Can. 199), and *P. betae* Keskin in sugar beet root cells and other hosts (F. Can. 200).

POLYPORUS Micheli ex Adans.
    Aphyllophorales   40 spp.
In recent years the genus *Polyporus* has been defined in a narrow sense and many species have been transferred to other genera. See related genera *Bondarzewia*, *Climacocystis*, *Dichomitus*, *Grifola*, *Inonotus*, *Laetiporus*, *Phaeolus*, *Pseudophaeolus*, *Postia*, and *Rigidoporus*. *Polyporus squamosus* Fr.:Fr. causes a heart rot of living trees. See also Aphyllophorales.

POLYSCYTALUM Riess
    Hyphomycetes   7 spp.
Ellis (1971, 1976) treats three species, including *Polyscytalum pustulans* (Owen & Wakefield) M. B. Ellis (syn. *Oospora pustulans* Owen & Wakefield), cause of skin spot of potato tubers. See Carmichael et al. (1980) for references to other species, most of which are saprophytic.

POLYTHRINCIUM See *Cymadothea*.

PORIA Pers.
    Aphyllophorales   23 spp.
Ginns, J. 1984. New names, new combinations and new synonymy in the Corticiaceae, Hymenochaetaceae, and Polyporaceae. Mycotaxon

PORIA - cont.
21:325-333. (discusses nomenclature of *Poria*, 23 *Perenniporia* spp. transferred to *Poria*)
Lowe, J. L. 1963. A synopsis of *Poria* and similar fungi from the tropical regions of the world. Mycologia 55:453-486. (key to 187 species of *Poria* in the broad sense, brief descriptions)
Lowe, J. L. 1966. Polyporaceae of North America. The genus *Poria*. State Univ. Coll. For. Syracuse Univ., Tech. Publ. 90:1-183. (key to 133 species of *Poria* in the broad sense, descriptions and illustrations)
The taxonomic status of *Poria* is in question. Most species have been transferred to other genera. Lowe's (1963, 1966) works are still useful for species identification although the nomenclature is outdated. See *Phellinus, Rigidoporus*, and Aphyllophorales.

POSTIA Fr.
    Aphyllophorales   25 spp.
Postia balsamea (Peck) Juelich (syn. *Polyporus balsameus* Peck) causes a butt rot of balsam fir and *P. amara* (Hedgc.) M. Larsen & Lombard (syn. *Polyporus amarus* Hedgc.) causes a heartwood rot or "pecky cedar" of living incense cedar. See Aphyllophorales.

POTEBNIAMYCES Smerlis
    Helotiales   2 spp.
DiCosmo, F., Nag Raj, T. R., and Kendrick, W. B. 1984. A revision of the Phacidiaceae and related anamorphs. Mycotaxon 21:1-234. (monographic treatment, restricts *Potebniamyces* to *P. pyri*, descriptions and illustrations)
Funk, A., and Smith, R. B. 1981. *Potebniamyces gallicola* n. sp. from dwarf mistletoe infections in western hemlock. Can. J. Bot. 59:1610-1612. (descriptions and illustrations)
DiCosmo et al. (1984) restrict *Potebniamyces* to *P. pyri* (Berk. & Broome) Dennis, anamorph *Phacidiopycnis piri* (Fuckel) Weindlmayr, the cause of apple bark canker. Species on conifers previously placed in this genus are transferred to *Phacidium* and their anamorphs are transferred to *Apostrasseria*.

PRAGMOPARA C. Massal.
    Helotiales   6 spp.
Groves, J. W. 1967. The genus *Pragmopara*. Can. J. Bot. 45:169-181. (key to six species, descriptions and illustrations)
Most species occur on conifers causing stem cankers associated with insects. Funk (1981) describes and illustrates *Pragmopara pini* Groves and *P. pithya* (Fr.) Groves, anamorph *Pragmopycnis pithya* Sutton & Funk.

PROSPODIUM Arth.
    Uredinales   40 spp.
Ferreira, F. A., and Hennen, J. F. 1986. The life cycle, pathology, and taxonomy of the rust, *Prospodium bicolor* sp. nov., on yellow ipe, *Tabebuia serratifolia*, in Brazil. Mycologia 78:795-803. (key to ten *Prospodium* spp. on *Tabebuia*, describes and illustrates *P. bicolor*)
See Uredinales, especially Cummins and Hiratsuka (1984).

PROTOMYCES Unger
    Taphrinales   10 spp.
Reddy, M. S., and Kramer, C. L. 1975. A taxonomic revision of the Protomycetales. Mycotaxon 3:1-50. (treats five genera, key to ten *Protomyces* spp. based on host, descriptions and illustrations)
Species of *Protomyces* cause the formation of galls on various plant parts of Apiaceae and Asteraceae.

PSEUDOCERCOSPORA Speg.
    Hyphomycetes   226 spp.
Deighton, F. C. 1976. Studies on *Cercospora* and allied genera. VI. *Pseudocercospora* Speg., *Pantospora* Cif. and *Cercoseptoria* Petr. Mycol. Pap. 140:1-168. (defines *Pseudocercospora*, describes and illustrates 54 species, transfers an additional 171 names into *Pseudocercospora*)
Deighton, F. C. 1979. Studies on *Cercospora* and allied genera. VII. New species and redispositions. Mycol. Pap. 144:1-56. (treats ten *Pseudocercospora* spp.)
Yen, J.-M. 1978. Etude sur les champignons parasites du Sud-Est Asiatique. 33. Les *Cercospora* de Formose. V. Les *Pseudocercospora*. Bull. Trimest. Soc. Mycol. Fr. 94:385-389. (in French, briefly describes and illustrates 13 species)
This genus is a segregate of *Cercospora*. Ellis (1971, 1976) includes five species based on the work of Deighton. Sivanesan (1984) provides a key to 11 *Mycosphaerella* spp. with *Pseudocercospora* anamorphs. *Pseudocercospora abelmoschi* (Ellis & Everh.) Deighton causes a leaf spot or blight of okra (CMI 625, 678); *P. cruenta* (Sacc.) Deighton causes leaf spot of cowpea (CMI 463 as *Cercospora cruenta*); *P. fuligena* (Roldan) Deighton causes leaf spot of tomato (CMI 465 as *Cercospora fuligena*); and *P. musae* (Zimmermann) Deighton causes Sigatoka of banana (CMI 414 under *Mycosphaerella musicola* as *Cercospora musae*). See *Cercospora*.

PSEUDOCERCOSPORELLA Deighton
    Hyphomycetes   10 spp.
Cunningham, P. C. 1981. Occurrence, role and pathogenic traits of a distinct pathotype of *Pseudocercosporella herpotrichoides*. Trans. Br. Mycol. Soc. 76:3-15. (defines pathogenic types from

PSEUDOCERCOSPORELLA - cont.
various grasses)
Nirenberg, H. I. 1981. [Differentiation of *Pseudocercosporella*
strains causing foot rot diseases of cereals 1. Morphology.]
Z. Pflanzenkr. Pflanzensch. 88:241-248. (in German with English
summary, differentiates two varieties of *P. herpotrichoides* and
two additional species)
Pseudocercosporella herpotrichoides (Fron) Deighton causes eyespot of
cereals (CMI 386). Sivanesan (1984) includes two species with
Mycosphaerella teleomorphs. See *Cercosporella* and *Cercospora*.

PSEUDOCOCHLIOBOLUS Tsuda, Ueyama & Nishihara
    Pleosporales   4 spp.
Tsuda, M., and Ueyama, A. 1985. Two new *Pseudocochliobolus* and a new
species of *Curvularia*. Trans. Mycol. Soc. Jpn. 26:321-330.
(descriptions and illustrations)
See the related genus *Cochliobolus*, particularly Alcorn (1983).

PSEUDOPERONOSPORA Rostovzev
    Peronosporales   8 spp.
Constantinescu, O. 1985. Notes on *Pseudoperonospora*. Mycotaxon
24:301-311. (discusses three species, distinguishes two species
on *Urtica*)
Waterhouse, G. M., and Brothers, M. P. 1981. The taxonomy of
Pseudoperonospora. Mycol. Pap. 148:1-28. (describes and
illustrates eight species, discusses the synonym
Peronoplasmopara)
Pseudoperonospora (syn. *Peronoplasmopara*) is similar to *Peronospora*
and *Plasmopara* as discussed by Waterhouse and Brothers (1981).
Pseudoperonospora cubensis (Berk. & M. A. Curtis) Rostovzev causes
downy mildew of Cucurbitaceae (CMI 457), and *P. humuli* (Miyabe &
Takah) G. W. Wilson causes hop downy mildew. See also
Peronosporales.

PSEUDOPEZICULA Korf
    Helotiales   2 spp.
Korf, R. P., Pearson, R. C., Zhuang, W.-Y., and Dubos, B. 1986.
Pseudopezicula (Helotiales, Peziculoideae), a new discomycete
genus for pathogens causing an angular leaf scorch disease of
grapes ("Rotbrenner"). Mycotaxon 26:457-471. (describes and
illustrates a new genus with two species)
Pseudopezicula tracheiphila (H. Muell.-Thurg.) Korf & Zhuang causes a
disease of grapes. The anamorph is *Phialophora tracheiphila* (Sacc. &
Sacc.) Korf.

PSEUDOPEZIZA Fuckel
    Helotiales   3 spp.
Schuepp, H. 1959. Untersuchungen ueber *Pseudopezizoideae* sensu

PSEUDOPEZIZA - cont.
Nannfeldt. Phytopathol. Z. 36:213-269. (in German, key to three *Pseudopeziza* spp., descriptions and illustrations)
Schuepp's (1959) monograph is the most comprehensive treatment available. Diseases include: *Pseudopeziza medicaginis* (Lib.) Sacc., leaf spot of lucerne (CMI 637); *P. singularis* (Peck) Peck ex J. J. Davis, on leaves of *Ranunculus* spp. (F. Can. 229); and *P. trifolii* (Fr.) Fuckel, leaf spot of clovers (CMI 636). See *Blumeriella*.

PSEUDOPHACIDIUM P. Karst.
    Rhytismatales   6 spp.
DiCosmo, F., Nag Raj, T. R., and Kendrick, W. B. 1984. A revision of the Phacidiaceae and related anamorphs. Mycotaxon 21:1-234. (monographic treatment, reviews described species, describes and illustrates *P. ledi*)
Egger, M. C. 1968. Morphologie und Biologie von *Pseudophacidium*-Arten (Ascomycetes). Sydowia 20:288-328. (in German with English summary, key to four species, descriptions and illustrations)
Funk, A. 1980. *Pseudophacidium garmanii* n. sp., on interior spruce in British Columbia. Can. J. Bot. 58:2447-2449. (description and illustration)
DiCosmo et al. (1984) discuss the generic concept, review the species and describe and illustrate *Pseudophacidium ledi* (Albertini & Schwein.:Fr.) P. Karst. on *Ledum*. They disagree with the concepts of Egger (1968) and Arx and Mueller (1954). A monographic treatment of the genus is needed. Anamorphs have been placed in *Myxofusicoccum*.

PSEUDOPHAEOLUS Ryvarden
    Aphyllophorales   1 sp.
Ofosu-Asiedu, A. 1975. A new disease of eucalypts in Ghana. Trans. Br. Mycol. Soc. 65:285-289. (description and illustration of *Pseudophaeolus*, *P. baudonii*, and the disease)
Westhuizen, G. C. A. van der 1973. *Polyporus baudoni* [sic] Pat. on *Eucalyptus* spp. in South Africa. Bothalia 11:143-151. (descriptions and illustrations of the fungus and disease)
Pseudophaeolus baudonii (Pat.) Ryvarden causes a root rot and decay of various tree species in Africa (CMI 442 as *Polyporus baudonii*). See Aphyllophorales, especially Ryvarden and Johansen (1980).

PSEUDOROBILLARDA Morelet
    Coelomycetes   5 spp.
Nag Raj, T. R., Morgan-Jones, G., and Kendrick, B. 1972. Genera coelomycetarum. IV. *Pseudorobillarda* gen. nov., a generic segregate of *Robillarda* Sacc. Can. J. Bot. 50:861-867. (describes and illustrates four species)
Nag Raj, T. R., Morgan-Jones, G., and Kendrick, B. 1973. *Pseudorobillarda* Nag Raj et al., a later homonym of *Pseudorobillarda* Morelet. Can. J. Bot. 51:688-689. (transfers

PSEUDOROBILLARDA - cont.
four species)
Uecker, F. A., and Kulik, M. M. 1986. *Pseudorobillarda sojae*, a new pycnidial coelomycete from soybean stems. Mycologia 78:449-453. (description and illustrations)
Sutton (1980) provides a key to four species with brief descriptions and illustrations.

PSEUDOSEPTORIA Speg.
Coelomycetes   5 spp.
Sutton (1980) reviews this genus and includes four species with brief descriptions and illustrations. Diseases include: *Pseudoseptoria bromigena* (Sacc.) Sutton, leaf spot of *Bromus inermis* (F. Can. 237); *P. donacis* (Pass.) Sutton, halo spot of grasses and cereals (CMI 400 as *Selenophoma donacis*); and *P. stomaticola* (Baeumler) Sutton, on various Poaceae (F. Can. 238).

PUCCINIA Pers.:Pers.
Uredinales   3000 spp.
Baum, B. L., and Savile, D. B. O. 1985. Rusts (Uredinales) of Triticeae: evolution and extent of coevolution, a cladistic analysis. Bot. J. Linn. Soc. 91:367-394. (reviews evolutionary trends, evaluates primitive and advanced characters for each species)
Hiratsuka, N. 1976. Microcyclic species of *Puccinia* in the Japanese Archipelago. Contributions to the rust-flora of Eastern Asia XI. Rep. Tottori Mycol. Inst. 14:1-77. (key to 83 species, descriptions and illustrations)
Hiratsuka, N. 1980. A taxonomic revision of the autoecious species of *Puccinia* parasitic on the Compositae in the Japanese Archipelago. Contributions to the rust flora of Eastern Asia. XIII. Rep. Tottori Mycol. Inst. 18:1-52. (key to 20 species, descriptions and illustrations)
Hiratsuka, N., and Hasebe, S. 1978. A taxonomic revision of the species of *Puccinia* parasitic on the Liliales (Liliaceae, Amaryllidaceae, Dioscoreaceae and Iridaceae) in the Japanese Archipelago. Contributions to the rust flora of Eastern Asia XII. Rep. Tottori Mycol. Inst. 16:1-36. (key to 26 species, descriptions and illustrations)
Hiratsuka, N., and Kaneko, S. 1968. A taxonomic revision of the species of *Puccinia* parasitic on the Umbelliferae in the Japanese Archipelago. Rep. Tottori Mycol. Inst. 6:74-110. (key to 24 species, descriptions and illustrations)
Hiratsuka, N., and Kaneko, S. 1973. A taxonomic revision of the species of *Puccinia* parasitic on the Polygonaceae in the Japanese Archipelago. Rep. Tottori Mycol. Inst. 10:99-140. (key to 22 taxa, descriptions and illustrations)
Hiratsuka, N., and Kaneko, S. 1983. A provisional list of *Puccinia*

PUCCINIA - cont.
species on the grasses in Japan. Rep. Tottori Mycol. Inst.
21:61-75. (list of 64 species with brief annotations)
Newton, A. C., Caten, C. E., and Johnson, R. 1985. Variation for
isozymes and double-stranded RNA among isolates of *Puccinia
striiformis* and two other cereal rusts. Plant Pathol.
34:235-247. (supports the separation of formae speciales *tritici*
and *hordei*)
Parmelee, J. A. 1986. The autoecious species of *Puccinia* on
Polemoniaceae in North America. Mycologia 78:454-468. (key to 11
species, descriptions and illustrations)
Parmelee, J. A., and Savile, D. B. O. 1981. Autoecious species of
Puccinia on Cichorieae in North America. Can. J. Bot.
59:1078-1101. (key to 17 species, descriptions and illustrations)
Savile, D. B. O. 1973. Revisions of the microcyclic *Puccinia* species
on Saxifragaceae. Can. J. Bot. 51:2347-2370. (descriptions and
illustrations of 17 taxa, distribution maps)
Savile, D. B. O. 1979. Fungi as aids in higher plant classification.
Bot. Rev. 45:377-503. (emphasizes rusts and their host
specificity, references to many other articles by this author)
Puccinia is a large genus that includes many important cereal
pathogens. Major references are available dealing with a specific
host range or geographical area. See Hawksworth et al. (1983) and
Uredinales. Important species include: *Puccinia allii* E. Rudolphi,
rust of *Allium* spp. (CMI 52); *P. anemones-virginianae* Schwein., on
Anemone spp. (F. Can. 262); *P. antirrhini* Dietel & Holw., rust of
antirrhinum (CMI 262); *P. apii* Desmaz., celery rust (CMI 284); *P.
arachidis* Speg., peanut or groundnut rust (CMI 53); *P. araliae* Ellis
& Everh., on *Panax trifolius* (F. Can. 222); *P. areolata* Dietel &
Holw. in Dietel, on *Caltha* spp. (F. Can. 269); *P. arnicalis* Peck, on
Arnica spp. (F. Can. 111); *P. asparagi* DC., asparagus rust (CMI 54);
P. balsamorhizae Peck, on *Balsamorhiza* spp. and *Wyethia* spp. (F.
Can. 89); *P. blyttiana* Lagerh., on *Ranunculus* spp. (F. Can. 263); *P.
cacabata* Arth. & Holw., cotton rust (CMI 294); *P. calthae* Link, on
Caltha palustris (F. Can. 270); *P. campanulae* Carmich. ex Berk., on
Campanula spp. (F. Can. 219); *P. canadensis* Arth., on *Viola
orbiculata* (F. Can. 56); *P. carthami* Corda, safflower rust (CMI
174); *P. chrysanthemi* Roze, black rust of *Chrysanthemum* (CMI 175);
P. codyi Savile, on *Smelowskia borealis* var. *borealis* (F. Can. 46);
P. conglomerata (Strauss) Roehl., on *Petasites* spp. (F. Can. 110);
P. cynodontis Lacr., leaf rust of bermuda grass (CMI 292); *P. dayi*
G. P. Clinton ex Peck, on *Lysimachia* spp. (F. Can. 204); *P.
distichlidis* Ellis & Everh., on Primulaceae and *Spartina* (F. Can.
205); *P. evadens* Harkn., rust of *Baccharis* spp. (CMI 620); *P.
fergussonii* Berk. & Broome, on *Viola* spp. (F. Can. 64); *P. gemella*
Dietel & Holw., on *Caltha leptosepala* (F. Can. 264); *P. glacieri*
Savile, on *Viola glabella* (F. Can. 78); *P. graminis* Pers.:Pers.,
stem rust of wheat and other Poaceae; *P. granularis* Kalchbr. &

PUCCINIA - cont.
Cooke, rust of pelargonium (CMI 263); *P. helianthi* Schwein.,
sunflower rust (CMI 55, F. Can. 95); *P. holboellii* (Hornem.) Rostr.,
on *Arabis*, *Draba*, and *Erysimium* (F. Can. 47); *P. horiana* Henn.,
white rust of chrysanthemum (CMI 176); *P. intermixta* Peck, on *Iva
axillaris* (F. Can. 96); *P. iridis* Rabenh., rust of iris (CMI 285);
P. karelica Tranz., on *Trientalis* spp. and *Carex* spp. (F. Can. 206);
P. kuehnii E. J. Butler, leaf rust of sugarcane (CMI 10); *P.
leveillei* Mont., rust of geranium (CMI 264); *P. limosae* Magnus, on
Lysimachia spp. and *Carex* spp. (F. Can. 207); *P. lobeliae* W. Gerard
ex Peck, on *Lobelia* spp. (F. Can. 220); *P. longipes* Lagerh., on
Vernonia altissima (F. Can. 97); *P. malvacearum* Mont., rust of
hollyhock (CMI 265, F. Can. 171); *P. melanocephala* Sydow & P. Sydow,
leaf rust of sugarcane (CMI 9 as *P. erianthi*); *P. menthae* Pers.,
mint rust (CMI 7); *P. oahuensis* Ellis & Everh., rust of *Digitaria*
spp. (CMI 516); *P. ornatula* Holw., on *Viola glabella* (F. Can. 77);
P. ortonii H. Jacks., on *Dodecatheon* spp. (F. Can. 208); *P.
parnassiae* Arth., on *Parnassia fimbriata* (F. Can. 19); *P.
pelargonii-zonalis* Doidge, rust of pelargonium (CMI 266); *P.
penicillariae* Speg., leaf rust of millet (CMI 6); *P. physalidis*
Peck, on *Physalis* spp. (F. Can. 187); *P. pittieriana* Henn., potato
and tomato rust (CMI 286); *P. polysora* Underwood, corn rust (CMI 4);
P. psidii Wint., guava rust (CMI 56); *P. pulsatillae* Kalchbr., on
Anemone spp. (F. Can. 265); *P. purpurea* Cooke, leaf rust of sorghum
(CMI 8); *P. recondita* Roberge, on *Secale* (F. Can. 310); *P.
schedonnardi* Kellerm. & Swingle, a minor leaf rust of cotton (CMI
293, F. Can. 172); *P. sherardiana* Koern., on *Sphaeralcea coccinea*
(F. Can. 173); *P. silphii* Schwein., on *Silphium* spp. (F. Can. 98);
P. sorghi Schwein., corn rust (CMI 3, F. Can. 302); *P. striiformis*
Westendorp, stripe rust of wheat and other cereals (CMI 291, F. Can.
250); *P. subnitens* Dietel, on Primulaceae and *Distichlis* (F. Can.
209); *P. thaliae* Dietel, rust of canna (CMI 267); *P. trautvetteriae*
Sydow & Holw., on *Trautvetteria carolinensis* var. *occidentalis* (F.
Can. 266); *P. treleasiana* Pazschke, on *Caltha leptosepala* (F. Can.
267); *P. triticina* Eriks., on *Triticum*, *Secale*, and *Aegilops* (F.
Can. 309); *P. tumidipes* Peck, on *Lycium* spp. (F. Can. 188); *P.
urbanis* Savile, on *Geum calthifollium* (F. Can. 18); *P. ustalis*
Berk., on *Ranunculus* spp. (F. Can. 268); *P. violae* (Schumach.) DC.
ssp. *americana* Savile, on *Viola* spp. (F. Can. 75); *P. volkartiana* E.
Fisch., on *Androsace* (F. Can. 210); and *P. xanthii* Schwein., on
Ambrosia spp. and *Xanthium* spp. (F. Can. 99).

PUCCINIASTRUM G. Otth
    Uredinales   37 spp.
Hiratsuka, N. 1958. Revision of Taxonomy of the Pucciniastreae, with
    Special Reference to Species of the Japanese Archipelago. Kasai,
    Tokyo. 167 pp. (key to 23 *Pucciniastrum* spp., lists hosts and
    distribution for each species)

PUCCINIASTRUM - cont.
Hiratsuka, Y. 1970. Identification and morphology of the aecial state of *Puccinastrum sparsum* in northwestern Canada. Can. J. Bot. 48:433-435. (descriptions, illustrations, hosts, geographic distribution)
Kaneko, S., and Hiratsuka, N. 1980. Fungi inhabiting on fagaceous trees II. Host alternation of the beech rust, *Pucciniastrum fagi*. Trans. Mycol. Soc. Jpn. 21:417-421. (describes and illustrates both states)
The anamorph of *Pucciniastrum* is placed in *Peridermium*. See *Puccinia, Thekopsora*, and Uredinales, especially Cummins and Hiratsuka (1983). *Pucciniastrum americanum* (Farl.) Arth. causes needle rust of white spruce and late leaf rust of raspberry (CMI 210).

PUCCINIOSIRA Lagerh.
    Uredinales   8 spp.
Buritica, P., and Hennen, J. F. 1980. Pucciniosireae (Uredinales, Pucciniaceae). Flora Neotropica 24:1-50. (key to genera, key to seven *Pucciniosira* spp., descriptions and illustrations)
See Uredinales.

PYCNOSTYSANUS   See *Briosia*.

PYRENOCHAETA De Not.
    Coelomycetes   10 spp.
Morgan-Jones, G., and White, J. F. 1983. Studies in the genus *Phoma*. III. *Paraphoma*, a new genus to accommodate *Phoma radicina*. Mycotaxon 18:57-65. (discusses the generic concept of *Pyrenochaeta*)
Schneider, R. 1984. The genus *Pyrenochaeta* De Not. Pages 513-524 in: Taxonomy of fungi. Part 2. C. V. Subramanian, ed. Amra Press, Madras. (discusses 75 *Pyrenochaeta* names, accepts 10 species, places *P. terrestris* in *Phoma*)
Schneider, R., and Schwarz, R. 1979. Die Gattung *Pyrenochaeta* De Notaris. Mitt. Biol. Bundes. Land- & Forst. 189:1-73. (in German, key to ten taxa, descriptions and illustrations)
Sutton (1980) discusses the generic concept and treats five species.
Diseases include: *Pyrenochaeta lycopersici* R. Schneider & Gerlach, brown root rot of tomato (CMI 398); *P. oryzae* Shirai ex Miyake, sheath blotch of rice (CMI 666); and *P. terrestris* (E. Hans.) Gorenz, Walker & Larson, pink root of onion (CMI 397). Teleomorphs are placed in *Herpotrichia* and *Neopeckia*. See *Dactuliophora*.

PYRENOPEZIZA Fuckel
    Helotiales   50 spp.
Gremmen, J. 1958. Taxonomical notes on mollisiaceous fungi-VI. The genus *Pyrenopeziza* Fuck. Fungus 28:37-46. (keys to sections and 45 species, descriptions)

PYRENOPEZIZA - cont.
Ilott, T. W., Ingram, D. S., and Rawlinson, C. J. 1984.
Heterothallism in *Pyrenopeziza brassicae*, cause of light leaf
spot of Brassicas. Trans. Br. Mycol. Soc. 82:477-483. (mating
types determined by two alleles at a single locus)
Maddock, S. E., and Ingram, D. S. 1981. Studies of the perfect stage
of the light leaf spot pathogen of Brassicas, *Pyrenopeziza
brassicae*. Trans. Br. Mycol. Soc. 77:207-210. (describes
developmental morphology)
Rawlinson, C. J., Sutton, B. C., and Muthyalu, G. 1978. Taxonomy and
biology of *Pyrenopeziza brassicae* sp. nov. (*Cylindrosporium
concentricum*), a pathogen of winter oilseed rape (*Brassica napus*
spp. *oleifera*). Trans. Br. Mycol. Soc. 71:425-439. (descriptions
and illustrations)
The anamorphs of *Pyrenopeziza* belong to *Cylindrosporium* and
Phialophora. *Pyrenopeziza brassicae* Sutton & Rawlinson causes light
spot of *Brassica* spp. (CMI 536 under *Cylindrosporium concentricum*).
See also the related genus *Leptotrochila*.

PYRENOPHORA Fr.
    Pleosporales   10 spp.
Ammon, H. U. 1963. Ueber einige Arten aus den Gattungen *Pyrenophora*
Fries und *Cochliobolus* Drechsler mit *Helminthosporium* als
Nebenfruchtform. Phytopathol. Z. 47:244-300. (keys to species,
descriptions and illlustrations)
Luttrell, E. S. 1977. Correlations between conidial and ascigerous
state characters in *Pyrenophora, Cochliobolus*, and *Setosphaeria*.
Rev. Mycol. 41:271-279. (proposes taxonomic disposition of
Helminthosporium-like species)
Shoemaker, R. A. 1966. A pleomorphic parasite of cereal seeds,
Pyrenophora semeniperda. Can. J. Bot. 44:1451-1456. (describes
and illustrates three anamorph states)
Smedegard-Petersen, V. 1971. *Pyrenophora teres* f. *maculata* f. nov.
and *Pyrenophora teres* f. *teres* on barley in Denmark. R. Vet.
Agric. Univ., Copenhagen, Yearb. 1971:124-144. (describes new
form, confirmed by inoculation experiments)
Smedegard-Petersen, V. 1978. Genetics of heterothallism in
Pyrenophora graminea and *Pyrenophora teres*. Trans. Br. Mycol.
Soc. 70:99-102. (mating types determined by a single pair of
alleles)
Additional references are included under the related genus *Pleospora*.
The anamorphs are placed in *Drechslera*. Sivanesan (1984) provides a
key to 12 species based on the anamorph. Species causing diseases
include: *Pyrenophora chaetomioides* Speg., leaf stripe and seedling
blight of oats (CMI 389 as *P. avenae*); *P. dictyoides* Paul & Parbery,
net blotch and leaf spot, primarily of fescues (CMI 493); *P.
graminea* Ito & Kuribayashi, leaf stripe of barley (CMI 388); *P.
teres* Drechs., net blotch of barley (CMI 390); and *P.*

PYRENOPHORA - cont.
tritici-repentis (Died.) Drechs., yellow leaf spot of cereals and grasses (CMI 494).

PYRICULARIA Sacc.
    Hyphomycetes   5 spp.
Hashioka, Y. 1971. Notes on *Pyricularia*. I. Three species parasitic to Musaceae, Cannaceae and Zingiberaceae. Trans. Mycol. Soc. Jpn. 12:126-135. (descriptions and illustrations)
Hashioka, Y. 1973. Notes on *Pyricularia*. II. Four species and one variety parasitic on Cyperaceae, Gramineae and Commelinaceae. Trans. Mycol. Soc. Jpn. 14:256-265. (descriptions and illustrations)
Ou, S. H. 1985. Rice Diseases. Second Edition. Commonwealth Mycological Institute, Kew, Surrey, England. 380 pp. (describes and illustrates the cause of rice blast disease, *P. oryzae*)
Rathaiah, Y. 1980. Leaf blast of turmeric. Plant Dis. 64:104-105. (describes and illustrates *P. curcumae*)
Yaegashi, H., and Udagawa, S. 1978. The taxonomical identity of the perfect state of *Pyricularia grisea* and its allies. Can. J. Bot. 56:180-183. (describes and illustrates the teleomorph *Magnaporthe grisea*, discusses relationship of various *Pyricularia* spp.)
Ellis (1971, 1976) describes and illustrates four species. Ou (1985) includes an extensive discussion of rice blast disease caused by *Pyricularia oryzae* Cavara (CMI 169). The teleomorph belongs in *Magnaporthe*.

PYTHIUM Pringsh.
    Peronosporales   87 spp.
Gerrettson-Cornell, L., and Simpson, J. 1984. Three new marine *Phytophthora* species from New South Wales. Mycotaxon 19:453-470. (key to related marine species including seven *Pythium* and 11 *Phytophthora* spp.)
Hendrix, F. F., Jr., and Campbell, W. D. 1983. Some pythiaceous fungi - new roles for old organisms. Pages 123-160 in: Zoosporic Plant Pathogens. A Modern Perspective. S. T. Buczaki, ed. Academic Press, New York. (reviews taxonomy, isolation techniques, pathology, and other aspects)
Mer, G. S., Verma, B. L., and Khulbe, R. D. 1984. Aquatic fungi of Kumaun Himalaya, India: *Pythium* Pringsheim. Sydowia 37:208-221. (key to 11 species, descriptions and illustrations)
Plaats-Niterink, A. J. van der 1981. Monograph of the genus *Pythium*. Stud. Mycol. 21:1-242. (key to 87 species, descriptions and illustrations)
Robertson, G. I. 1980. The genus *Pythium* in New Zealand. N. Z. J. Bot. 18:73-102. (key to 27 species, descriptions, illustrations, host list)
Waterhouse, G. M. 1967. Key to *Pythium* Pringsheim. Mycol. Pap.

PYTHIUM - cont.
109:1-15. (includes 89 species)
Waterhouse, G. M. 1968. The genus *Pythium* Pringsheim. Mycol. Pap.
110:1-71. (original descriptions, illustrations)
The recent monograph by van der Plaats-Niterink (1981) provides an excellent treatment of the genus and includes a key to all known species. Domsch et al. (1980) include ten species commonly occurring in soil and numerous references. Pathogenic species include: *Pythium aphanidermatum* (Edson) Fitzp., diseases of many hosts (CMI 36); *P. arrhenomanes* Drechs., seedling blight and root rot of Poaceae (CMI 39); *P. butleri* L. Subramanian, a variety of diseases on numerous hosts (CMI 37); *P. deliense* Meurs, damping off of seedlings and root rot of various hosts (CMI 116); *P. graminicola* L. Subramanian, seedling blight, collar and root rot of Poaceae (CMI 38); *P. intermedium* de Bary, damping off, foot and root rot of mainly ornamentals (CMI 40); *P. mamillatum* Meurs, damping off of seedlings and root rot of various hosts (CMI 117); *P. myriotylum* Drechs., various diseases on numerous hosts (CMI 118); *P. oligandrum* Drechs., various diseases on numerous hosts (CMI 119); and *P. splendens* Braun, various diseases on numerous hosts (CMI 120).

RAMICHLORIDIUM Stahel ex de Hoog
    Hyphomycetes   17 spp.
Hoog, G. S. de, Rahman, M. A., and Boekhout, T. 1983. *Ramichloridium, Veronaea* and *Stenella*: Generic delimitation, new combinations and two new species. Trans. Br. Mycol. Soc. 81:485-490. (key to 36 taxa in *Ramichloridium* and related genera, describes and illustrates *R. pini*)
De Hoog et al. (1983) provide a comprehensive treatment of *Ramichloridium. Ramichloridium pini* de Hoog & Rahman causes a dieback of *Pinus contorta* in Scotland.

RAMULARIA Unger
    Hyphomycetes   300 spp.
Gunnerbeck, E. 1967. *Ramularia* and related fungi on Phanerogams in Uppland (Sweden). Sven. Bot. Tidskr. 61:126-138. (lists over 100 *Ramularia* and similar species and their hosts)
Poelt, J., and Fritz-Schroeder, J. 1983. *Ramularia* und verwandte Pilze in der Steiermark (eine erste Uebersicht). Mitt. Naturwiss. Ver. Steiermark 13:79-89. (in German, lists 70 species according to host)
Vaneb, S. 1974. Fungi of the genus *Ramularia* Sacc. in the Central Balkan Range (Teteven Mountain). Izv. Bot. Inst. Bulg. Akad. Nauk. 25:145-180. (in Russian, lists 46 species)
Despite the above references, a monograph is needed. Sivanesan (1984) includes a key to eight *Mycosphaerella* species with *Ramularia* anamorphs. *Ramularia* species cause leaf spot diseases including: *Ramularia alba* (Dowson) Nannf., white blight or *Cladosporium* blight

RAMULARIA - cont.
of sweet pea (CMI 869); *R. anomala* Peck, leaf spot of *Polygonum* spp. (CMI 858); *R. astragali* Ellis & Holw., leaf spot of *Astragalus carolinianus* (CMI 861); *R. bistortae* Fuckel, leaf spot of *Polygonum bistorta* (CMI 854); *R. brunnea* Peck, leaf spot or white spot of strawberry (CMI 708 under *Mycosphaerella fragariae*); *R. coalescens* (J. J. Davis) Pirozynski, on leaves of *Ribes* spp. (F. Can. 23); *R. decipiens* Ellis & Everh., leaf spot of *Rumex* spp. (CMI 852); *R. deusta* (Fuckel) Karakulin, leaf spot of *Lathyrus* and *Lotus* spp. (CMI 868); *R. galegae* Sacc., leaf spot of *Galega officinalis* (CMI 864); *R. gossypii* (Speg.) Cif., grey mildew of cotton (CMI 520); *R. oxyria-digynae* Gjaerum, leaf spot of *Oxyria digyna* (CMI 857); *R. pakistanica* Shakil, Khan & Kamal, leaf spot of *Polygonum dentatus* (CMI 860); *R. polygoni* Pandotra & Ganguly, leaf spot of *Polygonum amplexicaule* (CMI 859); *R. pratensis* Sacc., leaf spot of *Rumex* spp. (CMI 853); *R. psolareae* Ellis & Everh., leaf spot of *Psoralea machrostachya* (CMI 867); *R. rigidula* (Delacr.) Nannf., leaf spot of *Polygonum aviculare* (CMI 855); *R. rubella* (Bonord.) Nannf., leaf spot of *Rumex* spp. (CMI 851); *R. rufomaculans* Peck, leaf spot of *Polygonum* spp. (CMI 856); *R. sphaeroidea* Sacc., leaf spot of *Lotus* spp. (CMI 862); *R. trifolii* Jaap, leaf spot of *Trifolium rybergii* (CMI 866); *R. vallisumbrosae* Cavara, *Narcissus* white mold; and *R. winteri* Thuem., on leaves of *Ononis* spp. (CMI 865).

RAMULISPORA Miura
    Hyphomycetes   4 spp.
Rawla, G. S. 1973. *Gloeocercospora* and *Ramulispora* in India. Trans. Br. Mycol. Soc. 60:283-292. (describes and illustrates three *Ramulispora* spp.)
Diseases include *Ramulispora sorghi* (Ellis & Everh.) Olive & Lefebvre, leaf spot, sooty stripe of *Sorghum* spp. (CMI 585), and *R. sorghicola* E. Harris, leaf spot of *Sorghum* spp. (CMI 586).

RAVENELIA Berk.
    Uredinales   150 spp.
Baxter, J. W. 1980. A study of three African species of *Ravenelia* on *Cassia*. Mycologia 72:840-842. (describes one species with two synonyms)
Tyagi, R. N. S., and Prasad, N. 1972. The monographic studies on genus *Ravenelia* occurring in Rajasthan. Indian J. Plant Pathol. 2:108-135. (key to 11 species, descriptions and illustrations)
Most of these autoecious rusts occur on legumes and are included in Cummins (1978) and other references under Uredinales.

RHABDOCLINE Sydow
    Rhytismatales   3 spp.
Parker, A. K., and Reid, J. 1969. The genus *Rhabdocline* Syd. Can. J. Bot. 47:1533-1545. (key to two species and their five subspecies,

RHABDOCLINE - cont.
descriptions and illustrations)
Sherwood-Pike, M., Stone, J. K., and Carroll, G. C. 1986. *Rhabdocline parkeri*, a ubiquitous foliar endophyte of Douglas-fir. Can. J. Bot. 64:1849-1855. (describes and illustrates *R. parkeri* and its anamorph *Meria parkeri*, presents tables comparing three *Rhabdocline* spp. and two *Meria* spp.)
These species occur on conifer needles. Funk (1985) includes a key to five taxa with descriptions and illustrations. *Rhabdocline pseudotsugae* Sydow ssp. *pseudotsugae* causes needlecast of *Pseudotsuga* spp. (CMI 651). Anamorphs are placed in *Meria* and *Rhabdogloeum*.

RHIZINA Fr.:Fr.
    Pezizales   1 sp.
Tylutki, E. E. 1979. Mushrooms of Idaho and the Pacific Northwest. Discomycetes. University Press of Idaho, Moscow. 133 pp. (key to species in the family Helvellaceae, description and illustration)
Rhizina undulata Fr.:Fr. has been associated with decline of conifers (CMI 324, F. Can. 16).

RHIZOCTONIA DC.
    Agonomycetes   100 spp.
Adams, G. C., Jr., and Butler, E. E. 1979. Serological relationships among anastomosis groups of *Rhizoctonia solani*. Phytopathology 69:629-633. (presents a rapid, simple method for determining serological groups)
Burpee, L. L., Sanders, P. L., Cole, H., Jr., and Sherwood, R. T. 1980. Anastomosis groups among isolates of *Ceratobasidium cornigerum* and related fungi. Mycologia 72:689-701. (defines anastomosis groups within binucleate *Rhizoctonia solani*-like fungi)
Martin, B. 1987. Rapid tentative identification of *Rhizoctonia* spp. associated with diseased turfgrasses. Plant Dis. 71:47-49. (describes technique to determine number of nuclei per hyphal cell using fluorescence microscopy and a DAPI probe)
Murray, D. I. L. 1982. A modified procedure for fruiting *Rhizoctonia solani* on agar. Trans. Br. Mycol. Soc. 79:129-135. (illustrates the teleomorph *Thanatephorus cucumeris*)
Ogoshi, A. 1984. [Studies on the taxonomy of the genus *Rhizoctonia*.] Ann. Phytopathol. Soc. Jpn. 50:307-309. (in Japanese)
Ogoshi, A., Oniki, M., Araki, T., and Ui, T. 1983. Studies on the anastomosis groups of binucleate *Rhizoctonia* and their perfect state. J. Fac. Agric., Hokkaido Univ., Japan 61:244-260. (lists pathogenic species of *Rhizoctonia* and *Ceratobasidium* for each anastomosis group)
Ogoshi, A., Oniki, M., Sakai, R., and Ui, T. 1979. Anastomosis

RHIZOCTONIA - cont.
grouping among isolates of binucleate *Rhizoctonia*. Trans. Mycol. Soc. Jpn. 20:33-39. (applies hyphal anastomosis techniques for identification of binucleate *Rhizoctonia* spp.)
Oniki, M., Ogoshi, A., Araki, T., Sakai, R., and Tanaka, S. 1985. [The perfect state of *Rhizoctonia oryzae* and *R. zeae*, and the anastomosis groups of *Waitea circinata*.] Trans. Mycol. Soc. Jpn. 26:189-198. (in Japanese with English summary, taxonomy of *R. zeae* and *Waitea circinata*, teleomorph of both *R. zeae* and *R. oryzae*, illustrations)
Parmeter, J. R., Jr., and Whitney, H. S. 1970. Taxonomy and nomenclature of the imperfect state. Pages 7-19 in: *Rhizoctonia solani*: Biology and Pathology. J. R. Parmeter, Jr., ed. University of California Press, Berkeley. (reviews *R. solani* and similar species)
Tu, C. C., and Kimbrough, J. W. 1975. Morphology, development and cytochemistry of the hyphae and sclerotia of species in the *Rhizoctonia* complex. Can. J. Bot. 53:2282-2296. (discusses morphological and cytological characteristics of seven genera)
Tu, C. C., and Kimbrough, J. W. 1978. Systematics and phylogeny of fungi in the *Rhizoctonia* complex. Bot. Gaz. 139:454-466. (compares related genera)
Rhizoctonia includes fungi which produce neither asexual nor sexual structures, rather form only mycelium and sclerotia. Some isolates develop sexual structures that belong to a number of basidiomycetous genera. See *Ceratobasidium, Thanatephorus*, and *Waitea*. Taxonomic concepts are presented in Parmeter and Whitney (1970) and Tu and Kimbrough (1978). Anastomosis groups have been defined and can be used to characterize isolates. Diseases caused by *Rhizoctonia* species include: *Rhizoctonia carotae* Rader, crater rot of carrot in cold storage (CMI 408); *R. oryzae-sativae* (Sawada) Mordue, sheath spot of rice (CMI 409); *R. solani* Kuehn, various diseases on many host plants (CMI 406 under *Thanatephorus cucumeris*); and *R. tuliparum* Whetzel & Arth., grey bulb rot (CMI 407). See also the anamorph *Macrophomina*.

RHIZOPUS Ehrenb.
Mucorales   10 spp.
Dabinett, P. E., and Wellman, A. M. 1973. Numerical taxonomy of the genus *Rhizopus*. Can. J. Bot. 51:2053-2064. (defines seven groups)
Ellis, D. H. 1981. Sporangiophore ornamentation of thermophilic *Rhizopus* species and some allied genera. Mycologia 73:511-523. (presents table comparing sporangiospores of nine species)
Ellis, J. J. 1985. Species and varieties in the *Rhizopus arrhizus-Rhizopus oryzae* group as indicated by their DNA complementarity. Mycologia 77:243-247. (proposes synonymy based on DNA relatedness)

RHIZOPUS - cont.
Ellis, J. J. 1986. Species and varieties in the *Rhizopus microsporus* group as indicated by their DNA complementarity. Mycologia 78:508-510. (key to eight taxa)
Schipper, M. A. A. 1984. A revision of the genus *Rhizopus*. I. The *Rhizopus stolonifer*-group and *Rhizopus oryzae*. Stud. Mycol. 25:1-19. (key to groups and four taxa in the *R. stolonifer*-group, descriptions and illustrations)
Schipper, M. A. A., and Stalpers, J. A. 1984. A revision of the genus *Rhizopus*. II. The *Rhizopus microsporus*-group. Stud. Mycol. 25:20-34. (key to five taxa, descriptions and illustrations)
Several good monographs are available. Domsch et al. (1980) provide a key to seven species, descriptions of two species, and numerous references. Diseases caused by *Rhizopus* species include: *Rhizopus microsporus* van Tieghem, animal pathogen (CMI 108, CMI 523); *R. oryzae* Went & Geerl., plant and animal pathogen (CMI 109, CMI 525); *R. rhizopodiformis* (Cohn) Zopf, human and animal pathogen (CMI 522); *R. sexualis* (G. Sm.) Callen, soft rot of strawberries and other soft fruits (CMI 526); and *R. stolonifer* (Ehrenb.:Fr.) Vuill. (syn. *R. nigricans* Ehrenb.), rot of various plant parts (CMI 110, CMI 524).

RHIZOSPHAERA L. Mangin & Hariot
    Coelomycetes   4 spp.
Martinez, A. T., and Ramirez, C. 1983. *Rhizosphaera oudemansii* (Sphaeropsidales) associated with a needle cast of Spanish *Abies pinsapo*. Mycopathologia 83:175-182. (describes a *Hormonema*-state)
Sutton (1980) presents a key to four species with brief descriptions and illustrations; Funk (1985) treats three species with descriptions and illustrations. The teleomorph is *Phaeocryptopus*. These fungi cause needle diseases on conifers including *Rhizosphaera kalkhoffii* Bubak, needle blight of pine and spruce (CMI 656), and *R. pini* (Corda) Maubl., needle blight of fir (CMI 657).

RHYNCHOSPHAERIA See *Lepteutypa*.

RHYNCHOSPORIUM Heinsen ex Frank
    Hyphomycetes   2 spp.
Most *Rhynchosporium* species have been transferred to *Microdochium*. *Rhynchosporium secalis* (Oudem.) J. J. Davis causes barley and rye leaf blotch and leaf scald (CMI 387).

RHYTISMA Fr.:Fr.
    Rhytismatales   20 spp.
Cannon, P. F., and Minter, D. W. 1986. The Rhytismataceae of the Indian subcontinent. Mycol. Pap. 155:1-123. (key to genera of Rhytismataceae, key to seven *Rhytisma* spp., descriptions and illustrations)

RHYTISMA - cont.
Duravetz, J. S., and Morgan-Jones, J. F. 1971. Ascocarp development in *Rhytisma acerinum* and *R. punctatum*. Can. J. Bot. 49:1267-1272. (compares these species)
Hudler, G. W., Banik, M. T., and Miller, S. G. 1987. Unusual epidemic of tar spot on Norway maple in upstate New York. Plant Dis. 71:65-68. (describes and illustrates the fungus and disease, compares the pathogen with *R. acerinum*)
Funk (1985) provides descriptions and illustrations of three species. Diseases caused by *Rhytisma* species include: *Rhytisma acerinum* (Pers.:Fr.) Fr., tar spot of *Acer* spp. (CMI 791); *R. punctatum* (Pers.:Fr.)Fr., tar spot of *Acer* spp.; and *R. salicinum* (Pers.:Fr.) Fr., tar spot of *Salix*. Sutton (1980) includes the anamorph of *R. acerinum* in *Melasmia*.

RIGIDOPORUS Murrill
    Aphyllophorales   14 spp.
Ryvarden and Johansen (1980) provide a key to six species. Some of the pathogenic species include: *Rigidoporus lignosus* (Klotzsch) Imazeki, white root rot of *Hevea* and other tropical genera (CMI 198); *R. ulmarius* (Sowerby:Fr.) Imazeki, elm butt rot, also reported on other trees (CMI 199); *R. vinctus* (Berk.) Ryv., root rot on various hosts (CMI 322 as *Poria hypobrunnea*); and *R. zonalis* (Berk.) Imazeki, white pocket rot on various hosts (CMI 200). See Aphyllophorales.

ROESTELIA Rebent.
    Uredinales   14 spp.
Roestelia is the form-genus for the aecial form of *Gymnosporangium* spp. *Roestelia brucensis* Parmelee occurs on *Juniperus horizontalis* (F. Can. 140). See *Gymnosporangium* and Uredinales.

ROSELLINIA De Not.
    Sphaeriales   100 spp.
Dargan, J. S., and Thind, K. S. 1979. Xylariaceae of India-VII. The genus *Rosellinia* in the Northwest Himalayas. Mycologia 71:1010-1023. (key to ten species)
Francis, S. M. 1985. *Rosellinia necatrix* - fact or fiction? Sydowia 38:75-86. (discussion of *R. necatrix* and three other *Rosellinia* spp., descriptions and illustrations)
Martin, P. 1967. Studies in the Xylariaceae-II. *Rosellinia* and the Primocinerea section of *Hypoxylon*. J. S. Afr. Bot. 33:315-328. (transfers all *Rosellinia* spp. to other genera, later authors disagree, key to *Rosellinia* species placed in other genera)
Petrini, L., and Petrini, O. 1985. Xylariaceous fungi as endophytes. Sydowia 38:216-234. (keys to European xylariaceous endophytes based on cultural characters, includes descriptions of three *Rosellinia* anamorphs placed in *Nodulisporium, Rhinocladiella*, and

ROSELLINIA - cont.
Sporothrix)
Saccas, A. M. 1956. Les *Rosellinia* des Cafeiers en Oubangui-Chari. Agron. Trop. (Paris) 11:551-595. (in French, descriptions and illustrations, comparative tables)
No comprehensive account of this genus exists. Several root rot diseases are caused by *Rosellinia* species including: *Rosellinia aquila* (Fr.) De Not., on mulberry; *R. arcuata* Petch, on tea (CMI 353); *R. bunodes* (Berk. & Broome) Sacc., causing black root rot of tropical woody hosts (CMI 351); *R. necatrix* Prill., on apple and grapes (CMI 352); *R. pepo* Pat., on various woody hosts (CMI 354); and *R. quercina* R. Hartig, on oak.

SAROCLADIUM Gams & D. Hawksworth
    Hyphomycetes   2 spp.
Gams, W., and Hawksworth, D. L. 1975. The identity of *Acrocylindrium oryzae* Sawada and a similar fungus causing sheath-rot of rice. Kavaka 3:57-61. (describes and illustrates two species)
Ngala, G. N. 1983. *Sarocladium attenuatum* as one of the causes of rice grain spotting in Nigeria. Plant Pathol. 32:289-293. (discusses diseases)
Sarocladium attenuatum Gams & D. Hawksworth causes sheath rot of rice (CMI 674) as does *S. oryzae* (Sawada) Gams & D. Hawksworth (CMI 673).

SAWADAEA Miyabe ex Sawada
    Erysiphales   6 spp.
Zheng, R.-y., and Chen, G.-q. 1980. [Taxonomic studies on the genus *Sawadaea* in China. I. Recognition of the genus *Sawadaea* and the new species and new combination on Aceraceae and Hippocastanaceae.] Acta Microbiol. Sin. 20:35-44. (in Chinese with English summary, key to six species, descriptions, illustrations)
This genus is a segregate of *Uncinula*.

SCHIZOTHYRIUM Desmaz.
    Dothideales   12 spp.
Mueller and Arx (1962) provide descriptions and illustrations for 12 species. These fungi cause fly speck on fruit, for example *Schizothyrium pomi* (Mont.) Arx (syn. *Leptothyrium pomi* (Mont.) Sacc.), fly speck of apple and pear.

SCINIATOSPORIUM   See *Stigmina*.

SCIRRHIA Nitschke ex Fuckel
    Dothideales   8 spp.
No monographs are available. Barr (1972) includes six species with descriptions and illustrations. Sivanesan (1984) and Funk (1985)

SCIRRHIA - cont.
describe and illustrate *Mycosphaerella pini* Rostr. (as *Scirrhia pini* Funk & A. Parker), the cause of *Dothistroma* blight of pines (CMI 368). *Mycosphaerella dearnessii* Barr is now considered the correct name for *Scirrhia acicola* (Dearn.) Siggers, the cause of *Lecanosticta* or brown spot needle blight of pines (CMI 367). See *Mycosphaerella*, particularly Evans (1984).

SCLERODERRIS See *Ascocalyx*.

SCLEROPHOMA See *Sydowia*.

SCLEROPHTHORA Thirumalachar, C. G. Shaw & Narasimhan
    Peronosporales   4 spp.
Payak, M. M., Renfro, B. L., and Lal, S. 1970. Downy mildew diseases incited by *Sclerophthora*. Indian Phytopathol. 23:183-193. (key to five taxa, discussion)
See Peronosporales.

SCLEROSPORA J. Schroet.
    Peronosporales   16 spp.
Shaw, C. G. 1978. *Peronosclerospora* species and other downy mildews of the Gramineae. Mycologia 70:594-604. (transfers most *Sclerospora* species to *Peronosclerospora*)
Westhuizen, G. C. A. van der 1977. Downy mildew fungi of maize and sorghum in South Africa. Phytophylactica 9:83-89. (discusses *S. graminicola* and *Peronosclerospora sorghi* as *Sclerospora sorghi*)
Sclerospora graminicola (Sacc.) J. Schroet. causes downy mildew of pearl millet (CMI 452, CMI 770). Most species have been transferred to *Peronosclerospora*. See also Peronosporales.

SCLEROTINIA Fuckel
    Helotiales   3 spp.
Cruickshank, R. H. 1983. Distinction between *Sclerotinia* species by their pectic zymograms. Trans. Br. Mycol. Soc. 80:117-119. (finds zymograms useful for rapid identification)
Kohn, L. M. 1979a. A monographic revision of the genus *Sclerotinia*. Mycotaxon 9:365-444. (key to sclerotium-forming genera of Sclerotiniaceae, key to three *Sclerotinia* spp., descriptions and illustrations; discusses current disposition of 259 names previously placed in *Sclerotinia*)
Kohn, L. M. 1979b. Delimitation of the economically important plant pathogenic *Sclerotinia* species. Phytopathology 69:881-890. (keys to three *Sclerotinia* spp. and related genera of plant pathogens)
Pyykko, M., and Hamet-ahti, L. 1980. *Sclerotinia pirolae*: sclerotial ontogeny and occurrence in Finland. Karstenia 20:28-32. (description and illustrations)
Sharma, M. P. 1983. The genus *Sclerotinia* Fuckel in India.

SCLEROTINIA - cont.
Biovigyanam 9:105-108. (key to seven species with brief descriptions of three species, does not agree with other authors in generic concept of *Sclerotinia*)
Tariq, V.-N., Gutteridge, C. S., and Jeffries, P. 1985. Comparative studies of cultural and biochemical characteristics used for distinguishing species within *Sclerotinia*. Trans. Br. Mycol. Soc. 84:381-397. (suggests that *S. sclerotiorum* and *S. trifoliorum* are subspecies)
Willetts, H. J., and Wong, A.-L. 1980. The biology of *Sclerotinia sclerotiorum, S. trifoliorum*, and *S. minor* with emphasis on specific nomenclature. Bot. Rev. 46:101-165. (a comprehensive review)
Wong, A.-L., and Willetts, H. J. 1973. Electrophoretic studies of soluble proteins and enzymes of *Sclerotinia* species. Trans. Br. Mycol. Soc. 61:167-178. (finds protein and enzymes patterns useful in separating species)
Wong, A.-L., and Willetts, H. J. 1975. Electrophoretic studies of Australasian, North American and European isolates of *Sclerotinia sclerotiorum* and related species. J. Gen. Microbiol. 90:355-359. (defines three species based on these studies)
Wong, A.-L., and Willetts, H. J. 1979. Cytology of *Sclerotinia sclerotiorum* and related species. J. Gen. Microbiol. 112:29-34. (nuclear characteristics and chromosome numbers of three species)
Several major studies support the acceptance of three species in *Sclerotinia*, namely: *Sclerotinia minor* Jagger, pathogenic on a wide range of herbaceous plants in several families including the Asteraceae, Brassicaceae, and Fabaceae; *S. sclerotiorum* (Lib.) de Bary, pathogenic on over 350 species of herbaceous plants in 60 families (CMI 513); and *S. trifoliorum* Eriks., occurring almost exclusively on forage legumes and on uncultivated leguminous plants (Kohn, 1979a, 1979b). Although not belonging to *Sclerotinia*, *S. homeocarpa* F. T. Bennett, the cause of dollar spot of turf (CMI 618), has not been placed in another genus. See *Botryotinia*, *Monilinia*, and *Stromatinia*.

SCLEROTIUM Tode:Fr.
Agonomycetes   100 spp.
Georgy, N. I., and Coley-Smith, J. R. 1982. Variation in morphology of *Sclerotium cepivorum* sclerotia. Trans. Br. Mycol. Soc. 79:534-536. (includes descriptions and illustrations of variability)
Insell, J. P., Huner, N. P. A., Newsted, W. J., and Huystee, R. B. van 1985. Light microscopic and polypeptide analyses of sclerotia from mesophilic and psychrophilic pathogenic fungi. Can. J. Bot. 63:2305-2310. (compares sclerotia of *Sclerotinia sclerotiorum*, *Sclerotium rolfsii*, *Botrytis cinerea*, and *Myrosclerotinia borealis*)

SCLEROTIUM - cont.
Mordue, J. E. M. 1983. Dolipore septa in *Sclerotium hydrophilum*. Trans. Br. Mycol. Soc. 81:654-655. (illustrates septum)
Ou, S. H. 1985. Rice Diseases. Second Edition. Commonwealth Mycological Institute, Kew, Surrey, England. 380 pp. (descriptions and illustrations of a number of sclerotial fungi pathogenic on rice)
Punja, Z. K. 1985. The biology, ecology, and control of *Sclerotium rolfsii*. Ann. Rev. Phytopathol. 23:97-127. (includes brief description)
Punja, Z. K., Grogan, R. G., and Adams, G. C., Jr. 1982. Influence of nutrition, environment, and the isolate, on basidiocarp formation, development, and structure in *Athelia (Sclerotium) rolfsii*. Mycologia 74:917-926. (includes description and illustrations)
Punter, D., Reid, J., and Hopkin, A. A. 1984. Notes on sclerotium-forming fungi from *Zizania aquatica* (wildrice) and other hosts. Mycologia 76:722-732. (describes and illustrates *Sclerotium oryzae*, the synanamorph *Nakataea sigmoidea*, and the teleomorph *Magnaporthe salvinii*)
Schoen, J. F. 1983. Identification of seed-like structures: a taxonomic review of sclerotial-forming fungi. Seed Sci. Technol. 11:639-650. (reviews major groups that form sclerotia)
Willetts, H. J. 1972. The morphogenesis and possible evolutionary origins of fungal sclerotia. Biol. Rev. 47:515-536. (summarizes types of sclerotia and the fungi that form them)
Fungi are placed in the genus *Sclerotium* because they produce only sterile mycelium and sclerotia. Identification is difficult without additional characters. Known teleomorphs belong to either Ascomycetes or Basidiomycetes. Ou (1985) provides an account of some aquatic, sclerotial-forming fungi. See *Athelia, Ceratobasidium, Magnaporthe, Rhizoctonia*, and *Sclerotinia*. Important disease-causing species include: *Sclerotium cepivorum* Berk., causing white rot of onions (CMI 512); *S. oryzae* Cattaneo, sclerotial state of *Magnaporthe salvinii* (Cattaneo) Krause & R. Webster, causing stem rot of rice (CMI 344 as *Leptosphaeria salvinii*); *S. rolfsii* Sacc., stem rot of legumes and many other economically important hosts (CMI 410 under *Athelia rolfsii* as *Corticium rolfsii*); and *S. tuliparum* Kleb., causing grey bulb rot of tulip.

SCOPELLA See *Maravalia*.

SCOPULARIOPSIS Bainier
    Hyphomycetes   12 spp.
Morton, F. J., and Smith, G. 1963. The genera *Scopulariopsis* Bainier, *Microascus* Zukal, and *Doratomyces* Corda. Mycol. Pap. 86:1-96. (key to 18 *Scopulariopsis* spp., descriptions and illustrations)
Domsch et al. (1980) include five common species of this generally

SCOPULARIOPSIS - cont.
non-pathogenic genus. *Scopulariopsis brevicaulis* (Sacc.) Bainier is saprophytic, rarely causing onychomycosis in humans (CMI 100), and *S. fimicola* (Costantina & Matr.) Arnaud & Mart. occurs as white plaster mould of mushroom beds. The teleomorph is *Microascus*.

SCYTINOSTROMA Donk
 Aphyllophorales 11 spp.
Rattan, S. S. 1974. *Scytinostroma* in India with notes on extralimital species. Trans. Br. Mycol. Soc. 63:1-12. (key to 11 species, descriptions and illustrations)
White, L. T. 1951. Studies of Canadian Thelephoraceae VIII. *Corticium galactinum* (Fr.) Burt. Can. J. Bot. 29:279-296. (reviews pathology, describes and illustrates the fungus, presents results of cultural and interfertility studies)
Scytinostroma galactinum (Fr.) Donk (syn. *Corticium galactinum* (Fr.) Burt) causes a white root rot of apples, and a root and butt rot of a number of hardwoods and conifers.

SEIMATOSPORIUM Corda
 Coelomycetes 38 spp.
Brockmann, I. (1975)1976. Untersuchungen ueber die Gattung *Discostroma* Clements (Ascomycetes). Sydowia 28:275-338. (in French with English summary, key to nine *Seimatosporium* and *Sporocadus* spp., descriptions and illustrations including teleomorph)
Funk, A. 1985. The anamorph of *Leciographa gallicola*. Can. J. Bot. 63:365. (describes and illustrates *Seimatosporium etheridgei*)
Nag Raj, T. R. 1986. Redisposals and redescriptions in the *Monochaetia-Seiridium*, *Pestalotia-Pestalotiopsis* complexes. V. *Monochaetia alnea* and *M. berberidicola*. Mycotaxon 26:187-198. (describes *Seimatosporium alnea*, transfers *Monochaetia berberidicola* to *Seimatosporium*, descriptions and illustrations)
Pirozynski, K. A., and Shoemaker, R. A. 1970. *Seimatosporium* leaf spot of *Ledum* and *Rhododendron*. Can. J. Bot. 48:2199-2203. (describes and illustrates *S. lichenicola* and *S. rhododendri*)
Shoemaker, R. A. 1964. *Seimatosporium* (= *Cryptostictus*) parasites of *Rosa, Vitis*, and *Cornus*. Can. J. Bot. 42:411-421. (key to four species, descriptions and illustrations)
Shoemaker, R. A., and Mueller, E. 1964. Generic correlations and concepts: *Clathridium* (=*Griphosphaeria*) and *Seimatosporium* (=*Sporocadus*). Can. J. Bot. 42:403-410. (describes two *Clathridium* spp. with *Seimatosporium* anamorphs)
Swart, H. J., and Griffiths, P. A. 1974. Australian leaf-inhabiting fungi. V. Two species of *Seimatosporium* on *Eucalyptus*. Trans. Br. Mycol. Soc. 62:359-366. (describes and illustrates *S. brevilatum* and *S. fusisporum*)
Sutton (1980) presents a key to 38 species with brief descriptions

SEIMATOSPORIUM - cont.
and illustrations. The teleomorphs are placed in *Clathridium*, *Discostroma*, *Leciographa*, and *Paradidymella*.

SEIRIDIUM Nees:Fr.
Coelomycetes   14 spp.
Boesewinkel, H. J. 1983. New records of the three fungi causing Cypress canker in New Zealand, *Seiridium cupressi* (Guba) comb. nov. and *S. cardinale* on *Cupressocyparis* and *S. unicorne* on *Cryptomeria* and *Cupressus*. Trans. Br. Mycol. Soc. 80:544-547. (recognizes three species causing Cypress canker, descriptions, illustrations, presents table comparing the three species)
Nag Raj, T. R., and Kendrick, B. 1985. *Ellurema* gen. nov., with notes on *Lepteutypa cisticola* and *Seiridium canariense*. Sydowia 38:178-193. (transfers *Adea canariensis* to *Seiridium*, description and illustrations)
Sasaki, K., and Kobayashi, T. 1974. Resinous canker disease of Cupressaceae caused by *Monochaetia unicornis* (Cke. et Ell.) Sacc. (I) - The causal fungus and its pathogenicity. Bull. Gov. For. Exp. Stn. (Jpn.) 271:27-38. (in Japanese with English summary, description, illustration, pathogenicity, this species now placed in *Seimatosporium*)
Sutton (1980) presents a key to 12 *Seiridium* spp. with brief descriptions and illustrations. The teleomorphs are placed in *Blogiascospora* and *Lepteutypa*. *Seiridium cardinale* (Wagener) Sutton & I. A. S. Gibson causes *Coryneum* canker of *Cupressus* (CMI 326), and *S. cupressii* (Guba) Boesewinkel causes cypress canker (CMI 325 under *Lepteutypa cupressi* as *Rhynchosphaeria cupressi*).

SELENOPHOMA Maire
Coelomycetes   5 spp.
Latterell, F. M., Rossi, A. E., and Trujillo, E. E. 1986. A previously undescribed *Selenophoma* leaf spot of maize in Colombia. Plant Dis. 70:472-474. (description and illustrations, first report of a *Selenophoma* sp. on maize)
Sutton (1980) includes a key to five species with brief descriptions and illustrations. The teleomorphs are placed in *Discosphaerina* (Sivanesan, 1984). See *Pseudoseptoria*.

SEPTOCYTA Petr.
Coelomycetes   1 sp.
This monotypic genus is included in Sutton (1980). *Septocyta ruborum* (Lib.) Petrak causes purple blotch or stem spot of blackberries (CMI 667).

SEPTOGLOEUM Sacc.
Coelomycetes   3 spp.
Sutton, B. C., and Pollack, F. G. 1974. Microfungi on *Cercocarpus*.

SEPTOGLOEUM - cont.
Mycopathol. Mycol. Appl. 52:331-351. (describes the type of *Septogloeum*, compares it with *Phloeospora*, describes and illustrates two species)
Sutton, B. C., and Webster, J. 1984. *Septogloeum japonicum* and *Marssonina pakistanica* spp. nov., Coelomycetes with *Tricellula*-like conidia. Trans. Br. Mycol. Soc. 83:59-64. (on living leaves of *Euonymous*, descriptions and illustrations)
Sutton (1980) includes a key to two species with brief descriptions and illustrations.

SEPTORIA Sacc.
  Coelomycetes   1000 spp.
Cejp, K., and Dolejs, K. 1967. Rare species of the genus *Septoria* from Czechoslovakia. Ceska Mykol. 21:213-219. (descriptions of fifteen species, illustrations)
Cejp, K., and Jechova, V. 1967. Beitrag zur Kenntnis einiger Tschechoslowakischen Arten der Gattung *Septoria* Fries. Sb. Nar. Muz. Praze, Rada B 23:101-123. (in German with Czechoslovakian summary, key to related genera and 86 species, brief descriptions and illustrations)
Constantinescu, O. 1984. Taxonomic revision of *Septoria*-like fungi parasitic on Betulaceae. Trans. Br. Mycol. Soc. 83:383-398. (key to six species, descriptions and illustrations, list of twelve excluded taxa)
Jorstad, I. 1965. *Septoria* and septorioid fungi on Dicotyledons in Norway. Skr. Nor. Vidensk.-Akad. Kl. I. Mat. Naturvidensk. Kl. 22:1-110. (includes 110 species of *Septoria* and related genera with brief descriptions, lists by host)
Jorstad, I. 1967. *Septoria* and related fungi on Gramineae in Norway. Skr. Nor. Vidensk.-Akad. Kl. I. Mat. Naturvidensk. Kl. 24:1-63. (keys to 41 species in *Septoria* and related genera, brief descriptions)
Makela, K. 1977. *Septoria* and *Selenophoma* species on Gramineae in Finland. Ann. Agric. Fenn. 16:256-276. (descriptions and illustrations of ten *Septoria* spp.)
Punithalingam, E. 1981. New microfungi from cereals and grasses. II. Nova Hedw. 34:67-95. (key to two *Septoria* spp. on *Setaria*, descriptions and illustrations)
Punithalingam, E., and Wheeler, B. E. J. 1965. *Septoria* spp. occurring on species of *Chrysanthemum*. Trans. Br. Mycol. Soc. 48:423-439. (compares four species, descriptions and illustrations)
Radulescu, E., Negru, A., and Docea, E. 1973. Septoriozele diu Romania. Edit. Acad. Rep. Soc. Rom. Lei., Bucharest. 325 pp. (in Czechoslovakian, lists 399 species according to host, brief descriptions and illustrations)
Richardson, M. J., and Noble, M. 1970. *Septoria* species on cereals -

SEPTORIA - cont.
a note to aid their identification. Plant Pathol. 19:159-163. (compares similar species, illustrations)
Sprague, R. 1950. *Septoria* disease of Gramineae in western United States. Oreg. State Monogr., Stud. Bot. 6:1-151. (key to related genera and *Septoria* spp., describes and illustrates twenty species, outdated but still a useful monograph)
Sukapure, R. S., and Thirumalachar, M. J. 1963. Studies on some *Septoria* species from India I. Sydowia 17:1-11. (includes 18 species with brief descriptions and a table comparing four species on *Artemisia*)
Teterevnikova-Babayan, D. N. 1976. [A critical survey of *Septoria* species parasitizing Salicaceae I. *Septoria* species inhabiting willow.] Biol. Zh. Arm. 29:3-11. (in Russian)
Teterevnikova-Babayan, D. N. 1976. [A critical survey of *Septoria* species parasitizing Salicaceae II. *Septoria* species on poplar.] Biol. Zh. Arm. 29:53-61. (in Russian)
Teterevnikova-Babayan, D. N. 1985. [*Septoria* leaf spot causal agents on stone fruits in the USSR and abroad and some of their ecological characters.] Mycol. & Phytopathol. 19:299-303. (in Russian, describes nine species, references to other articles by this author)
Although some regional or host-limited treatments of *Septoria* are available, identification of *Septoria* species is difficult and is usually based on host. The teleomorphs belong to *Leptosphaeria, Mycosphaerella, Phaeosphaeria,* and *Sphaerulina*. Some diseases caused by *Septoria* species are: *Septoria adanensis* Petr., leaf spot of *Chrysanthemum* (CMI 136); *S. apiicola* Speg., celery leaf spot (CMI 88); *S. avenae* f. sp. *avenae* Frank, speckled blotch of oats (CMI 312 under *Phaeosphaeria avenaria* f. sp. *avenaria* as *Leptosphaeria*); *S. azaleae* Vogl., azalea leaf scorch; *S. bromi* Sacc., on *Bromus* spp. (F. Can. 241); *S. cannabis* (Lasch) Sacc., white leaf spot of hemp (CMI 668); *S. carthami* Murashkinskij, white leaf spot of safflower (CMI 669); *S. chrysanthemella* Sacc., on *Chrysanthemum* spp. (CMI 137); *S. cucurbitacearum* Sacc., leaf spot of cucurbits (CMI 740); *S. elymi* Ellis & Everh., mostly on *Agropyron* and *Elymus* spp. (F. Can. 242); *S. glycines* Hemmi, brown spot of soybean (CMI 339); *S. helianthi* Ellis & Kellerm., leaf spot of sunflower (CMI 276); *S. humuli* Westendorp, leaf spot of hop (CMI 829); *S. lactucae* Pass., leaf spot of lettuce (CMI 335); *S. leucanthemi* Sacc. & Speg., leaf spot of *Chrysanthemum* spp. (CMI 138); *S. linicola* (Speg.) Garassini, Pasmo disease of flax (CMI 709 under *Mycosphaerella linicola*); *S. lycopersici* Speg., tomato leaf spot (CMI 89); *S. nodorum* Berk., wheat glume blotch (CMI 86 under *Phaeosphaeria nodorum* as *Leptosphaeria nodorum*; F. Can. 240 as *Stagonospora nodorum*); *S. obesa* Sydow, leaf spot of *Chrysanthemum* (CMI 139); *S. passerinii* Sacc., speckled leaf blotch of barley (CMI 277, F. Can. 243); *S. passifloricola* Punithalingam, diseases of *Passiflora* spp. (CMI 670);

SEPTORIA - cont.
S. socia Pass., leaf spot of *Chrysanthemum leucanthemum* (CMI 140); *S. tritici* Roberge, speckled leaf blotch of wheat (CMI 90, F. Can. 244); and *S. vignae* Henn., leaf spot of cowpea (CMI 830).

SETOSPHAERIA Leonard & Suggs
    Pleosporales   7 spp.
Alcorn, J. L. 1986. A new homothallic *Setosphaeria* species and its *Exserohilum* anamorph. Trans. Br. Mycol. Soc. 86:313-317. (description and illustrations, isolated from *Dactyloctenium*)
Leonard, K. J., and Suggs, E. G. 1974. *Setosphaeria prolata*, the ascigerous state of *Exserohilum prolatum*. Mycologia 66:281-297.
    (describes new teleomorph and new anamorph genera, lists species) Sivanesan (1984) provides a key to seven species, some previously *Trichometaspheria* species, with brief descriptions and illustrations. Anamorphs of *Setosphaeria* are placed in *Exserohilum*. *Setosphaeria rostrata* Leonard causes foot rot of wheat and other Poaceae (CMI 587). The cause of northern leaf blight of maize and sorghum is *S. turcica* (Luttrell) Leonard & Suggs (CMI 304 as *Trichometaspheria turcica*) with anamorph *Exserohilum turcicum* (Pass.) Leonard & Suggs.

SIROCOCCUS G. Preuss
    Coelomycetes   2 spp.
Sutherland, J. R., Lock, W., and Farris, S. H. 1981. *Sirococcus* blight: a seed-borne disease of container-grown spruce seedlings in coastal British Columbia forest nurseries. Can. J. Bot. 59:559-562. (describes detection of *S. strobilinus* in seeds)
Sutton (1980) briefly describes and illustrates two species. Funk (1985) includes *S. strobilinus* G. Preuss (syn. *Discella strobilina* (Desmaz.) Dietel) which causes shoot and leaf blight of various conifers.

SOLEELLA Darker
    Rhytismatales   1 sp.
Soleella striiformis (Darker) Darker causes needle cast of *Pinus torreyana* (CMI 800).

SOROSPORIUM F. Rudolphi
    Ustilaginales   1 sp.
Sorosporium is treated as a monotypic genus by Vanky (1985) who includes the species previously described on various Caryophyllaceae in *S. saponariae* F. Rudolphi.

SPHACELIA Lev.
    Hyphomycetes   5 spp.
Loveless, A. R. 1964. Use of the honeydew state in the identification of ergot species. Trans. Br. Mycol. Soc. 47:205-213.

SPHACELIA - cont.
(illustrates thirteen groups based on conidial morphology)
Mantle, P. G. 1968. Studies on *Sphacelia sorghi* McRae, an ergot of *Sorghum vulgare* Pers. Ann. Appl. Biol. 62:443-449. (compares sphacelial and sclerotial states)
Rykard, D. M., Luttrell, E. S., and Bacon, C. W. 1984. Conidiogenesis and conidiomata in the Clavicipitoideae. Mycologia 76:1095-1103. (compares *Ephelis* and *Sphacelia* states of clavicipitaceous genera)
Sphacelia includes the anamorphs of *Claviceps* species. Loveless (1964) compares the conidial morphology of *Sphacelia* spp. and presents his findings in a comparative table from which it is possible to identify species based on conidia. See *Claviceps* and *Epichloe*.

SPHACELOMA de Bary
    Coelomycetes   180 spp.
Stevenson, J. A. 1971. An account of fungus exsiccati containing material from the Americas. Nova Hedw. Beih. 36:1-563. (lists Jenkins' publications)
Sutton, B. C., and Pollack, F. G. 1973. *Gloeosporium cercocarpi* and *Sphaceloma cercocarpi*. Mycologia 65:1125-1134. (describes and illustrates three *Sphaceloma* spp.)
Sphaceloma includes the anamorphs of *Elsinoe* species which cause scab and anthracnose diseases of numerous hosts. A. E. Jenkins worked extensively on these genera and issued specimens as exsiccatae. See Stevenson (1971) for a list of her publications. No comprehensive treatment exists for *Sphaceloma* or *Elsinoe*. *Sphaceloma fawcettii* var. *scabiosa* (McAlpine & Tryon) Jenkins causes Tryon's scab or Australian citrus scab (CMI 437).

SPHACELOTHECA de Bary
    Ustilaginales   30 spp.
Langdon, R. F. N., and Fullerton, R. A. 1978. The genus *Sphacelotheca* (Ustilaginales): criteria for its delimitation and the consequences thereof. Mycotaxon 6:421-456. (delimits genus, describes and illustrates five species of *Sphacelotheca* and *Sporisorium*)
Sphacelotheca is a cosmopolitan genus occurring on Polygonaceae. Langdon and Fullerton (1978) place *Sphacelotheca* species on Poaceae in *Sporisorium* including some well-known smuts of sorghum and millet. See *Sporisorium, Ustilago*, and Ustilaginales.

SPHAEROPSIS Sacc.
    Coelomycetes   30 spp.
Petrak, F., and Sydow, H. 1927. Die Gattungen der Pyrenomyzeten, Sphaeropsideen, und Melanconieen. Beih. Rep. Spec. Nov. Regni Veg. 42:1-551. (in German, outdated but disposes of many names,

SPHAEROPSIS - cont.
host index)
Punithalingam, E. 1969. Studies on Sphaeropsidales in culture. Mycol. Pap. 119:1-24. (describes *Neodeightonia subglobosa*, now considered a *Botryosphaeria*, the teleomorph of *Sphaeropsis subglobosa* on bamboo)
Rodriguez, R., and Melendez, P. L. 1984. Occurrence of *Sphaeropsis* knot on citron (*Citrus medica* L.) in Puerto Rico. J. Agric. Univ. P. R. 68:179-183. (describes the disease and its causal organism)
Wang, C.-g., Blanchette, R. A., Jackson, W. A., and Palmer, M. A. 1985. Differences in conidial morphology among isolates of *Sphaeropsis sapinea*. Plant Dis. 69:838-841. (correlates characteristics of type A and B isolates)
No modern treatment exists for this genus; see Sutton (1980). *Sphaeropsis sapinea* (Fr.) Dyko & Sutton causes diseases of *Pinus* and other conifers (CMI 273 as *Diplodia pinea*) and *S. tumefaciens* Hedges causes knot of lime and other *Citrus* spp. (CMI 278).

SPHAEROSTILBE   See *Nectria*.

SPHAEROTHECA Lev.
   Erysiphales   6 spp.
Ballantyne, B. 1975. Powdery mildew on Cucurbitaceae: Identity, distribution, host range and sources of resistance. Proc. Linn. Soc. N. S. W. 99:100-120. (synopsis of characteristics and distribution of powdery mildews on cucurbits, discussion of occurrence of *S. fuligena* and comparision with *Erysiphe cichoracearum*)
Boesewinkel, H. J. 1979. Differences between the conidial states of *Podosphaera tridactyla* and *Sphaerotheca pannosa*. Ann. Phytopathol. 11:525-527. (descriptions and illustrations)
Braun, U. 1984. Taxonomic notes on some powdery mildews. (IV). Mycotaxon 20:483-489. (describes and illustrates two *Sphaerotheca* spp.)
Braun, U. 1985. Miscellaneous notes on the genus *Sphaerotheca*. I. Zbl. Mikrobiol. 140:161-170. (describes eight new taxa, discusses three additional species)
Braun, U. 1985. Miscellaneous notes on the genus *Sphaerotheca*. II. Zbl. Mikrobiol. 140:237-246. (describes four new taxa, key to 22 *Sphaerotheca* spp. in North America)
Junell, L. 1966. A revision of *Sphaerotheca fuliginea* ([Schlecht.] Fr.) Poll. s. lat. Sven. Bot. Tidskr. 60:365-392. (key to 12 *Sphaerotheca* spp., descriptions)
Reifschneider, F. J. B., Boiteux, L. S., and Occhiena, E. M. 1985. Powdery mildew of melon (*Cucumis melo*) caused by *Sphaerotheca fuliginea* in Brazil. Plant Dis. 69:1069-1070. (compares with *Erysiphe cichoracearum*)

SPHAEROTHECA - cont.
As for other genera in the Erysiphales, this genus is often treated geographically or by host. Species include: *Sphaerotheca fuliginea* (Schlechtend.:Fr.) Pollacci, on numerous genera of Asteraceae, Cucurbitaceae, and Scrophulariaceae (CMI 159); *S. macularis* (Wallr.:Fr.) Lind, hop and strawberry mildew (CMI 188); *S. mors-uvae* (Schwein.) Berk. & M. A. Curtis, American gooseberry mildew (CMI 254); and *S. pannosa* (Wallr.:Fr.) Lev., rose mildew (CMI 189). See *Erysiphe* and Erysiphales.

SPHAERULINA Sacc.
  Dothideales 40 spp.
Boerema, G. H. 1963. [Leaf scorch and surface canker (dieback) of rose rootstocks and other species roses caused by *Sphaerulina rehmiana* (stat. con. *Septoria rosae*).] Neth. J. Plant Pathol. 69:76-103. (in Dutch with English summary, descriptions and illustrations of morphology and pathology)
Members of this genus cause leaf spot diseases on a variety of hosts. Barr (1972) treats six species; Sivanesan (1984) treats three species. The anamorphs are placed in *Cercospora* and *Septoria*. Diseases caused by these fungi include: *Sphaerulina rehmiana* Jaap, rose leaf scorch; *S. rubi* Desmaz. & Wilcox, raspberry leaf spot; and *S. taxi* (Cooke) Massee, on yew.

SPILOCAEA Fr.
  Hyphomycetes 6 spp.
Ellis (1971, 1976) includes six species. *Spilocaea* species are usually treated with their *Venturia* teleomorphs.

SPONGOSPORA Brunchorst
  Plasmodiophorales 3 spp.
Flett, S. P. 1983. A technique for detection of *Spongospora subterranea* in soil. Trans. Br. Mycol. Soc. 81:424-425. (uses tomato seedlings as bait)
Karling, J. S. 1968. The Plasmodiophorales. Hafner Publishing, New York. 256 pp. (a comprehensive monograph covering all aspects of these fungi, their diseases and control, key to four *Spongospora* taxa, descriptions and illustrations)
Karling (1968) includes a key to taxa with descriptions and illustrations. *Spongospora subterranea* (Wallr.) Lagerh. f. sp. *subterranea* Toml. causes powdery scab of potato (CMI 477).

SPORISORIUM Ehrenb. ex Link
  Ustilaginales 9 spp.
Deml, G. 1983. Untersuchungen an Heterobasidiomyceten, Teil 32 ueber die Brandpilze von *Hyparrhenia hirta* (L.) Stapf. I. *Sporisorium transfissum* (Tul.) G. Deml comb. nov. Z. Mykol. 49:171-178. (in German with English summary, description and illustration)

SPORISORIUM - cont.
Frederiksen, R. A., ed. 1986. Compendium of Sorghum Diseases. American Phytopathological Society, St. Paul, MN. 82 pp. (includes three species, one as *Sphacelotheca*, descriptions and illustrations)
Langdon, R. F. N., and Fullerton, R. A. 1978. The genus *Sphacelotheca* (Ustilaginales): criteria for its delimitation and the consequences thereof. Mycotaxon 6:421-456. (delimits genus, describes and illustrates five species of *Sphacelotheca* and *Sporisorium*)
The genus includes species occurring on Poaceae that were previously placed in *Sphacelotheca*. Vanky (1985) includes a key to nine *Sporisorium* spp., with descriptions and illustrations. Important diseases include: *Sporisorium cruentum* (Kuehn) Vanky, loose smut of sorghum (CMI 71 as *Sphacelotheca cruenta*); *Sporisorium destruens* (Schlectend.) Vanky, head smut of millet (CMI 72 as *Sphacelotheca destruens*); *Sporisorium reilianum* (Kuehn) Langdon & Fullerton, head smut of sorghum (CMI 73 as *Sphacelotheca reiliana*; but see Vanky, 1985 under *Sporisorium holci-sorghi*); and *Sporisorium sorghi* Ehrenb., covered kernel smut of sorghum (CMI 74 as *Sphacelotheca sorghi*). See Ustilaginales.

SPORONEMA Desmaz.
    Coelomycetes   15 spp.
Sutton (1980) discusses this genus and describes *Sporonema phacidioides* Desmaz., the anamorph of *Leptotrochila medicaginis* (Fuckel) Schuepp, found on *Medicago* spp.

SPOROTHRIX Hektoen & Perkins ex Nicot & Mariat
    Hyphomycetes   23 spp.
Hoog, G. S. de 1974. The genera *Blastobotrys, Sporothrix, Calcarisporium*, and *Calcarisporiella* gen. nov. Stud. Mycol. 7:1-84. (descriptions and illustrations of over 20 *Sporothrix* and *Ophiostoma* spp., key to 28 species in five genera)
Sporothrix is one of the anamorphs of *Ophiostoma*.

STACHYBOTRYS Corda
    Hyphomycetes   12 spp.
Jong, S. C., and Davis, E. E. 1976. Contribution to the knowledge of *Stachybotrys* and *Memnoniella* in culture. Mycotaxon 3:409-485. (keys to 11 *Stachybotrys* spp. and two *Memnoniella* spp., descriptions and illustrations)
Morgan-Jones, G., and Karr, G. W., Jr. 1976. Notes on Hyphomycetes. XVI. A new species of *Stachybotrys*. Mycotaxon 4:510-512. (describes and illustrates *S. zeae* on maize)
Stachybotrys and the related genus *Memnoniella* are ubiquitous saprobes commonly found in soil and many species are capable of degrading cellulose. Strains of *S. chartarum* (Ehrenb.) S. J. Hughes

STACHYBOTRYS - cont.
produce mycotoxins responsible for stachybotryotoxicosis. Ellis
(1971, 1976) describes and illustrates ten species.

STAGONOSPORA (Sacc.) Sacc.
  Coelomycetes 200 spp.
Castellani, E., and Germano, G. 1977. Le Stagonosporae graminicole.
 Ann. Fac. Sci. Agrar., Univ. Torino 10:1-135. (in French, keys to
 77 *Stagonospora* spp. on grasses, descriptions and illustrations)
Hsieh, W. H. 1979. The causal organism of sugarcane leaf blight.
 Mycologia 71:892-898. (describes *Leptosphaeria taiwanensis* and
 its anamorph *S. tainanensis*)
Kobayashi, T. 1972. Notes on new or little-known fungi inhabiting
 woody plants in Japan III. Trans. Mycol. Soc. Jpn. 13:22-33.
 (compares morphological characteristics of *Stagonospora* species
 on leguminous trees, Japanese woody plants, and *Maackia*.)
Sutton (1980) includes a key to nine *Stagonospora* spp. on *Carex* and
Phragmites with brief descriptions and illustrations. The
teleomorphs are placed in *Leptosphaeria* and *Phaeosphaeria*. Diseases
include *Stagonospora curtisii* (Berk.) Sacc. causing *Narcissus* leaf
scorch, and *S. sacchari* Lo & Ling causing leaf scorch of sugarcane
(CMI 776). See *Septoria*.

STEGONSPORIUM Corda
  Coelomycetes 2 spp.
Warmelo, K. T. van, and Sutton, B. C. 1981. Coelomycetes VII.
 Stegonsporium. Mycol. Pap. 145:1-45. (key to two species,
 descriptions and illustrations, disposition of excluded species)
Stegonsporium acerinum Peck occurs on bark and branches of *Acer
saccharum* (F. Can. 102) and *S. pyriforme* (Hoffm.) Corda is found on
bark of *Acer* spp. (F. Can. 103 as *S. ovatum*).

STEGOPHORA Sydow & P. Sydow
  Diaporthales 1 sp.
McGranahan, G. H., and Smalley, E. B. 1984. Conidial morphology,
 axenic grouth, and sporulation of *Stegophora ulmea.*
 Phytopathology 74:1300-1303. (describes the anamorph with
 macroconidia as *Gloeosporium ulmicolum* and microconidia as
 Cylindrosporella ulmeum)
Barr (1978) presents a thorough discussion of this genus including
Stegophora ulmea (Schwein.:Fr.) Sydow & P. Sydow (syn. *Gnomonia
ulmea* (Schwein.:Fr.) Thuem.).

STEMPHYLIUM Wallr.
  Hyphomycetes 20 spp.
Chau, K. F., and Alvarez, A. M. 1983. Postharvest fruit rot of papaya
 caused by *Stemphylium lycopersici.* Plant Dis. 67:1279-1281.
 (includes description of fungus)

STEMPHYLIUM - cont.
Irwin, J. A. G. 1984. Etiology of a new *Stemphylium*-incited leaf disease of alfalfa in Australia. Plant Dis. 68:531-532. (morphologically close to *S. botryosum* and *S. vesicarium*)
Lamprecht, S. C., Baxter, A. P., and Thompson, A. H. 1984. *Stemphylium vesicarium* on *Medicago* spp. in South Africa. Phytophylactica 16:73-75. (describes and illustrates anamorph and its teleomorph *Pleospora allii*)
Simmons, E. G. 1967. Typification of *Alternaria, Stemphylium*, and *Ulocladium*. Mycologia 59:67-92. (defines genera)
Simmons, E. G. 1969. Perfect states of *Stemphylium*. Mycologia 61:1-26. (describes and illustrates six species)
Simmons, E. G. 1985. Perfect states of *Stemphylium*. II. Sydowia 38:284-293. (descriptions and illustrations of three holomorphs from *Medicago sativa, Pleospora herbarum* and *Stemphylium botryosum* not a holomorph, elucidation of their respective genetic connections)
Ellis (1971, 1976) includes several species, with descriptions and illustrations. Diseases caused by these fungi include: *Stemphylium botryosum* Wallr., leaf spots and net blotch of field and broad beans (CMI 150 under *Pleospora herbarum*, F. Can. 232); *S. lycopersici* (Enjoji) Yamamoto, *Stemphylium* leaf blight of tomato (CMI 471); *S. sarciniforme* (Cavara) Wiltshire, target spot of clover (CMI 671, F. Can. 233); and *S. solani* Weber, grey leaf spot of tomato, potato, and other Solanaceae (CMI 472).

STENELLA Sydow
    Hyphomycetes   13 spp.
Deighton, F. C. 1979. Studies on *Cercospora* and allied genera. VII. New species and redispositions. Mycol. Pap. 144:1-56. (defines genus, discusses five *Stenella* spp.)
Hoog, G. S. de, Rahman, M. A., and Boekhout, T. 1983. *Ramichloridium, Veronaea* and *Stenella*: generic delimitation, new combinations and two new species. Trans. Br. Mycol. Soc. 81:485-490. (describes and illustrates two *Stenella* spp., key to 37 taxa in this and related genera)
Mulder, J. L. 1975. Notes on *Stenella*. Trans. Br. Mycol. Soc. 65:514-517. (describes five species, illustrates one species)
Mulder, J. L. 1982. New species and combinations in *Stenella*. Trans. Br. Mycol. Soc. 79:469-478. (describes and illustrates six species)
Ellis (1971, 1976) includes eleven species of *Stenella*, with brief descriptions and illustrations.

STENOCARPELLA Sydow
    Coelomycetes   2 spp.
Sutton, B. C. 1964. Coelomycetes III. *Annellolacinia* gen. nov., *Aristostoma, Phaeocytostroma, Seimatosporium*, etc. Mycol. Pap.

STENOCARPELLA - cont.
97:1-42. (description, illustrations, discussion of *Stenocarpella maydis* and *Stenocarpella zeae* as *Diplodia* spp.)
Sutton (1980) includes a brief description and illustration of two species, both of which occur on corn. *Stenocarpella maydis* (Berk.) Sutton causes stalk rot, white ear rot, and seedling blight of maize (CMI 84 as *Diplodia maydis*) and *Stenocarpella macrospora* (Earle) Sutton causes dry rot of ears and stalks of maize (CMI 83 as *Diplodia macrospora*, syn. *S. zeae* Sydow).

STEREUM J. Hill ex Pers.
    Aphyllophorales   25 spp.
Boidin, J. 1960. Le Genre *Stereum* Pers. S. L. au Congo Belge. Bull. Jard. Bot. Brux. 30:283-355. (in French, key to 85 species in five subgenera)
Jahn, H. 1971. Stereoide pilze in Europa (Stereaceae Pil. emend. Parm. u.a., Hymenochaete). Westf. Pilzbriefe 8:69-176. (in German, key to stereoid genera, treats 43 European species of *Stereum* and related genera, descriptions and illustrations)
Lentz, P. L. 1955. *Stereum* and allied genera of fungi in the Upper Mississippi Valley. U.S. Dep. Agric., Agric. Monogr. 24. 74 pp. (key to 26 species in *Stereum* and related genera, descriptions and illustrations, nomenclature outdated)
Talbot, P. H. B. 1954. The genus *Stereum* in South Africa. Bothalia 6:303-332. (key to 22 *Stereum* spp., illustrations, discusses 67 names)
Welden, A. L. 1971. An essay on *Stereum*. Mycologia 63:790-799. (key to eleven taxa, defines four subcomplexes in North America)
No monographic treatment of *Stereum* and related genera is yet available. *Stereum gausapatum* (Fr.:Fr.) Fr. and *S. sanguinolentum* (Albertini & Schwein.:Fr.) Fr. cause heart rot of *Quercus* spp. and conifers, respectively. *Xylobolus* species, formerly *Stereum* species, cause heart rots. See *Amylostereum, Chondrostereum, Cystostereum*, and Aphyllophorales.

STIGMELLA Lev.
    Coelomycetes   2 spp.
Hughes, S. J. 1952. Studies on microfungi. XIV. *Stigmella, Stigmina, Camptomeris, Polythrincium* and *Fusicladiella*. Mycol. Pap. 49:1-25. (describes two *Stigmella* spp., disposes of excluded names)
Sutton (1980) includes *Stigmella effigurata* (Schwein.) S. J. Hughes causing lesions on oak leaves.

STIGMINA Sacc.
    Hyphomycetes   35 spp.
Morgan-Jones, G., and Kendrick, B. 1972. Notes on Hyphomycetes. III. Redisposition of six species of *Exosporium*. Can. J. Bot.

STIGMINA - cont.
50:1817-1824. (transfers three *Exosporium* spp. to *Stigmina*, descriptions and illustrations)
Sutton, B. C. 1972. '*Sciniatosporium*' Kalchbr. Trans. Br. Mycol. Soc. 58:164-167. (rejects *Sciniatosporium* for *Stigmina* and transfers 12 species to *Stigmina*)
Ellis (1971, 1976) includes numerous species. *Stigmina carpophila* (Lev.) M. B. Ellis is the cause of peach shot hole, *S. dothideoides* (Ellis & Everh.) M. B. Ellis occurs on *Gaillardia* and *Shepherdia* (F. Can. 212), and *Stigmina robusta* (Cooke & Ellis) Sutton is found on bark of *Populus* spp. (F. Can. 167).

STRASSERIA Bres. & Sacc.
    Coelomycetes   8 spp.
Parmelee, J. A., and Cauchon, R. 1979. *Strasseria* on Pinaceae in Canada. Can. J. Bot. 57:1660-1662. (describes and illustrates *S. geniculata*, lists 11 additional species in the genus)
Schwarz, M. R., and Boone, D. M. 1983. Black rot of cranberry caused by *Strasseria oxycocci*. Plant Dis. 67:31-32. (compares with *Ceuthospora lunata*)
Sutton (1980) includes a brief description and illustration of *Strasseria geniculata* (Berk. & Broome) Hoehn., the cause of black rot of apple and other diseases.

STROMATINIA (Boud.) Boud.
    Helotiales   6 spp.
Darvas, J. M. 1984. The occurrence of *Stromatinia gladioli* in South Africa. Phytophylactica 16:255. (describes the cause of dry rot of *Gladiolus* corms)
Stromatinia is related to *Sclerotinia* and is in need of monographic work. See Kohn (1979a, 1979b) under *Sclerotinia*. *Stromatinia gladioli* (Drayton) Whetzel occurs on *Gladiolus* corms and *Stromatinia panacis* (Rankin) Kohn causes a black rot of ginseng tubers.

SYDOWIA Bres.
    Dothideales   5 spp.
Froidevaux, L. 1972. Contribution a l'etude des Dothioracees (Ascomycetes). Nova Hedw. 23:679-734. (in French, keys to genera and species based on host and macroscopic characters, describes and illustrates two *Sydowia* spp.)
Barr (1972) includes five species with descriptions and illustrations. The anamorph is *Sclerophoma*. *Sydowia polyspora* (Bref. & F. Tavel) E. Mueller and its anamorph *Sclerophoma pythiophila* (Corda) Hoehn. causes pine leaf blight and dieback (CMI 228; Sutton, 1980). Funk (1981) presents a table comparing this species with two other *Sclerophoma* species on conifers.

SYNCHYTRIUM de Bary & Woronin
    Chytridiales   121 spp.
Karling, J. S. 1964. *Synchytrium*. Academic Press, New York. 470 pp. (key, descriptions and illustrations of all species, information on biology)
Langerfeld, E. 1984. *Synchytrium endobioticum* (Schilb.) Perc. Zusammenfassende Darstellung des Erregers des Kartoffelkrebses anhand von Literatureberichten. Mitt. Biol. Bundesanst. Land-Forstwirtsch 219:1-142. (in German with English summary, describes morphology and biology, host list)
Lenne, J. M. 1985. *Synchytrium desmodii*, cause of wart disease of the tropical pasture legume *Desmodium ovalifolium* in Colombia. Plant Dis. 69:806-808. (describes fungus briefly)
Sparrow, F. K. 1973. Chytridiomycetes, Hyphochytridiomycetes. Pages 85-110 in: The Fungi Vol. IVB. G. C. Ainsworth, F. K. Sparrow, and A. S. Sussman, eds. Academic Press, New York. (keys to taxa)
Synchytrium is a large, cosmopolitan genus of plant parasites. Karling (1964) presents a comprehensive monograph that includes keys to species as well as descriptions, illustrations, a host index, and much additional information. Diseases caused by these fungi include: *Synchytrium endobioticum* (Schilberszky) Percival, potato wart (CMI 755); *S. lagenariae* Mhatre & Mundkur, gall disease of cucurbits (CMI 756); *S. macrosporum* Karling, lavender-red galls on many hosts (CMI 757); *S. phaseoli* Weston, galls on Fabaceae (CMI 758); *S. phaseoli-radiati* Sinha & Gupta, galls on Fabaceae (CMI 759); and *S. psophocarpi* (Racib.) Gaeumann, false rust of winged bean (CMI 760).

TALAROMYCES C. Benjamin
    Eurotiales   16 spp.
Pitt, J. I. 1979. The Genus *Penicillium* and its Teleomorphic States *Eupenicillium* and *Talaromyces*. Academic Press, New York. 634 pp. (synoptic key to 16 *Talaromyces* spp. with *Penicillium* anamorphs, descriptions and illustrations)
Stolk, A. C., and Samson, R. A. 1971. Studies on *Talaromyces* and related genera I. *Hamigera* gen. nov. and *Byssochlamys*. Persoonia 6:341-357. (describes the type species of *Talaromyces* and related genera)
Talaromyces is well-documented by Pitt (1979). See the anamorphs *Paecilomyces* and *Penicillium*.

TAPHRINA Fr.:Fr.
    Taphrinales   95 spp.
Gjaerum, H. B. 1964. The genus *Taphrina* Fr. in Norway. Nytt Magn. Bot. 11:5-26. (discusses 21 species)
Mix, A. J. 1949. A monograph of the genus *Taphrina*. Univ. Kansas Sci. Bull. 33:3-167. (descriptions and illustrations of 98 species, arranged by host family)
Mix, A. J. 1954. Additions and emendations to a monograph of the

TAPHRINA - cont.
genus *Taphrina*. Trans. Kansas Acad. Sci. 57:55-65. (comments on 32 species)
Snider, R. D., and Kramer, C. L. 1974. An electrophoretic protein analysis and numerical taxonomic study of the genus *Taphrina*. Mycologia 66:754-772. (confirms grouping by host family and ascus morphology)
Mix (1949) provides the only comprehensive account of this genus. Funk (1985) presents a table comparing eleven species that occur on western trees. Species causing diseases include: *Taphrina betulina* Rostr., witches broom of birch; *T. bullata* (Berk.) Tul., pear leaf blister; *T. caerulascens* (Desmaz. & Mont.) Tul., oak leaf curl; *T. cerasi* (Fuckel) Sadebeck, witches broom of cherry; *T. deformans* (Berk.) Tul., peach leaf curl (CMI 711); *T. insititiae* (Sadebeck) Johan., witches broom of plum; *T. maculans* E. J. Butler, leaf spot of turmeric (CMI 507); *T. minor* Sadebeck, cherry leaf curl; *T. populina* Fr., poplar leaf blister; *T. pruni* Tul., pocket plum (CMI 713); and *T. wiesneri* (Rathay) Mix, witches broom and leaf curl of cherry and apricot (CMI 712).

THANATEPHORUS Donk
 Tulasnellales   4 spp.
Talbot, P. H. B. 1965. Studies of "*Pellicularia*" and associated genera of hymenomycetes. Persoonia 3:371-406. (keys to related genera and three *Thanatephorus* spp.)
Talbot, P. H. B. 1970. Taxonomy and nomenclature of the perfect state. Pages 20-31 in: *Rhizoctonia solani*: Biology and Pathology. J. R. Parmeter, Jr., ed. University of California Press, Berkeley. (discusses taxonomic relationship to *Ceratobasidium*)
Warcup, J. H., and Talbot, P. H. B. 1967. Perfect states of Rhizoctonias associated with orchids. New Phytol. 66:631-641. (describes and illustrates three species)
Thanatephorus cucumeris (Frank) Donk is the teleomorph of *Rhizoctonia solani* Kuehn, a fungus that causes serious diseases on a wide variety of hosts (CMI 406). See *Rhizoctonia*.

THECAPHORA Fingerh.
 Ustilaginales   15 spp.
Harada, Y. 1983. Material for the smut flora of Japan I. Trans. Mycol. Soc. Jpn. 24:299-306. (reviews *Thecaphora* spp. on Fabaceae, descriptions and illustrations)
Zambettakis, C., and Joly, P. 1975. Application de traitements numeriques a la systematique des Ustilaginales. III. Le genre *Thecaphora*. Bull. Trimest. Soc. Mycol. Fr. 91:71-88. (in French, lists characteristics of 39 species)
See *Angiosorus* and Ustilaginales.

THEKOPSORA Magnus
    Uredinales   15 spp.
Hiratsuka, N. 1958. Revision of Taxonomy of the Puccinastreae, with
    Special Reference to Species of the Japanese Archipelago. Kasai,
    Tokyo. 167 pp. (key to 13 *Thekopsora* spp., lists hosts and
    distribution for each species)
Hiratsuka, Y., and Sato, S. 1976. Species of *Thekopsora* on *Tsuga*.
    Trans. Mycol. Soc. Jpn. 17:543-548. (in Japanese, host list)
Sato, S., and Katsuya, K. 1979. Heteroecism of two rust fungi on the
    needles of *Tsuga diversifolia* and *T. sieboldii*. Trans. Mycol.
    Soc. Jpn. 20:1-4. (illustrations of aecia and aeciospores of two
    species)
Cummins and Hiratsuka (1983) consider *Thekopsora* to be a synonym of
Pucciniastrum.

THERRYA Sacc.
    Rhytismatales   5 spp.
Funk, A. 1980. New *Therrya* species parasitic on western conifers.
    Can. J. Bot. 58:1291-1294. (descriptions and illustrations of
    three species)
Reid, J., and Cain, R. F. 1961. The genus *Therrya*. Can. J. Bot.
    39:1117-1129. (describes and illustrates two species)
Therrya species occur on coniferous bark, often associated with
cankers. Funk (1981) treats three species.

THIELAVIOPSIS Went
    Hyphomycetes   2 spp.
Ellis (1971) includes descriptions and illustrations of both species.
The teleomorph belongs to *Ceratocystis* and the synanamorph is placed
in *Chalara*. *Thielaviopsis basicola* (Berk. & Broome) Ferraris is the
cause of black root rot in tobacco and other crop plants (CMI 170).
Carmichael in Carmichael et al. (1980) places this species in
Trichocladium.

THYRONECTRIA Sacc.
    Hypocreales   20 spp.
Bedker, P. J., and Wingfield, M. J. 1983. Taxonomy of three
    canker-causing fungi of honey locust in the United States.
    Trans. Br. Mycol. Soc. 81:179-183. (compares *Nectria cinnabarina*,
    anamorph *Tubercularia vulgaris*, with *Thyronectria*
    austroamericana, anamorph *Gyrostroma austroamericana*)
Seeler, E. V., Jr. 1940. A monographic study of the genus
    Thyronectria. J. Arnold Arbor., Harv. Univ. 21:429-460. (key to
    16 species, descriptions and illustrations)
Subramanian, C. V., and Bhat, D. J. 1984. Developmental morphology of
    Ascomycetes. XII. *Thyronectria pseudotrichia*. Crypt. Mycol.
    5:307-321. (developmental study, discussion of *Thyronectria*
    perithecial centrum, illustrations)

THYRONECTRIA - cont.
The monograph by Seeler (1940) is still useful in the identification of *Thyronectria* species. The anamorphs are placed in *Gyrostroma*.

TIAROSPORELLA   See *Darkera*.

TILLETIA Tul. & C. Tul.
  Ustilaginales   76 spp.
Banowetz, G. M., Trione, E. J., and Krygier, B. B. 1984.
  Immunological comparisons of teliospores of two wheat bunt fungi,
  Tilletia species, using monoclonal antibodies and antisera.
  Mycologia 76:51-62. (no qualitative distinctions were found)
Hess, W. M., and Trione, E. J. 1986. Use of electron microscopy to
  characterize teliospores of *Tilletia caries* and *T. controversa*.
  Plant Dis. 70:458-460. (could not be used to identify species due
  to inconsistent morphological differences)
Hoffman, J. A. 1982. Bunt of wheat. Plant Dis. 66:979-986. (compares
  T. caries, T. controversa, and *T. foetida*)
Stockwell, V. O., and Trione, E. J. 1986. Distinguishing teliospores
  of *Tilletia controversa* from those of *T. caries* by fluorescence
  microscopy. Plant Dis. 70:924-926. (describes rapid, sensitive
  method for distinguishing teliospores)
Weber, G., and Schauz, K. 1985. Characterization of spore protein
  patterns in *Tilletia controversa* and *Tilletia caries* with gel
  electrophoretic methods. Z. Pflanzenkr. Pflanzensch. 92:600-605.
  (finds species-related differences in protein patterns)
Zogg, H. 1972. Die *Tilletia*-Streifenbrandkrankheiten der Graeser.
  Phytopathol. Z. 74:218-229. (in German, key to seven *Tilletia*
  stripe smuts on grasses, descriptions and illustrations)
Zogg, H. 1983. *Tilletia sabaudiae*, a new smut fungus (Tilletiales)
  and some observations on the gelatinoid sheath of the *Tilletia*
  species. Bot. Helv. 93:91-98. (description and illustrations)
Duran and Fischer (1961) provide an account of *Tilletia* species,
including several species now placed in *Neovossia*. Mordue and
Ainsworth (1984) include descriptions and a key to seven species.
This cosmopolitan genus contains a number of important pathogens:
Tilletia controversa Kuehn, dwarf bunt of winter wheat (CMI 746, F.
Can. 33); *T. holci* (Westendorp) J. Schroet., covered smut on *Holcus*
and *Anthoxanthum* (CMI 747); *T. laevis* Kuehn, common bunt of wheat
(CMI 720 as *T. foetida*); *T. lolii* Auersw., covered smut on ryegrass
(CMI 804); *T. sphaerococca* (Rabenh.) A. Fisch. v. Waldh., *Agrostis*
smut (CMI 805); and *T. tritici* (Bjerk.) Wint., common bunt of wheat
(CMI 719 as *T. caries*). See *Neovossia* and Ustilaginales.

TILLETIOPSIS Derx
  Hyphomycetes   7 spp.
Gokhale, A. A. 1972. Studies on the genus *Tilletiopsis*. Nova Hedw.
  23:795-809. (includes key to seven species)

TILLETIOPSIS - cont.
Nyland, G. 1950. The genus *Tilletiopsis.* Mycologia 42:487-496. (describes and illustrates two species)
Tubaki, K. 1952. Studies on the Sporobolomycetaceae in Japan: I. On *Tilletiopsis.* Nagaoa 1:26-31. (key to five taxa based on colony color, later authors suggest that this character is variable)
Yamazaki, M., Goto, S., and Komagata, K. 1985. Taxonomical studies of the genus *Tilletiopsis* on physiological properties and electrophoretic comparision of enzymes. Trans. Mycol. Soc. Jpn. 26:13-22. (seven species with distinct enzyme patterns)
These yeast-like fungi are anamorphs of basidiomycetes and are difficult to distinguish by their morphological or physiological properties alone. See the similar genus *Itersonilia.*

TOLYPOSPORIUM Woronin ex J. Schroet.
    Ustilaginales   27 spp.
Zambettakis, C., and Joly, P. 1973. Application de traitements numeriques a la systematique des Ustilaginales. II. Le genre *Tolyposporium.* Bull. Trimest. Soc. Mycol. Fr. 89:83-97. (in French, lists characteristics of species)
Species include *Tolyposporium ehrenbergii* (Kuehn) Pat., long smut of sorghum (CMI 76). Vanky (1977, 1986 under *Moesziomyces*) places several species in the segregate genus *Moesziomyces.* See Ustilaginales.

TRACHYSPHAERA Tabor & Bunting
    Peronosporales   1 sp.
Trachysphaera fructigena Tabor & Bunting causes mealy pod of cacao and fruit rot of coffee and banana (CMI 229). See Peronosporales.

TRACHYSPORA Fuckel
    Uredinales   4 spp.
Gjaerum, H. B., and Cummins, G. B. 1982. Rust fungi (Uredinales) on East African *Alchemilla.* Mycotaxon 15:420-424. (key to four *Trachyspora* spp.)
Henderson, D. M. 1973. Studies in the morphology of fungal spores: *Trachyspora intrusa.* Rep. Tottori Mycol. Inst. 10:163-168. (developmental morphology of uredinial aeciospores and of teliospores)
Species of *Trachyspora* occur on *Alchemilla* (Rosaceae). See Uredinales.

TRANZSCHELIA Arth.
    Uredinales   15 spp.
Bennell, A. P., and Henderson, D. M. 1978. Urediniospore and teliospore development in *Tranzschelia* (Uredinales). Trans. Br. Mycol. Soc. 71:271-278. (compares *T. anemones* and *T. discolor* with *T. pruni-spinosae*)

TRANZSCHELIA - cont.
Bolkan, H. A., Ogawa, J. M., Michailides, T. J., and Kable, P. F. 1985. Physiological specialization in *Tranzschelia discolor*. Plant Dis. 69:485-486. (defines three formae speciales)
Laundon, G. F. 1975. Taxonomy and nomenclature notes on Uredinales. Mycotaxon 3:133-161. (discusses *T. discolor*)
Linfield, C., and Price, D. 1983. Host range of plum anemone rust, *Tranzschelia discolor*. Trans. Br. Mycol. Soc. 80:19-21. (infection only found on *Prunus domestica*)
Species causing diseases include *Tranzschelia pruni-spinosae* (Pers.) Dietel var. *discolor* (Fuckel) Dunegan, peach rust (CMI 287), and *T. pruni-spinosae* var. *pruni-spinosae*, rust of plum (CMI 288). See Uredinales, particularly Cummins and Hiratsuka (1983).

TRECHISPORA   See *Phymatotrichopsis*.

TRICHOCLADIUM   See *Thielaviopsis*.

TRICHODERMA Pers.:Fr.
    Hyphomycetes   9 spp.
Bissett, J. 1984. A revision of the genus *Trichoderma*. I. Section *Longibrachiatum* sect. nov. Can. J. Bot. 62:924-931. (key to six species, descriptions and illustrations)
Doi, N., and Doi, Y. 1979. Notes on *Trichoderma* and its allies. 1. A list of teleomorphic species with *Trichoderma* or its allied anamorphs hitherto known. Bull. Nat. Sci. Mus., Ser. B. (Tokyo) 5:117-123. (includes 75 teleomorph names)
Doi, N., and Doi, Y. 1986. Notes on *Trichoderma* and its allies 4. A list of specific names proposed for the genus *Trichoderma*. Bull. Nat. Sci. Mus., Ser. B. (Tokyo) 12:1-15. (lists 63 names in *Trichoderma* and 14 names in related genera)
Komatsu, M. 1976. [Studies on *Hypocrea, Trichoderma* and allied fungi antagonistic to shiitake, *Lentinus edodes* (Berk.) Sing.]. Rep. Tottori Mycol. Inst. 13:1-113. (in Japanese with English summary, excellent illustrations of 19 *Hypocrea* and *Trichoderma* spp.)
Rifai, M. A. 1969. A revision of the genus *Trichoderma*. Mycol. Pap. 116:1-56. (key to nine species, descriptions and illustrations)
Vajna, L. 1983. *Trichoderma* species in Hungary. Acto Phytopathol. Acad. Sci. Hung. 18:291-301. (includes five species, follows Rifai)
Webster, J., and Rifai, M. A. 1968. Culture studies on *Hypocrea* and *Trichoderma*. IV. *Hypocrea pilulifera* sp. nov. Trans. Br. Mycol. Soc. 51:511-514. (descriptions and illustrations)
Despite several major references, identification of *Trichoderma* species remains difficult. Domsch et al. (1980) include a key to eight species, descriptions, illustrations, and numerous references. The teleomorph is *Hypocrea*.

TRICHOMETASPHAERIA   See *Setosphaeria*.

TRICHOSCYPHELLA   See *Lachnellula*.

TRICHOTHECIUM Link:Fr.
    Hyphomycetes   5 spp.
Madelin, M. F. 1966. *Trichothecium acridiorum* (Trabut) comb. nov. on red locusts. Trans. Br. Mycol. Soc. 49:275-288. (description, illustrations, biology)
Meyer, J. 1958. Appareil conidien de *Trichothecium roseum* Lk. ex Fr., *Cylindrocarpon congoensis* nov. sp. et *Arthrobotrys stilbacea* nov. sp. Bull. Trimest. Soc. Mycol. Fr. 74:236-248. (in French, descriptions and illustrations)
Rifai, M. A., and Cooke, R. C. 1966. Studies on some didymosporous genera of nematode-trapping hyphomycetes. Trans. Br. Mycol. Soc. 49:147-168. (discusses generic delimitation of *Trichothecium* and accepted species, describes and illustrates four species)
Domsch et al. (1980) include *Trichothecium roseum* (Pers.:Fr.) Link, the cause of pink rot of apples and pink pod rot of beans.

TRIMMATOSTROMA Corda
    Hyphomycetes   9 spp.
Gadgil, P. D., and Dick, M. 1983. Fungi Eucalyptorum Novazelandiae: *Septoria pulcherrima* sp. nov. and *Trimmatostroma bifarium* sp. nov. N. Z. J. Bot. 21:49-52. (descriptions and illustrations)
Sutton, B. C., and Ganapathi, A. 1978. *Trimmatostroma excentricum* sp. nov. on *Eucalyptus* from New Zealand and Fiji. N. Z. J. Bot. 16:529-533. (describes and illustrates *T. excentricum*, compares with similar taxa, discusses conidial development)
Ellis (1971, 1976) includes five species with descriptions and illustrations. Most species are not pathogenic, although a few cause leaf spots.

TUBAKIA Sutton
    Coelomycetes   5 spp.
Kobayashi, T., Horie, H., and Sasaki, K. 1979. Notes on new or little-known fungi inhabiting woody plants in Japan. IX. Trans. Mycol. Soc. Jpn. 20:325-337. (compares *T. subglobosa* and *T. dryina* causing sooty leaf blotch of fagaceous plants)
Sutton, B. C. 1973. *Tubakia* nom. nov. Trans. Br. Mycol. Soc. 60:164-165. (nomenclature)
Yokoyama, T., and Tubaki, K. 1971. Cultural and taxonomical studies on the genus *Actinopelte*. Inst. Ferment., Osaka, Res. Commun. 5:43-77. (describes and illustrates five *Actinopelte* spp., now placed in *Tubakia*, comparative table, cultural characteristics and taxonomic discussion)
Sutton (1973) established the name *Tubakia* for *Actinopelte* and transferred the five species discussed by Yokoyama and Tubaki

TUBAKIA - cont.
(1971). *Tubakia subglobosa* (Yokayama & Tubaki) Sutton and *T. dryina* (Sacc.) Sutton cause sooty leaf blotch of fagaceous plants. *Tubakia japonica* (Sacc.) Sutton causes chestnut leaf spot in Japan.

TUBERCULARIA Tode:Fr.
    Hyphomycetes   25 spp.
Seifert, K. 1985. A monograph of *Stilbella* and some allied hyphomycetes. Stud. Mycol. 27:1-235. (key to seven *Tubercularia* spp., descriptions and illustrations including teleomorphs)
This genus contains anamorphs of hypocrealean fungi. *Tubercularia vulgaris* Tode:Fr. is the anamorph of *Nectria cinnabarina* Tode:Fr., the coral spot fungus on many hosts (CMI 531 under *N. cinnabarina*).

TYPHULA Fr.:Fr.
    Aphyllophorales   63 spp.
Berthier, J. 1976. Monographie des *Typhula* Fr., *Pistillaria* Fr. et genres voisins. Numero Special, Bull. Soc. Linn. Lyon 45:1-213. (in French, a comprehensive monograph)
Bruehl, G. W., and Cunfer, B. M. 1975. *Typhula* species pathogenic to wheat in the Pacific Northwest. Phytopathology 65:755-760. (differentiates three species, *T. incarnata*, *T. idahoensis*, and *T. ishikariensis*, compares these with other *Typhula* spp.)
Species of *Typhula* are similar to *Pistillaria* but form sclerotia. Many *Typhula* species are pathogenic to cereals under snow causing "snow mold", especially of winter wheat.

ULOCLADIUM G. Preuss
    Hyphomycetes   9 spp.
Simmons, E. G. 1967. Typification of *Alternaria, Stemphylium,* and *Ulocladium*. Mycologia 59:67-92. (key to nine *Ulocladium* spp., descriptions and illustrations)
Ellis (1976) presents a key to seven species with brief descriptions and illustrations.

UNCINULA Lev.
    Erysiphales   20 spp.
Braun, U. 1983. Taxonomic notes on some powdery mildews. (II). Mycotaxon 16:425-428. (two new *Uncinula* taxa, descriptions and illustrations)
Braun, U. 1984. Taxonomic notes on some powdery mildews. (IV). Mycotaxon 20:483-489. (reviews *Uncinula* spp. on Sterculiaceae and Aceraceae)
Chen, Z.-x., Gao, R.-x., Luo, S.-b., and Liu, B.-c. 1984. [New species of powdery mildews from Wuyishan.] Acta Mycol. Sin. 3:75-80. (in Chinese with English summary, describes and illustrates *U. hydrangeae*)
Nomura, Y., and Tanda, S. 1983. [Notes on powdery mildew fungus of

UNCINULA - cont.
Uncinula on *Acer* in Japan.] Trans. Mycol. Soc. Jpn. 24:197-200. (in Japanese with English summary, morphological characteristics of *U. aduncoides* on various *Acer* spp.)
Pirozynski, K. A. 1965. African species of *Uncinula*. Pages 2-23 in: F. C. Deighton and K. A. Pirozynski. 1965. Microfungi. I. Mycol. Pap. 101:1-43. (key to 12 *Uncinula* spp. based on fungus and host, descriptions and illustrations)
Sharma, A. K. 1985. A new species of *Uncinula* (Erysiphaceae) from Kashmir. Curr. Sci. 54:237-239. (description, illustrations, comparison with *U. adunca*)
Zheng, R.-y., and Chen, G.-q. 1977. [Taxonomic studies on the genus *Uncinula* in China. I. Discussion on *Uncinula sinensis* Tai et Wei.] Acta Microbiol. Sin. 17:189-197. (in Chinese with English summary, describes and illustrates three species on *Acer, Alchornea*, and *Sophora*)
Uncinula bicornis (Fr.) Lev. (syn. *U. aceris* (DC.) Sacc.) causes powdery mildew of sycamore and *Acer* (CMI 190), and *U. necator* (Schwein.) Burr causes vine (*Vitis*) mildew (CMI 160). See *Erysiphe, Oidium*, and Erysiphales.

UNCINULIELLA Zheng & Chen
  Erysiphales 3 spp.
Zheng, R.-y., and Chen, G.-q. 1979. [Taxonomic studies on the genus *Uncinuliella* of China. 1. The establishment of *Uncinuliella* gen. nov. and identification of the Chinese and Japanese species.] Acta Microbiol. Sin. 19:280-291. (in Chinese with English summary, describes new genus with two species on *Rosa*)
Zheng, R.-y., and Chen, G.-q. 1982. [Taxonomic studies on the genus *Uncinuliella* of China. I. *U. australiana* (McAlp.) comb. nov.] Acta Bot. Yunnanica 4:363-366. (in Chinese with English summary, key to three *Uncinuliella* spp.)
See Erysiphales.

UREDO Pers.:Pers.
  Uredinales 500 spp.
This is a form-genus for Uredinales that produce only urediniospores. Diseases caused by *Uredo* spp. include: *Uredo beringiana* (Tranz.) Parmelee, on *Gentiana glauca* (F. Can. 297); *U. cajani* Sydow, rust of pigeon pea (CMI 590); and *U. ficina* Juel, *Ficus* rust (CMI 289). See Uredinales.

UROCYSTIS Rabenh. ex Fuckel
  Ustilaginales 60 spp.
Sampson, M. G., and Watson, A. K. 1985. Host specificity of *Urocystis agropyri* isolated from *Agropyron repens*. Can. J. Plant Pathol. 7:52-54. (infects only this host)
Ulvinen, T. 1980. *Urocystis carcinodes* discovered in Finland.

UROCYSTIS - cont.
Karstenia 20:16-18. (*Urocystis* spp. of *Cimicifuga* and *Actaea*)
Some important species include: *Urocystis agropyri* (G. Preuss) J.
Schroet., flag smut of wheat, barley, and other grasses (CMI 716);
U. anemones (Pers.) Wint., Anemone and Ranunculus smut (CMI 806); *U. cepulae* Frost, smut of onion (CMI 298); *U. gladiolicola* Ainsworth, Gladiolus smut (CMI 807); and *U. occulta* Rabenh., stripe or flag smut of rye (CMI 808). See *Ustilago* and Ustilaginales.

UROHENDERSONIA Speg.
    Coelomycetes   5 spp.
Nag Raj, T. R., and Kendrick, B. 1971. Genera coelomycetarum. I. *Urohendersonia*. Can. J. Bot. 49:1853-1862. (key to five species, descriptions and illustrations)
Sutton (1980) includes a key to five species with brief descriptions and illustrations based on Nag Raj and Kendrick (1971). Members of the genus cause leaf lesions on a variety of hosts.

UROMYCES (Link) Unger
    Uredinales   600 spp.
Guyot, A. L. 1957. Les Rouilles des Legumineuses. Paul Lechevalier, Paris. 647 pp. (in French, tables comparing *Uromyces* spp. according to host, descriptions and illustrations)
Hiratsuka, N. 1973. Revision of taxonomy of the genus *Uromyces* in the Japanese Archipelago. Contributions to the rust-flora of Eastern Asia X. Rep. Tottori Mycol. Inst. 10:1-98. (key to 86 species, descriptions)
Laberry, R., Lozano, J. C., and Buritica, P. 1984. Estudio taxonomico de especies del genero *Uromyces* en yuca (*Manihot* spp.). Fitopath. Bras. 9:525-536. (in Spanish with English summary, key to six species, descriptions and illustrations)
Monoson, H. L., and Schlesser, P. E. 1980. *Uromyces* on New-World Convolvulaceae. Mycologia 72:817-820. (key to three species, descriptions and illustrations)
Uromyces is a large, cosmopolitan genus of rusts that occurs commonly on leguminous plants (Cummins, 1978). Although similar to *Puccinia*, *Uromyces* species differ in having one-celled teliospores. Several comprehensive references are available; in addition, *Uromyces* species are included in works on rust fungi based on host or geographical region. Mordue and Ainsworth (1984) provide a key to 17 British species with descriptions. See *Puccinia* and Uredinales.
Important species include: *Uromyces acuminatus* Arth., on *Dodecatheon, Lysimachia,* and *Spartina* (F. Can. 203); *U. aloes* (Cooke) Magnus, rust of aloe (CMI 268); *U. appendiculatus* (Pers.) Unger, bean rust (CMI 57); *U. betae* Kickx, beet rust (CMI 177); *U. ciceris-arietini* Jacz., chickpea rust (CMI 178); *U. decoratus* Sydow, rust of Sunn hemp (*Crotalaria*) (CMI 179); *U. dianthi* (Pers.) Niessl, carnation rust (CMI 180); *U. dolicholi* Arth., rust of rhynchosia

UROMYCES - cont.
(CMI 269); *U. eugentianae* Cummins, on *Gentianella amarella* (F. Can. 298); *U. fabae* (Pers.) de Bary, on *Vicia*; *U. geranii* (DC.) Fr., rust of geranium (CMI 270); *U. lapponicus* Lagerh. var. *laponicus*, on *Astragalus* spp. (F. Can. 26a); *U. mucunae* Rabenh., velvet bean rust (CMI 290); *U. musae* Henn., rust of *Musa* spp. (CMI 295); *U. phacae-frigidae* (Wahl.) Hariot, on *Astragalus* spp. (F. Can. 25); *U. pisi-sativi* (Pers.) Liro, pea, vetch, and broad bean rust (CMI 58); *U. punctatus* J. Schroet., on *Astragalus*, *Euphorbia*, and *Oxytropis* (F. Can. 24); *U. striatus* J. Schroet., lucerne and sweet clover rust (CMI 59); *U. transversalis* (Thuem.) Wint., rust of gladiolus; *U. trifolii* (R. Hedw.) Lev., on *Trifolium*; and *U. viciae-fabae* (Pers.) J. Schroet., broad bean rust (CMI 60).

UROPHLYCTIS See *Physoderma*.

UROPYXIS J. Schroet.
    Uredinales   13 spp.
Baxter, J. W. 1959. A monograph of the genus *Uropyxis*. Mycologia 51:210-226. (key to thirteen species, descriptions and illustrations)
Lindquist, J. C. 1967. Las especies argentinas de *Uropyxis* (Uredinae). Uredineana 6:283-287. (key to four species, descriptions and illustrations)
Most *Uropyxis* species occur on legumes. See Cummins (1978) and Uredinales.

USTILAGINOIDEA Bref.
    Hyphomycetes   5 spp.
Ellis (1971) includes two species. *Ustilaginoidea virens* (Cooke) Takahashi causes false smut of rice (CMI 299).

USTILAGO (Pers.) Roussel
    Ustilaginales   300 spp.
Kim, W. K., Rohringer, R., and Nielsen, J. 1984. Comparison of polypeptides in *Ustilago* spp. pathogenic on wheat, barley, and oats: a chemotaxonomic study. Can. J. Bot. 62:1431-1437. (using isoelectric focusing finds *U. nigra, U. hordei,* and *U. avenae* to be one taxon; *U. tritici* and *U. nuda* distinct)
Nielsen, J. 1985. *Ustilago* spp. pathogenic on *Aegilops*. II. *Ustilago tritici*. Can. J. Bot. 63:765-771. (confirms synonymous species)
Terrell, E. E., and Batra, L. R. 1982. *Zizania latifolia* and *Ustilago esculenta*, a grass-fungus association. Econ. Bot. 36:274-285. (description and illustration of fungus)
Zambettakis, C. 1973. Recherches sur les charbons des Arundinelleae. Rev. Mycol. 38:67-90. (in French, lists six *Ustilago* spp., brief descriptions and illustrations)
Species of *Ustilago* are important pathogens of cereals and other

USTILAGO - cont.
Poaceae. Many references on smut fungi are available mostly based on host or geographic area. Hawksworth et al. (1983) include an extensive list of general and regional references under Ustilaginales. Important species include: *Ustilago avenae* (Pers.) Rostr., loose smut of oats, barley, and tall oat grass (CMI 279); *U. bullata* Berk., head smut of grasses (CMI 718); *U. crameri* Koern., head smut of millet (CMI 78); *U. cynodontis* (Pass.) Henn., Bermuda grass smut (CMI 297); *U. hordei* (Pers.) Lagerh., covered smut of barley (CMI 749); *U. hypodytes* (Schlechtend.:Fr.) Fr., stem smut of grasses (CMI 809); *U. nuda* (Jensen) Rostr., loose smut of wheat, barley, oats, and rye (CMI 280); *U. pinguiculae* Rostr., on *Pinguicula* spp. (F. Can. 198); *U. scitaminea* Sydow, sugarcane smut (CMI 80); *U. striiformis* (Westendorp) Niessl, stripe smut of grasses (CMI 717); *U. vaillantii* Tul. & C. Tul., anther smut of many small, spring-flowering Liliaceae (CMI 810); *U. violacea* (Pers.) Roussel, anther smut of carnation (CMI 750); and *U. zeae* (Beckm.) Unger, maize smut (CMI 79 as *U. maydis*). See Ustilaginales.

USTULINA Tul. & C. Tul.
    Sphaeriales   6 spp.
Dennis, R. W. G. 1963. Hypoxyloideae of Congo. Bull. Jard. Bot. Brux. 33:317-343. (includes *U. deusta* as *Hypoxylon*)
Ustulina deusta (Hoffm.:Fr.) Lind causes butt, stump, and root rot of various woody hosts (CMI 360). See the related genus *Hypoxylon*.

VALSA Fr.
    Diaporthales   60 spp.
Gaiova, V. P. 1985. [Systematics, morphology and biology of the fungus genus *Valsa* Fr. s. l. and its anamorph genus *Cytospora* Ehr.:Fr.] Ukr. Bot. Zh. 42(1):86-94. (in Russian with English summary, defines related taxa and anamorphs, lists teleomorph-anamorph connections, hosts, references)
Kobayashi, T. 1970. Taxonomic studies of Japanese Diaporthaceae with special reference to their life-histories. Bull. Gov. For. Exp. Stn. (Jpn.) 226:1-242. (keys to genera and species of Diaporthaceae, key to nine *Valsa* spp., descriptions and illustrations)
Spielman, L. J. 1985. A monograph of *Valsa* on hardwoods in North America. Can. J. Bot. 63:1355-1378. (describes and illustrates six *Valsa* spp. with their anamorphs, key to related genera and species, lists all *Cytospora* and *Valsa* epithets and their disposition)
Urban, Z. 1958. Revise ceskolovenskych zastpcu rodu *Valsa*, *Leucostoma*, a *Valsella*. Rozpr. Cesk. Akad. Ved., Rada Mat. Prir. Ved. 68:1-101. (in Czechoslovakian, keys to teleomorphs, detailed descriptions of both teleomorphs and anamorphs, host lists)
No comprehensive monograph exists. Spielman (1985) represents a major

VALSA - cont.
contribution and cites many references to previous work. See also Barr (1978) and the anamorph *Cytospora*. *Valsa eugeniae* Nutman & Roberts causes sudden death of clove (CMI 230).

VARARIA P. Karst.
    Aphyllophorales   15 spp.
Boidin, J., and Lanquetin, P. 1975. *Vararia* subgenus *Vararia* (Basidiomycetes: Lachnocladiaceae): Etude speciale des especes d'Afrique intertropicale. Bull. Trimest. Soc. Mycol. Fr. 91:457-513. (in French, key to subgenera and 32 species, descriptions and illustrations)
Boidin, J., and Lanquetin, P. 1977. Les genres *Dichostereum* et *Vararia* en Guadeloupe (Basidiomycetes, Lachnocladiaceae). Mycotaxon 6:277-336. (in French with English summary, key to 12 species occurring in Central America, descriptions and illustrations)
Boidin, J., Lanquetin, P., Terra, P., and Gomez, C. E. 1976. *Vararia* subg. *Vararia* (Basidiomycetes Lachnocladiaceae), deuxieme partie: caracteres culturaux. Bull. Trimest. Soc. Mycol. Fr. 92:247-277. (in French with English summary, key to 11 species, descriptions and illustrations)
Gilbertson, R. L. 1965. Some species of *Vararia* from temperate North America. Michigan Acad. Sci., Arts, and Letters Pap. 50:161-184. (key to nine species, descriptions and illustrations)
Pascoe, I. G., Washington, W. S., and Guy, G. 1984. White root rot of raspberry in Victoria is caused by a *Vararia* species. Trans. Br. Mycol. Soc. 82:723-726. (describes first pathogenic *Vararia* species, not named)
Welden, A. L. 1965. West Indian species of *Vararia* with notes on extralimital species. Mycologia 57:502-520. (key to eight species, descriptions and illustrations)
Pascoe et al. (1984) describe a pathogenic *Vararia* species, causing white root rot of raspberry. Other *Vararia* species are saprophytic, mostly causing wood decay. See Aphyllophorales.

VENTURIA Sacc.
    Pleosporales   53 spp.
Barr, M. E. 1968. The Venturiaceae in North America. Can. J. Bot. 46:799-864. (keys to 13 genera and 80 taxa, key to 34 *Venturia* spp., descriptions, illustrations, host index)
Morelet, M. 1985. Les *Venturia* des peupliers de la section *Leuce* I. Taxinomie. Crypt. Mycol. 6:101-117. (in French with English summary, recognizes ten taxa on poplars, key, descriptions and illustrations)
Sivanesan, A. 1977. The taxonomy and pathology of *Venturia* species. Bibl. Mycol. 59:1-139. (key to 52 *Venturia* spp., descriptions and illustrations, list of doubtful and excluded species, host index)

VENTURIA - cont.
Sivanesan, A. 1985. *Venturia ribis*: a new species of ascomycete. Karstenia 25:50-52. (describes and illustrates the first *Venturia* on *Ribes*, compares five species)
Barr (1968) and Sivanesan (1977, 1984) are useful in identifying *Venturia* species. The anamorphs belong to *Cladosporium, Fusicladium, Pollaccia*, and *Spilocaea*. See *Monographella* for "*Venturia*" *cucumerina*. Species causing diseases include: *Venturia adusta* (Fuckel) E. Mueller, on *Epilobium latifolium* (F. Can. 194); *V. asperata* Samuels & Sivanesan, on *Malus* sp. (F. Can. 291); *V. canadensis* Barr, on *Rumex acetosella* (F. Can. 182); *V. carpophila* E. E. Fisher, peach freckle (CMI 402); *V. cerasi* Aderhold, cherry scab (CMI 706); *V. chlorospora* (Ces.) P. Karst., willow scab and canker (F. Can. 225); *V. inaequalis* (Cooke) Wint. and the anamorph, *Spilocaea pomi* Fr., apple scab (CMI 401, F. Can. 35); *V. macularis* (Fr.) E. Mueller & Arx, scab of poplars (CMI 403); *V. minuta* Barr, on overwintered leaves of *Salix* sp. (F. Can. 223); *V. pirina* Aderhold, pear scab (CMI 404, F. Can. 36); *V. populina* (Vuill.) L. Fabricius, blight or dieback of poplar (CMI 483); *V. rumicis* (Desmaz.) Wint., leaf spot of *Rumex* sp. (CMI 405, F. Can. 181); *V. saliciperda* Nuesch, willow scab (CMI 482; F. Can. 247); and *V. subcutanea* Dearn., on overwintered leaves of *Salix* sp. (F. Can. 224).

VERONAEA Cif. & Montem.
    Hyphomycetes   8 spp.
Hoog, G. S. de, Rahman, M. A., and Boekhout, T. 1983. *Ramichloridium, Veronaea* and *Stenella*: generic delimitation, new combinations and two new species. Trans. Br. Mycol. Soc. 81:485-490. (discusses generic concepts, key to 37 taxa in *Veronaea* and related genera)
Ellis (1976) includes a key to eight species. *Veronaea musae* M. B. Ellis causes minute brown or black spots on living leaves of *Musa*.

VERRUCALVUS M. W. Dick & P. T. W. Wong
    Peronosporales   1 sp.
Dick, M. W., Wong, P. T. W., and Clark, G. 1984. The identity of the oomycete causing 'Kikuyu Yellows', with a reclassification of the downy mildews. Bot. J. Linn. Soc. 89:171-197. (describes and illustrates new genus and species)
This unusual fungus caused a reevaluation of the traditional concept of the Oomycetes and suggests a polyphyletic origin of the group.

VERRUCISPORA D. E. Shaw & Alcorn
    Hyphomycetes   1 sp.
Shaw, D. E., and Alcorn, J. L. 1967. The genus *Verrucispora* gen. nov. (Fungi Imperfecti) on Proteaceae in New Guinea and Queensland. Linn. Soc. N. S. W. 92:171-173. (descriptions and illustration)

VERRUCISPORA - cont.
Ellis (1971) includes *Verrucispora proteacearum* D. E. Shaw & Alcorn.

VERTICICLADIELLA   See *Leptographium*.

VERTICILLIUM Nees
    Hyphomycetes   40 spp.
Gams, W. 1971. *Cephalosporium*-artige Schimmelpilze (Hyphomycetes). Gustav Fischer Verlag, New York. 262 pp. (in German, descriptions and illustrations of 20 *Verticillium* spp.; summary, glossary, and keys in English)
Gams, W., and Zaayen, A. van 1982. Contribution to the taxonomy and pathogenicity of fungicolous *Verticillium* species. I. Taxonomy. Neth. J. Plant Pathol. 88:57-78. (key to sections and nine species, descriptions and illustrations)
Hastie, A. C., and Heale, J. B. 1984. Genetics of *Verticillium*. Phytopathol. Mediterr. 23:130-162. (genetic study of five pathogenic species including the basis of morphological characters)
Isaac, I. 1967. Speciation in *Verticillium*. Ann. Rev. Phytopathol. 5:201-222. (reviews taxonomy and biology)
Pegg, G. F. 1974. *Verticillium* diseases. Rev. Plant Pathol. 53:157-182. (reviews all aspects of the diseases)
Smith, H. C. 1965. The morphology of *Verticilllium albo-atrum*, *V. dahliae*, and *V. tricorpus*.   N. Z. J. Agric. Res. 8:450-478. (differentiates three species)
The taxonomy of *Verticillium* species is not yet settled. Five important plant pathogenic species have been studied in detail (Hastie & Heale, 1984) but a number of soil-borne species need to be delimited. Domsch et al. (1980) include a key to four species, with descriptions, illustrations, and numerous references. The teleomorphs belong in *Cordyceps*, *Nectria*, and *Torrubiella*. *Verticillium* species include: *Verticillium albo-atrum* Reinke & Berthier, wilt of many hosts (CMI 255); *V. dahliae* Kleb., wilt disease of many plants (CMI 256); *V. fungicola* (G. Preuss) Hasseb., on cultivated mushrooms (CMI 498); *V. lecanii* (Zimmermann) Viegas, on insects (CMI 610); *V. nigrescens* Pethybr., usually saprophytic or weakly pathogenic (CMI 257); *V. nubilum* Pethybr., weakly pathogenic causing twisted sprout of potato (CMI 258); *V. psalliotae* Treschow, on cultivated mushrooms (CMI 497); *V. theobromae* (Turconi) E. Mason & S. J. Hughes, cigar end of banana (CMI 259); and *V. tricorpus* Isaac, wilt of tomato (CMI 260).

VIZELLA Sacc.
    Dothideales   7 spp.
Sivanesan, A. 1973. New species of *Vizella* on *Pycnanthus*. Trans. Br. Mycol. Soc. 60:586-588. (describes and illustrates one species)
Swart, H. J. 1971. Australian leaf-inhabiting fungi. I. Two species

VIZELLA - cont.
of *Vizella*. Trans. Br. Mycol. Soc. 57:455-464. (describes and illustrates two species)
Swart, H. J. 1975. Australian leaf-inhabiting fungi. VII. Further studies in *Vizella*. Trans. Br. Mycol. Soc. 64:301-306. (describes and illustrates two species)
Wyk, P. S. van, Marasas, W. F. O., and Hattingh, M. J. 1976. Morphology and taxonomy of *Vizella interrupta* (Ascomycetes: Vizellaceae). Trans. Br. Mycol. Soc. 66:489-494. (considers *Entopeltis* a synonym of *Vizella*, description and illustrations of *V. interrupta*, discussion of generic characters)
These fungi are leaf parasites, most commonly recorded from Australia and South Africa. Sivanesan (1984) provides a key to three species.

VOLUTELLA Tode:Fr.
    Hyphomycetes   20 spp.
Bezerra, J. L. 1963. Studies on *Pseudonectria rousseliana*. Acta Bot. Neerl. 12:58-63. (describes and illustrates *V. buxi* and its teleomorph *Pseudonectria rousseliana*)
Chilton, J. E. 1954. *Volutella* species on alfalfa. Mycologia 46:800-809. (describes and illustrates four taxa)
Kobayashi, T. 1980. Notes on the Philippine fungi parasitic to woody plants (3). Trans. Mycol. Soc. Jpn. 21:311-319. (describes *V. pini-caribaeae*, reviews four other *Volutella* spp. on pines)
Samuels, G. J. 1977. *Nectria consors* and its *Volutella* conidial state. Mycologia 69:255-262. (reviews *Volutella* species, description and illustration)
These fungi are anamorphs of *Nectria* and *Pseudonectria*. Domsch et al. (1980) present a comparative table to six species with a description and illustrations of *Volutella ciliata* Albertini & Schwein.:Fr.

WAITEA Warcup & Talbot
    Tulasnellales   1 sp.
Narayanaswamy, T., and Venkata, Rao A. 1984. *Waitea circinata*-a new fungus causing sheath blight on rice. Curr. Sci. 53:874. (symptoms similar to those caused by *Rhizoctonia solani*)
Oniki, M., Ogoshi, A., Araki, T., Sakai, R., and Tanaka, S. 1985. [The perfect state of *Rhizoctonia oryzae* and *R. zeae*, and the anastomosis groups of *Waitea circinata*.] Trans. Mycol. Soc. Jpn. 26:189-198. (in Japanese with English summary, defines anastomosis groups for *Waitea circinata*, description and illustration)
Warcup, J. H., and Talbot, P. H. B. 1962. Ecology and identity of mycelia isolated from soil. Trans. Br. Mycol. Soc. 45:495-518. (describes and illustrates *Waitea circinata*, a species similar to *Rhizoctonia* and *Thanatephorus* species)
Waitea circinata Warcup & Talbot causes diseases on a number of hosts and has a *Rhizoctonia* anamorph.

WETTSTEININA Hoehn.
  Dothideales   15 spp.
Walker, J. 1984. *Wettsteinina phyllodiorum* (McAlp.) comb. nov. on *Acacia* in Australia. Trans. Br. Mycol. Soc. 83:705-709. (forms spots on phyllodes, description and illustration)
Barr (1972) provides a key to 13 species with descriptions and illustrations. Most species are saprophytic.

WOJNOWICIA Sacc.
  Coelomycetes   2 spp.
Sutton, B. C. 1975. *Wojnowicia* and *Angiopomopsis*. Ceska Mykol. 29:97-104. (describes and illustrates two *Wojnowicia* spp., disposes of two excluded species)
Wojnowicia hirta Sacc., the correct name for *Hendersonia hirta* J. Schroet. and *H. graminis* McAlpine, causes foot rot or root rot of cereals and grasses (CMI 773). Sutton (1975, 1980) treats two species.

XYLARIA J. Hill ex Schrank
  Sphaeriales   100 spp.
Bertault, R. 1984. Xylaires d'Europe et d'Afrique du Nord. Bull. Trimest. Soc. Mycol. Fr. 100:139-175. (in French with English summary, key to 45 species, descriptions and illustrations)
Dennis, R. W. G. 1956. Some Xylarias of tropical America. Kew Bull. 3:401-444. (describes and illustrates 51 species)
Dennis, R. W. G. 1958. Some Xylosphaeras of tropical Africa. Rev. Biol. 1:175-208. (describes and illustrates 31 species)
Dennis, R. W. G. 1961. Xylarioideae and Thamnomycetoideae of Congo. Bull. Jard. Bot. Brux. 31:109-154. (key to genera, key to 25 *Xylaria* spp., descriptions and illustrations)
Dennis, R. W. G. 1974. Xylariaceae from Papua and New Guinea. Numero Special, Bull. Soc. Linn. Lyon 43:127-138. (describes and illustrates 12 *Xylaria* spp.)
Greenhalgh, G. N., and Roe, G. M. 1984. Conidial structure in *Xylaria* and related genera. Pages 365-378 in: Taxonomy of Fungi. Part 2. C. V. Subramanian, ed. Amra Press, Madras. (describes anamorphs of 11 *Xylaria* spp. and other xylariaceous fungi)
Joly, P. 1968. Elements de la flore mycologique du Viet-Nam (troiseme contribution: a propos de quelques Xylarias). Rev. Mycol. 33:155-207. (in French, describes and illustrates 21 species)
Martin, P. 1970. Studies in the Xylariaceae: VIII. *Xylaria* and its allies. J. S. Afr. Bot. 36:73-138. (key to 22 species, descriptions and illustrations)
Perez-Silva, E. 1975. El Genero *Xylaria* (Pyrenomycetes) en Mexico, I. Bol. Soc. Mex. Micol. 9:31-52. (in Spanish, key to 12 species, descriptions and illustrations)
Petrini, L., and Petrini, O. 1985. Xylariaceous fungi as endophytes.

XYLARIA - cont.
Sydowia 38:216-234. (keys to European xylariaceous endophytes based on cultural characters, key to ten anamorph genera, keys to genera and species of eight anamorph genera, descriptions of 22 cultures including three *Xylaria* spp.)
Rogers, J. D. 1983. *Xylaria bulbosa*, *Xylaria curta* and *Xylaria longipes* in continental United States. Mycologia 75:457-467. (descriptions and illustrations including anamorphs)
Rogers, J. D. 1984. *Xylaria acuta*, *Xylaria cornu-damae*, and *Xylaria mali* in continental United States. Mycologia 76:23-33. (descriptions and illustrations including anamorphs, also *X. adscendens*)
Rogers, J. D. 1984. *Xylaria cubensis* and its anamorph *Xylocoremium flabelliforme*, *Xylaria allantoidea*, and *Xylaria poitei* in continental United States. Mycologia 76:912-923. (descriptions and illustrations)
Rogers, J. D. 1985. Anamorphs of *Xylaria*: Taxonomic considerations. Sydowia 38:255-262. (divides genus into four sections based on anamorph characters)
Rogers, J. D. 1986. Provisional keys to *Xylaria* species in continental United States. Mycotaxon 26:85-97. (key to 30 species based on substrate and morphological characters, listed also by distinctive morphological characters)
Rogers, J. D., and Callan, B. E. 1986. *Xylaria poitei*: Stromata, cultural description, and structure of conidia and ascospores. Mycotaxon 26:287-296. (descriptions and illustrations of teleomorph, anamorph, and cultural characteristics)
Rogers, J. D., and Callan, B. E. 1986. *Xylaria polymorpha* and its allies in continental United States. Mycologia 78:391-400. (key to seven species, descriptions and illustrations)
No comprehensive monograph of this large genus exists. Rogers (1986) is useful for the continental United States. *Xylaria* species causing diseases include: *Xylaria digitata* (L.:Fr.) Grev., root rot of hardwoods; *X. hypoxylon* (L.:Fr.) Grev., the candle-snuff fungus, black root rot of apple; *X. polymorpha* (Pers.:Fr.) Grev., dead man's finger, decay of dead hardwoods (CMI 355); and *X. vaporaria* Berk. (syn. *X. pendunculata* (Dickson) Fr.), an invader of mushroom beds.

XYLOBOLUS   See *Stereum*.

ZIMMERMANNIELLA Henn.
Polystigmatales   1 sp.
Lim, T. K., and Khoo, K. C. 1983. Crusty leaf spot disease of mango. Pertanika 6:12-14. (describes *Z. trispora*)
Lim, T. K., and Khoo, K. C. 1983. *Zimmermanniella trispora*, a leaf parasite of mango in Malayasia. Plant Dis. 67:1389. (describes the disease and the fungus)
This tropical pathogen is thoroughly described and illustrated.

ZYTHIA Fr.
Coelomycetes   25 spp.
This genus is generally unstudied. Sutton (1980) includes two species as synonyms of *Zythiostroma*. *Zythia fragariae* Laibach causes a disease of strawberry (CMI 737 under *Gnomonia comari*).

ZYTHIOSTROMA Hoehn.
Coelomycetes   3 spp.
No monograph of this genus exists. Sutton (1980) includes two species that are anamorphs of *Nectria* and *Scoleconectria* species.

ZYXIPHORA Sutton
Hyphomycetes   1 sp.
Sutton, B. C. 1980. Synnematous fungi II. *Zyxiphora gorakhpuensis* n. gen. et sp., causing leaf lesions on *Streblus asper*. Kavaka 8:55-57. (description and illustration)
This monotypic genus is thoroughly described and illustrated.

INDEX TO AUTHORS

Aa, H. A. van der *Guignardia* 1973; *Herpotrichia* 1975; *Leptodothiorella* 1973; *Phoma* 1979, 1980; *Phyllosticta* 1971, 1973.
Abdel-Azim, O. F. *Marasmiellus* 1970.
Abe, Y. *Hypoxylon* 1984.
Adam, P. *Botrytis* 1984.
Adams, G. C., Jr. *Athelia* 1982; *Rhizoctonia* 1979; *Sclerotium* 1982.
Ahmad, S. *Diaporthopsis* 1955.
Ainsworth, G. C. General 1973, 1983; *Ustilaginales* 1984.
Al-Musallam, A. *Aspergillus* 1980.
Alcorn, J. L. *Bipolaris* 1981, 1982, 1983; *Cochliobolus* 1982, 1983; *Drechslera* 1983; *Exserohilum* 1983, 1986; *Helminthosporium* 1983; *Mycoleptodiscus* 1985; *Neottiosporina* 1974, 1985; *Setosphaeria* 1986; *Verrucispora* 1967.
Alfenas, A. C. *Cryphonectria* 1986.
Alfieri, S. A., Jr. *Calonectria* 1972, 1983, 1986; *Cephaleuros* 1881; *Dothichiza* 1982; *Nectriella* 1979.
Allison, J. R. *Phoma* 1985.
Alvarez, A. M. *Stemphylium* 1983.
Ames, L. M. *Chaetomium* 1961.
Ammirati, J. F. *Discula* 1983.
Ammon, H. U. *Cochliobolus* 1963; *Pyrenophora* 1963.
Anahosur, K. H. *Microcyclus* (1970)1971.
Anderson, J. B. *Armillaria* 1978, 1979, 1980, 1986.
Anderson, T-H. General 1980.
Anderson, T. R. *Botryosporium* 1983.
Ann, P. J. *Phytophthora* 1985.
Aquilar, A. M. *Ceratocystis* 1984.
Araki, T. *Erythricium* 1985; *Rhizoctonia* 1983, 1985; *Waitea* 1985.
Arthur, J. C. Uredinales 1934.
Arx, J. A. von *Acantharia* 1984; Ascomycotina 1954, 1962, 1973, 1975; *Ascotricha* 1982; *Blumeriella* 1961; *Chaetomium* 1984, 1986; *Colletotrichum* 1957, 1970; *Discula* 1970; *Elsinoe* 1963; General 1981; *Geotrichum* 1977; *Gloeosporidiella* 1970; *Gloeosporium* 1970; *Herpotrichia* 1984; *Idriella* 1981; *Kabatiella* 1970; *Kabatina* 1966; *Magnaporthe* 1977; *Marssonina* 1970; *Microdochium* 1981, 1985; *Monilia* 1981; *Monographella* 1985; *Monostichella* 1970; *Mycosphaerella* 1949, 1983; *Myriangium* 1963; *Ovularia* 1983.
Arya, A. *Botryodiplodia* 1985.
Ashour, W. A. *Marasmiellus* 1970.
Austwick, P. K. C. *Mastigosporium* 1954.
Ayers, W. A. *Aphanomyces* 1974.
Babcock, C. E. *Paraphaeosphaeria* 1985.
Backhouse, D. *Botrytis* 1984.
Backus, M. P. *Entomosporium* 1966, 1967.
Bacon, C. W. *Acremonium* 1985; *Atkinsonella* 1984; *Balansia* 1984; *Claviceps* 1984; *Epichloe* 1984; *Myriogenospora* 1977, 1982; *Sphacelia* 1984.
Bagyanarayana, G. *Kuehneola* 1985.
Bai, H. C. *Bremia* 1985.
Baijal, U. *Blakeslea* 1968.
Baker, K. F. *Didymella* 1983.
Baker, R. E. D. *Diatractium* 1951.
Baker, R. F. *Phoma* 1985.
Bakshi, B. K. Aphyllophorales 1971; *Dicellomyces* 1976.

Ballantyne, B. *Sphaerotheca* 1975.
Bandoni, R. *Herpobasidium* 1984; *Insolibasidium* 1984.
Bandoni, R. J. *Graphiola* 1982.
Baniecki, J. F. *Phymatotrichopsis* 1969.
Banik, M. T. *Rhytisma* 1987.
Banowetz, G. M. *Tilletia* 1984.
Baral, H. O. *Lachnellula* 1984.
Barnard, E. L. *Calonectria* 1986.
Barnett, H. L. Deuteromycotina 1972.
Barr, D. J. S. *Polymyxa* 1979.
Barr, M. E. *Apiosporina* 1968; Ascomycotina 1972, 1978, 1979; *Atopospora* 1968; *Botryosphaeria* 1970; *Coleroa* 1986; *Endothia* 1983; *Herpotrichia* 1984; *Hormotheca* 1986; *Magnaporthe* 1977; *Massaria* 1979; *Neopeckia* 1984; *Pestalosphaeria* 1975; *Physalospora* 1970; *Venturia* 1968.
Barron, G. L. Deuteromycotina 1968; *Meria* 1977; *Paecilomyces* 1967.
Bartnicki-Garcia, S. *Phytophthora* 1983.
Batista, A. C. *Phyllosticta* 1952.
Batra, L. R. *Ciborinia* 1959, 1960; *Eremothecium* 1973; *Monilinia* 1979, 1983, 1986; *Nematospora* 1973; *Ustilago* 1982.
Baum, B. L. *Puccinia* 1985.
Baxter, A. P. *Colletotrichum* 1983, 1984, 1985; *Pleospora* 1984; *Stemphylium* 1984.
Baxter, J. W. *Cumminsiella* 1957; *Ravenelia* 1980; *Uropyxis* 1959.
Bedker, P. J. *Thyronectria* 1983.
Bedlan, G. *Phragmidium* 1984.
Bell, J. C. *Phytophthora* 1984.
Benjamin, C. R. *Aspergillus* 1955; *Biscogniauxia* 1971.
Bennell, A. P. *Tranzschelia* 1978.
Benny, G. L. *Corynelia* 1985; *Moniliophthora* 1978.
Benoit, M. A. *Curvularia* 1970.
Berge, G. *Gliocladium* 1963.
Bertagnole, C. L. *Leptographium* 1983.
Bertault, R. *Xylaria* 1984.
Berthelay, S. *Armillaria* 1981.
Berthier, J. *Typhula* 1976.
Bestagno, G. *Coniothyrium* (1958)1959.
Bezerra, J. L. *Volutella* 1963.
Bhagyanarayana, G. *Oidium* 1983.
Bhat, D. J. *Monographella* 1978; *Nectria* 1984; *Pleiochaeta* 1983; *Thyronectria* 1984.
Bhatt, G. C. *Dactylaria* 1968.
Biga, M. L. B. *Albugo* 1955; *Coniothyrium* (1958)1959.
Bissett, J. *Guignardia* 1986; *Phyllosticta* 1979; *Trichoderma* 1984.
Blanchard, R. O. *Nectria* 1981.
Blanchette, R. A. *Sphaeropsis* 1985.
Blanz, P. *Graphiola* 1982.
Blaser, P. *Eurotium* (1975)1976.
Bloss, H. E. *Phymatotrichopsis* 1969, 1973.
Boedijn, K. B. *Cercospora* 1961; *Cylindrocladium* 1950; *Eremothecium* 1960.
Boekhout, T. *Ramichloridium* 1983; *Stenella* 1983; *Veronaea* 1983.
Boerema, G. H. *Ascochyta* 1973, 1975; *Leptosphaeria* 1981; *Phoma* 1973, 1975, 1976, 1981; *Sphaerulina* 1963.
Boeswinkel, H. J. *Calonectria* 1982;

Cylindrocladiella 1982; Lepteutypa 1983; Monilinia 1970; Oidium 1977, 1979, 1980; Podosphaera 1979; Seiridium 1983; Sphaerotheca 1979.
Boidin, J. Aleurodiscus 1985; Stereum 1960; Vararia 1975, 1976, 1977.
Boiteux, L. S. Sphaerotheca 1985.
Bojovic-Cvetic, D. Phomopsis 1985.
Bolay, A. Gnomonia 1971.
Bolkan, H. A. Tranzschelia 1985.
Bollard, E. G. Mastigosporium 1950.
Bollen, G. J. Ascochyta 1975; Phoma 1975.
Bonar, L. Diaporthopsis 1966.
Bondartsev, A. S. Aphyllophorales 1953.
Boone, D. M. Phyllosticta 1982; Strasseria 1983.
Booth, C. Cylindrocarpon 1966, 1974, 1984; Fusarium 1971, 1975; Gibberella 1971, 1984; Microdochium 1981; Monographella 1981; Nectria 1959.
Booth, C., ed. General 1971.
Boothroyd, C. W. Phyllosticta 1973.
Bose, S. K. Acantharia 1965; Herpotrichia 1961.
Bove, F. J. Claviceps 1970.
Bowerman, C. A. Ciborinia 1955.
Brady, L. R. Claviceps 1962.
Brandenburger, W. General 1985.
Braun, U. Anthracoidea 1978; Erysiphe 1981, 1982, 1983, 1984, 1985; Leveillula 1980; Microsphaera 1981, 1982, 1983, 1984, 1985; Oidium 1980, 1982; Phyllactinia 1985; Podosphaera 1984; Sphaerotheca 1984, 1985; Uncinula 1983, 1984.
Breitenbach, J. Aphyllophorales 1986; Ascomycotina 1983.
Bridge, P. D. Monascus 1985; Penicillium 1984.
Brockmann, I. Discostroma 1975; Seimatosporium (1975)1976.
Bromfield, K. R. Phakopsora 1984.
Brothers, M. P. Pseudoperonospora 1981.
Brown, A. H. S. Paecilomyces 1957.
Brown, L. G. Cercospora 1976, 1977; Phaeoisariopsis 1976.
Brown, R. D., Jr. Endothia 1982.
Bruehl, G. W. Typhula 1975.
Buckley, N. G. Aureobasidium 1971.
Buczacki, S. T. Plasmodiophora 1973.
Burdsall, H. H., Jr. Cronartium 1977; Laetisaria 1979; Phyllosticta 1982.
Buritica, P. Pucciniosira 1980; Uromyces 1984.
Burpee, L. L. Ceratobasidium 1984; Rhizoctonia 1980.
Burr, T. J. Phomopsis 1982.
Burt, E. A. Aphyllophorales 1966.
Bushnell, W. R. Uredinales 1984.
Butin, H. Ceratocystis 1984; Cryptodiaporthe 1958; Kabatina 1976.
Butler, E. E. Cylindrocarpon 1981; Geotrichum 1972; Rhizoctonia 1979.
Butterfill, G. B. Ascomycotina 1969.
Byford, W. J. Phoma 1985.
Byrde, R. J. W. Monilinia 1977.
Cain, R. F. Ceratocystis 1961; Chaetomium 1969; Phacidium 1962; Therrya 1961.
Callan, B. E. Xylaria 1986.
Calvo, M. A. Ascotricha 1983.
Cameron, H. R. Anisogramma 1976.
Campbell, W. A. Cystostereum 1941.
Campbell, W. D. Pythium 1983.

Candoussau, F. Aleurodiscus 1985.
Cannon, P. F. Ascomycotina 1985; Chaetomium 1986; Cyclaneusma 1986; Epicoccum 1986; Hypoderma 1983, 1986; Lophodermium 1983, 1986; Neocosmospora 1984; Rhytisma 1986.
Carmichael, J. W. Briosia 1976; Deuteromycotina 1973, 1980; Geotrichum 1957, 1976; Oospora 1976.
Carranza-Morse, J. Fomitopsis 1986.
Carroll, G. Phaeocryptopus 1985.
Carroll, G. C. Rhabdocline 1986.
Castellani, E. Monochaetiella 1943; Stagonospora 1977.
Caten, C. E. Puccinia 1985.
Cauchon, R. Strasseria 1979.
Cejp, K. Ascochyta 1968; Septoria 1967.
Chahal, S. S. Claviceps 1985.
Chandra-Reddy, K. R. Discosia 1974, 1984.
Channon, A. G. Itersonilia 1963.
Chant, S. R. Briosia 1984.
Chapman, R. L. Cephaleuros 1985.
Chastagner, G. A. Botryotinia 1983.
Chau, K. F. Stemphylium 1983.
Chen, G.-q. Erysiphe 1981; Sawadaea 1980; Uncinula 1977; Uncinuliella 1979, 1982.
Chen, S.-c. Nectria 1984.
Chen, Z.-x. Uncinula 1984.
Cheng, X. Y. Bremia 1985.
Chereponova, P. Peronospora 1982.
Chesters, C. G. C. Hypoxylon 1968.
Chi, I. H. Agaricodochium 1983.
Chidambaram, P. Cochliobolus 1973; Drechslera 1973.
Child, M. Daldinia 1932.
Chilton, J. E. Volutella 1954.
Choobamroong, W. Cercospora 1980.
Chowdhry, P. N. Dactylaria 1982.
Christensen, M. Aspergillus 1978, 1982; Emericella 1978.
Christensen, M. J. Acremonium 1984.
Chu, D. Armillaria 1985.
Chuandao, L. Marssonina 1984.
Chuang, T. Y. Phyllosticta 1981.
Chupp, C. Cercospora 1953.
Ciferri, R. Coniothyrium (1958)1959.
Claflin, L. E. Dothiorella 1980.
Clark, G. Verrucalvus 1984.
Clayton, C. N. Dothichiza 1959.
Cline, M. N. Cristulariella 1983.
Cline, S. D. Cristulariella 1983.
Cobb, F. W., Jr. Leptographium 1986.
Cole, G. T. Acremonium 1985; Graphiola 1983; Phialophora 1973.
Cole, H., Jr. Rhizoctonia 1980.
Coley-Smith, J. R. Botrytis 1980; Sclerotium 1982.
Commonwealth Mycological Institute. General 1983, 1964-1985.
Conners, I. L. Deuteromycotina 1980.
Constantinescu, O. Bremiella 1979; Deightoniella 1983; Mycocentrospora 1978; Peronosporales 1983; Phloeosporella 1984; Pseudoperonospora 1985; Septoria 1984.
Conway, K. E. Botryosporium 1978.
Cook, R. J., eds. Fusarium 1981.
Cooke, R. C. Trichothecium 1966.
Cooke, W. B. Aureobasidium 1962.
Corbaz, R. Didymella 1957.

238

Corbetta, G. *Curvularia* 1965.
Corbin, J. B. *Monilinia* 1970; *Mycosphaerella* 1979.
Corlett, M. *Coleroa* 1986; *Didymella* 1981; *Hormotheca* 1986; *Lophophacidium* 1984.
Corner, E. J. H. *Ganoderma* 1983.
Cotter, H. V. T. *Nectria* 1981.
Cowan, R. S. General 1976-1985.
Crane, J. L. *Cristulariella* 1983; *Nakataea* 1979; *Pesotum* 1973.
Cristinzio, G. *Phytophthora* 1983.
Crivelli, P. *Idriella* 1983.
Crompton, J. G. *Cladosporium* 1984.
Croxall, H. E. *Diatrypella* 1950.
Cruickshank, R. H. *Sclerotinia* 1983.
Crute, I. R. *Bremia* 1981.
Cummins, G. B. *Cronartium* 1984; *Maravalia* 1950; *Physopella* 1958; *Trachyspora* 1982; Uredinales 1962, 1971, 1978, 1981, 1983.
Cunfer, B. M. *Typhula* 1975.
Cunningham, J. L. *Dicellomyces* 1976.
Cunningham, P. C. *Pseudocercosporella* 1981.
Dabinett, P. E. *Rhizopus* 1973.
Dale, W. T. *Diatractium* 1951.
Dantanarayana, D. M. *Phytophthora* 1984.
Dargan, J. S. *Daldinia* 1978; *Rosellinia* 1979.
Darker, G. D. *Davisomycella* 1967; *Hypoderma* 1932, 1967; *Hypodermella* 1967; *Lirula* 1967; *Lophodermium* 1932, 1967; *Lophomerum* 1967.
Darvas, J. M. *Stromatinia* 1984.
Das, B. K. *Cercospora* 1982.
Datnoff, L. E. *Dactuliophora* 1986.
Davidson, R. W. *Ceratocystis* 1971; *Cystostereum* 1941; *Leptographium* 1978; *Ophiostoma* 1968.
Davis, E. E. *Stachybotrys* 1976.
Davis, L. H. *Phoma* 1985.
Davis, R. D. *Leptosphaerulina* 1985.
Daykin, M. E. *Phomopsis* 1983.
Deacon, J. W. *Magnaporthe* 1983.
Dehorter, B. *Nectria* 1983.
Deighton, F. C. *Cercoseptoria* 1976, 1983; *Cercospora* 1959, 1967, 1969, 1971, 1974, 1976, 1979, 1983; *Cercosporella* 1973; *Cercosporidium* 1967; *Gloeocercospora* 1971; *Microdochium* 1972; *Mycocentrospora* 1971, 1983; *Mycovellosiella* 1974, 1979; *Nodulisporium* 1985; *Ovularia* 1984; *Passalora* 1967; *Phaeoramularia* 1976; *Pseudocercospora* 1976, 1979; *Stenella* 1979.
Delon, R. *Chalara* 1983.
Deml, G. *Graphiola* 1982; *Sporisorium* 1983.
Dennis, R. W. G. Ascomycotina 1978; *Botryotinia* 1956; *Hypoxylon* 1963; *Lachnellula* 1962; *Pezicula* 1974; *Pezizella* 1976; *Ustulina* 1963; *Xylaria* 1956, 1958, 1961, 1974.
Dhanvantari, B. N. *Leucostoma* 1982.
Dharne, C. G. *Lachnellula* 1964.
Dhingra, O. D. *Macrophomina* 1977.
DiCosmo, F. *Cryptosporella* 1981; *Cyclaneusma* 1983, 1984; *Darkera* 1984; *Foveostroma* 1978; *Harknessia* 1981; *Lophophacidium* 1984; *Monochaetiellopsis* 1977; *Myrothecium* 1980; *Phacidiopycnis* 1984; *Phacidium* 1984; *Potebniamyces* 1984; *Pseudophacidium* 1984.
Diamandis, S. *Elytroderma* 1979.
Dias, M. R. de Sousa Peronosporales 1976, 1982.
Dick, E. A. General 1971.
Dick, M. *Trimmatostroma* 1983.
Dick, M. W. *Aphanomyces* 1973; *Verrucalvus* 1984.

Diehl, W. W. *Atkinsonella* 1950; *Balansia* 1950.
Dilley, M. A. *Eutypa* 1983.
Dingley, J. M. *Nectria* 1951, 1957.
Dixon, G. R. *Bremia* 1981.
Docea, E. *Septoria* 1973.
Doebbeler, P. *Nectria* 1979.
Doguet, G. *Epichloe* 1960.
Doi, N. *Trichoderma* 1979, 1986.
Doi, Y. *Trichoderma* 1979, 1986.
Dolejs, K. *Septoria* 1967.
Domanski, S. Aphyllophorales 1965, 1967.
Domsch, K. H. General 1980.
Donk, M. A. Aphyllophorales 1974.
Dorenbosch, M. M. J. *Ascochyta* 1973; *Phoma* 1970, 1973.
Dorworth, C. E. *Ascocalyx* 1983.
Doyle, A. F. *Nectria* 1978.
Dreyfuss, M. *Chaetomium* (1975)1976.
Dubos, B. *Pseudopezicula* 1986.
Dumont, K. P. *Nectria* 1982.
Duran, R. Ustilaginales 1961, 1973.
Duravetz, J. S. *Rhytisma* 1971.
Durrieu, G. *Leveillula* 1984.
Eboh, D. O. *Chaconia* 1985; *Newinia* 1983; Uredinales 1986.
Eckblad, F.-E. *Biscogniauxia* 1978; *Cenangium* 1977; *Daldinia* 1969.
Edwards, R. L. *Biscogniauxia* 1985; *Hypoxylon* 1983.
Egger, M. C. *Pseudophacidium* 1968.
Eicker, A. *Colletotrichum* 1983, 1985; *Oidium* 1984.
El-Gholl, N. E. *Calonectria* 1983, 1986.
Elkins, J. R. *Endothia* 1986.
Ellett, C. W. *Mycocentrospora* 1974.
Ellis, D. H. *Rhizopus* 1981.
Ellis, J. J. Mucorales 1973; *Rhizopus* 1985, 1986.
Ellis, J. P. General 1985.
Ellis, M. B. *Annellophora* 1958; *Arthrinium* 1963; *Curvularia* 1966; *Deightoniella* 1957; Deuteromycotina 1971, 1976; General 1985; *Haplobasidion* 1957, 1972; *Helminthosporium* 1961; *Lacellina* 1957; *Lacellinopsis* 1957; *Periconiella* 1971.
Emert, G. H. *Endothia* 1982.
Engelbrecht, C. *Batcheloromyces* 1983.
Eriksson, J. Aphyllophorales 1973, 1975, 1976, 1978, 1981, 1984.
Eriksson, O. *Anthostomella* 1966; *Clathrospora* 1967; *Leptosphaeria* 1967; *Phaeosphaeria* 1967; *Pleospora* 1967.
Erselius, L. J. *Phytophthora* 1984.
Ershad, D. *Cylindrocarpon* 1970.
Erwin, D. C. *Phytophthora* 1983.
Evans, H. C. *Ascochytulina* 1985; *Cercoseptoria* 1984; *Cylindrocarpon* 1984; *Dothistroma* 1984; *Lecanosticta* 1984; *Moniliophthora* 1978, 1981; *Mycosphaerella* 1984; *Ocotomyces* 1985.
Farr, M. L. *Dimeriella* 1979.
Farris, S. H. *Siroccocus* 1981.
Fayret, J. *Gliocladium* 1963.
Fennell, D. I. *Aspergillus* 1965, 1973.
Ferreira, F. A. *Cryphonectria* 1986; *Prospodium* 1986.
Fiasson, J.-L. *Phellinus* 1984.
Ficke, W. *Cytospora* 1983.
Fielding, A. H. *Monilinia* 1977.
Figueras, K. J. *Chaetomium* 1986.

Fischer, G. W. *Ustilaginales* 1953, 1961.
Fisher, N. L. *Fusarium* 1983.
Fisher, P. J. *Leiosphaerella* 1983.
Fitt, B. D. L. *Gibellina* 1985.
Flack, N. J. *Nectria* 1977.
Flett, S. P. *Spongospora* 1983.
Forrester, R. I. *Hendersonia* 1981.
Fosberg, F. R. *Diaporthopsis* 1983.
Fowler, M. C. *Leptosphaeria* 1984.
Francis, S. M. *Anthostomella* 1975; *Hypoxylon* 1983; *Rosellinia* 1985.
Frederiksen, R. A., ed. *Sporisorium* 1986.
Freyer, K. von *Herpotrichia* 1975.
Fripp, Y. J. *Hendersonia* 1981.
Fritz-Schroeder, J. *Ramularia* 1983.
Froidevaux, L. *Dothiora* 1972; *Sydowia* 1972.
Fulkerson, J. F. *Botryosphaeria* 1960.
Fullerton, R. A. *Sphacelotheca* 1978; *Sporisorium* 1978.
Fulton, R. H., ed. *Hemileia* 1984.
Funk, A. *Atropellis* 1966; *Botryosphaeria* 1985; *Caliciopsis* 1963; *Dermea* 1976; *Endothiella* 1984; *Foveostroma* 1976; General 1981, 1985; *Potebniamyces* 1981; *Pseudophacidium* 1980; *Seimatosporium* 1985; *Therrya* 1980.
Gadgil, P. D. *Trimmatostroma* 1983.
Gaeumann, E. *Uredinales* 1959.
Gaiova, V. P. *Cytospora* 1984; *Valsa* 1985.
Galan, R. *Lachnellula* 1985.
Galea, V. J. *Microdochium* 1986.
Gallegos, H. L. *Uredinales* 1981.
Gambogi, P. *Leptographium* 1977; *Phoma* 1985.
Gams, W. *Acremonium* 1971, 1982; *Aspergillus* 1984; *Epichloe* 1982; General 1980; *Hymenella* 1971; *Microdochium* 1980; *Phialophora* 1976; *Sarocladium* 1975; *Verticillium* 1971, 1982.
Ganapathi, A. *Mycosphaerella* 1979; *Trimmatostroma* 1978.
Gao, R.-x. *Uncinula* 1984.
Gbaja, I. S. *Briosia* 1984.
Georgy, N. I. *Sclerotium* 1982.
Gerlach, W. *Cylindrocarpon* 1970; *Fusarium* 1982.
Germano, G. *Stagonospora* 1977.
Gerrettson-Cornell, L. *Phytophthora* 1985; *Pythium* 1984.
Giatgong, P. *Cercospora* 1980.
Gibbs, J. N. *Phomopsis* 1984.
Gibson, I. A. S. *Dothistroma* 1972.
Gilbertson, R. L. *Aphyllophorales* 1971, 1978, 1986; *Fomitopsis* 1986; *Inonotus* 1976; *Peronospora* 1973; *Phellinus* 1979; *Vararia* 1965.
Gilles, G. *Aleurodiscus* 1985.
Gilliam, M. S. *Dicellomyces* 1976; *Marasmius* 1976.
Gilman, J. C. *Diaporthe* 1959; *Diatrype* 1965.
Ginns, J. *Poria* 1984.
Gjaerum, H. B. *Monilinia* 1969; *Taphrina* 1964; *Trachyspora* 1982; *Uredinales* 1986.
Gladders, P. *Ascochyta* 1985.
Glawe, D. A. *Cryptosphaeria* 1984; *Diatrype* 1982, 1983, 1984; *Diatrypella* 1982, 1984, 1986; *Eutypa* 1982, 1983, 1984; *Eutypella* 1982, 1983, 1984; *Ophiovalsa* 1986.
Glynne, M. B. *Gibellina* 1985.
Goidanich, G. *Macrophomina* 1947.
Gokhale, A. A. *Tilletiopsis* 1972.
Gomez, C. E. *Vararia* 1976.
Goodding, L. N. *Atropellis* 1930.

Gopalkrishnan, K. S. *Hemileia* 1951.
Gorter, G. J. M. A. *Oidium* 1984.
Goto, S. *Tilletiopsis* 1985.
Gottwald, T. R. *Anisogramma* 1979; *Cladosporium* 1982.
Govindu, H. C. *Balansia* 1961, 1973.
Graham, J. H. *Leptosphaerulina* 1961.
Granmo, A. *Biscogniauxia* 1978.
Grasso, V. *Gymnosporangium* 1972.
Graves, A. A. *Monochaetia* 1971.
Gray, E. *Gloeotinia* 1954.
Greenhalgh, G. N. *Hypoxylon* 1968, 1973; *Nodulisporium* 1975; *Xylaria* 1984.
Gregory, N. M. *Armillaria* 1982.
Gregory, P. H. *Phytophthora* 1974.
Gregory, S. C. *Armillaria* 1985.
Gremmen, J. *Cenangium* 1953; *Drepanopeziza* 1965; *Marssonina* 1965; *Pyrenopeziza* 1958.
Griffin, G. J. *Endothia* 1986.
Griffin, H. D. *Ceratocystis* 1968.
Griffiths, D. A. *Pestalotia* 1974.
Griffiths, P. A. *Seimatosporium* 1974.
Grogan, R. G. *Athelia* 1982; *Sclerotium* 1982.
Grosclaude, C. *Cytospora* 1979.
Gross, H. L. *Echinodontium* 1964.
Grove, W. B. *Deuteromycotina* 1935, 1937.
Groves, J. W. *Alternaria* 1944; *Ascocalyx* 1968; *Botryotinia* 1963; *Ciborinia* 1955; *Dermea* 1946; *Godronia* 1965; *Pezicula* 1939, 1940; *Pragmopara* 1967.
Guarro, J. *Ascotricha* 1983; *Chaetomium* 1986.
Guba, E. F. *Monochaetia* 1961; *Pestalotia* 1961; *Pestalotiopsis* 1961.
Gueho, E. *Geotrichum* 1979, 1985.
Guikwad, Y. B. *Botryosphaeria* 1974.
Guillaumin, J. J. *Armillaria* 1981.
Gunnerbeck, E. *Mastigosporium* 1971; *Ramularia* 1967.
Guo, L. *Gymnosporangium* 1984.
Guo, Y.-l. *Cercosporidium* 1982.
Gupta, S. R. *Nectria* 1980.
Gutteridge, C. S. *Sclerotinia* 1985.
Guy, G. *Vararia* 1984.
Guyot, A. L. *Uromyces* 1957.
Gvritishvili, M. N. *Cytospora* 1982.
Hadlok, R. *Penicillium* 1976.
Hallett, I. C. *Microdochium* 1983; *Monographella* 1983.
Hamet-ahti, L. *Sclerotinia* 1980.
Hammett, K. R. W. *Oidium* 1977.
Hanlin, R. T. *Brasiliomyces* 1984.
Hanson, E. W. *Erysiphe* 1966.
Harada, Y. *Gymnosporangium* 1984; *Monilinia* 1977, 1984, 1986; *Thecaphora* 1983; *Ustilaginales* 1983.
Hardin, H. *Bipolaris* 1980; *Cochliobolus* 1980.
Hardison, J. R. *Gloeotinia* 1962.
Harr, J. *Clathrospora* 1970.
Harrington, T. C. *Leptographium* 1986.
Harris, O. C. *Microdochium* 1985.
Harrison, K. A. *Aphyllophorales* 1973.
Hartman, G. L. *Colletotrichum* 1986.
Hasebe, S. *Puccinia* 1978.
Hashioka, Y. *Pyricularia* 1971, 1973.
Hastie, A. C. *Verticillium* 1984.
Hattingh, M. J. *Vizella* 1976.
Hawksworth, D. L. *Ascomycotina* 1985; *Ascotricha* 1971; *Chaetomium* 1973; General 1974, 1983;

Monascus 1983, 1985; *Nectria* 1980; *Nectriella* 1982; *Neocosmospora* 1984; *Penicillium* 1984; *Sarocladium* 1975.
Heale, J. B. *Verticillium* 1984.
Heath, M. C. Uredinales 1979.
Hedjaroude, G. A. *Phaeosphaeria* 1968.
Henderson, D. M. *Trachyspora* 1973; *Tranzschelia* 1978; Uredinales 1966.
Hendrix, F. F., Jr. *Pythium* 1983.
Henk, M. C. *Cephaleuros* 1985.
Hennebert, G. L. *Botryotinia* 1963; *Botrytis* 1973; *Dactylaria* 1983; *Phymatotrichopsis* 1973.
Hennen, J. F. *Chaconia* 1983; *Cumminsiella* 1982; *Goplana* 1983; *Olivea* 1983; *Prospodium* 1986; *Pucciniosira* 1980.
Hermanides-Nijhof, E. J. *Aureobasidium* 1977; *Kabatiella* 1977; *Kabatina* 1977.
Hess, W. M. *Tilletia* 1986.
Hesseltine, C. W. *Gilbertella* 1960; Mucorales 1973.
Hijwegen, T. *Cordana* 1983; *Dactylaria* 1983.
Hinds, T. E. *Cryptosphaeria* 1981; *Libertella* 1981.
Hino, T. *Cercospora* 1978.
Hinton, D. M. *Acremonium* 1985.
Hiratsuka, N. *Melampsora* 1982, 1984; *Melampsoridium* 1958, 1981, 1982, 1983; *Phragmidium* 1980; *Puccinia* 1968, 1973, 1976, 1978, 1980, 1983; *Pucciniastrum* 1958, 1980; *Thekopsora* 1958; *Uromyces* 1973.
Hiratsuka, Y. *Cronartium* 1976; *Endocronartium* 1969; *Pucciniastrum* 1970; *Thekopsora* 1976; Uredinales 1983.
Hirsch, G. *Anthracoidea* 1978.
Hirschhorn, E. Ustilaginales 1986.
Hjortstam, K. Aphyllophorales 1978, 1981, 1984.
Ho, H. H. *Phytophthora* 1981, 1984.
Hobbs, T. W. *Phomopsis* 1985.
Hochapfel, H. *Ascochyta* 1936; *Diplodia* 1943.
Hocking, A. D. *Aspergillus* 1985; *Paecilomyces* 1985; *Penicillium* 1985.
Hodges, C. S., Jr. *Brasiliomyces* 1985; *Cryphonectria* 1986; *Microdochium* 1976; *Mycoleptodiscus* 1976.
Hoffman, J. A. *Tilletia* 1982.
Holcomb, G. E. *Cephaleuros* 1985, 1986.
Holliday, P. *Microcyclus* 1970.
Holm, K. *Phaeosphaeria* 1981.
Holm, L. Ascomycotina 1975; *Didymella* 1953; *Entodesmium* 1957; *Leptosphaeria* 1952, 1957; *Nodulosphaeria* 1957, 1961, 1963; *Ophiobolus* 1957; *Phaeosphaeria* 1981.
Holubova-Jechova, V. *Chalara* 1984; *Phialophora* 1976.
Hoog, G. S. de *Ceratocystis* 1974, 1984; *Cordana* 1983; *Dactylaria* 1983, 1985; *Dichotomophthora* 1983; *Myrothecium* 1983; *Ophiostoma* 1984; *Pithomyces* 1986; *Ramichloridium* 1983; *Sporothrix* 1974; *Stenella* 1983; *Veronaea* 1983.
Hopcroft, D. H. *Marssonina* 1983.
Hopkin, A. A. *Magnaporthe* 1984; *Sclerotium* 1984.
Hopkins, J. C. *Atropellis* 1961, 1963.
Horie, H. *Entomosporium* 1980; *Tubakia* 1979.
Horie, Y. *Eupenicillium* 1973.
Hornby, D. *Gibellina* 1985.
Hosagoudar, V. B. *Phyllachora* 1985.
Hosford, R. M., Jr. *Leptosphaeria* 1978.
Hsieh, W. H. *Stagonospora* 1979.

Hubbes, M. *Leucostoma* 1960; *Melampsora* 1983.
Hudler, G. W. *Rhytisma* 1987.
Hudson, H. J. *Apiospora* 1976; *Nigrospora* 1963.
Hughes, S. J. *Botryosporium* 1978; Deuteromycotina 1958; *Mastigosporium* 1951; *Pleiochaeta* 1951; *Stigmella* 1952.
Hugueney, R. *Aleurodiscus* 1985.
Humphrey, J. G. *Plasmodiophora* 1973.
Huner, N. P. A. *Sclerotium* 1985.
Hunt, J. *Ceratocystis* 1956.
Hunt, R. S. *Hypoderma* 1978; *Leptomelanconium* 1985; *Lirula* 1978; *Lophodermium* 1978; *Ploioderma* 1978.
Hunter, B. B. Deuteromycotina 1972.
Hunter, L. Deuteromycotina 1979.
Huystee, R. B. van *Sclerotium* 1985.
Ikediugwu, F. E. O. *Marasmius* 1984.
Illman, W. I. *Butlerelfia* 1980.
Ilott, T. W. *Pyrenopeziza* 1984.
Ingram, D. S. *Bremia* 1985; *Pyrenopeziza* 1981, 1984.
Insell, J. P. *Sclerotium* 1985.
Irwin, J. A. G. *Leptosphaerulina* 1985; *Stemphylium* 1984.
Isaac, I. *Verticillium* 1967.
Ito, K. *Endothia* 1956.
Ivory, M. H. *Dothistroma* 1967.
Iwatsu, T. *Cladosporium* 1984.
Jaarsveld, A. B. van *Oidium* 1984.
Jackson, H. S. *Ceratobasidium* 1949.
Jackson, N. *Laetisaria* 1983.
Jackson, W. A. *Sphaeropsis* 1985.
Jahn, H. *Phellinus* 1981; *Stereum* 1971.
Jarvis, W. R. *Botrytis* 1977, 1980; *Cladosporium* 1984.
Jechova, V. *Septoria* 1967.
Jeffries, P. *Sclerotinia* 1985.
Jeng, R. S. *Melampsora* 1983.
Jensen, J. D. *Melanconis* 1984; *Ophiovalsa* 1986.
Joffe, A. Z. *Fusarium* 1986.
Johansen, I. Aphyllophorales 1980.
Johnson, A. G. *Neovossia* 1952.
Johnson, A. L. S. *Armillaria* 1985.
Johnson, M. C. *Acremonium* 1985.
Johnson, R. *Puccinia* 1985.
Johnston, P. R. *Coccomyces* 1986; *Phoma* 1981.
Joly, P. *Alternaria* 1964; *Melanotaenium* 1972; *Thecaphora* 1975; *Tolyposporium* 1973; *Xylaria* 1968.
Jong, S. C. *Biscogniauxia* 1971; *Hypoxylon* 1972; *Nodulisporium* 1972; *Phaeoisariopsis* 1968; *Stachybotrys* 1976.
Jooste, W. J. *Leptographium* 1978.
Jorstad, I. *Septoria* 1965, 1967.
Juelich, W. Aphyllophorales 1980, 1984; *Athelia* 1972.
Junell, L. Erysiphales 1967; *Erysiphe* 1967; *Sphaerotheca* 1966.
Kable, P. F. *Tranzschelia* 1985.
Kakishima, M. *Ceraceopsora* 1984; *Nyssopsora* 1984; *Pileolaria* 1980, 1984; Ustilaginales 1982.
Kamal, A. *Mycovellosiella* 1985.
Kamat, M. N. *Phyllachora* 1978.
Kaneko, S. *Coleosporium* 1981; *Melampsora* 1982, 1984; *Melampsoridium* 1981, 1982, 1983; *Phragmidium* 1980; *Puccinia* 1968, 1973, 1983; *Pucciniastrum* 1980.

Kaosiri, T. *Phytophthora* 1980.
Kaplan, J. D. *Laetisaria* 1983.
Kar, A. K. *Cercospora* 1982; *Nectria* 1980.
Karakulin, B. P. Deuteromycotina 1950.
Karling, J. S. *Physoderma* 1950, 1977;
Plasmodiophora 1968; *Polymyxa* 1968;
Spongospora 1968; *Synchytrium* 1964.
Karr, G. W., Jr. *Stachybotrys* 1976.
Kasimatis, A. N. *Phomopsis* 1981.
Kastirr, U. *Cytospora* 1984.
Kaszonyi, S. *Blumeriella* 1966.
Katsuki, S. *Cercospora* 1965, 1982.
Katsuya, K. *Pileolaria* 1980; *Thekopsora* 1979.
Katumoto, K. *Cystotheca* 1973; *Dermatodothis* 1983; *Leiosphaerella* 1981; *Mycosphaerella* 1983.
Keane, P. J. *Aulographina* 1984; *Hendersonia* 1984; *Mycosphaerella* 1982, 1984.
Kendrick, B. *Chalara* 1975; Deuteromycotina 1979; *Dinemasporium* 1986; *Lepteutypa* 1985; *Myrothecium* 1980; *Pseudorobillarda* 1972, 1973; *Seiridium* 1985; *Stigmina* 1972; *Urohendersonia* 1971.
Kendrick, W. B. *Cyclaneusma* 1984; *Dactylaria* 1968, 1977; *Darkera* 1984; Deuteromycotina 1973, 1980; *Leptographium* 1962, 1965, 1980; *Lophophacidium* 1984; *Phacidiopycnis* 1984; *Phacidium* 1984; *Phialophora* 1973; *Potebniamyces* 1984; *Pseudophacidium* 1984.
Kennel, W. *Pezicula* 1984.
Kenneth, R. G. Peronosporales 1984.
Kern, F. D. *Gymnosporangium* 1964, 1973.
Kern, H. *Leucostoma* 1961.
Kessler, K. J., Jr. *Mycosphaerella* 1984, 1985.
Kesteren, H. A. van *Phoma* 1979, 1980, 1981; *Phyllosticta* 1971.
Khanna, A. *Neovossia* 1966, 1968.
Khoo, K. C. *Zimmermanniella* 1983.
Khulbe, R. D. *Pythium* 1984.
Kiffer, E. *Chalara* 1983.
Kile, G. A. *Armillaria* 1982, 1983.
Kim, W. K. *Ustilago* 1984.
Kimbrough, J. W. *Athelia* 1978; *Calonectria* 1986; *Corynelia* 1985; *Rhizoctonia* 1975, 1978.
Kirk, P. M. *Blakeslea* 1984; *Chalara* 1984; *Choanephora* 1984; *Cladosporium* 1984.
Kisimova-Horovitz, L. *Graphiola* 1982.
Kliejunas, J. T. *Phoma* 1985.
Knox-Davies, P. S. *Batcheloromyces* 1983, 1985.
Ko, W. H. *Phytophthora* 1985.
Kobayashi, T. *Ascocalyx* 1984; Ascomycotina 1970; *Cercospora* 1982; *Cytoplea* 1965; *Diaporthe* 1970; *Endothia* 1956; *Entomosporium* 1980; *Linocarpon* 1970; *Melanconis* 1971; *Phomatospora* 1982, 1983; *Plagiosphaera* 1982; *Seiridium* 1974; *Stagonospora* 1972; *Tubakia* 1979; *Valsa* 1970; *Volutella* 1980.
Kohn, L. M. *Cenangium* 1976; *Ciborinia* 1984; *Monilia* 1979; *Sclerotinia* 1979a, 1979b.
Kojwang, H. O. *Linospora* 1984.
Komagata, K. *Tilletiopsis* 1985.
Komatsu, M. *Trichoderma* 1976.
Kondo, H. *Ascocalyx* 1984.
Koponen, H. *Leptosphaeria* 1975; *Paraphaeosphaeria* 1975; *Phaeosphaeria* 1975.
Korf, R. P. Ascomycotina 1972, 1973; *Cenangium* 1976; *Ciborinia* 1959; *Itersonilia* 1960; *Monilia* 1979; *Pseudopezicula* 1986.

Korhonen, K. *Armillaria* 1978, 1980.
Kraenzlin, F. Aphyllophorales 1986; Ascomycotina 1983.
Kramer, C. L. *Protomyces* 1975; *Taphrina* 1974.
Krause, R. A. *Magnaporthe* 1972.
Krygier, B. B. *Tilletia* 1984.
Kuhlman, E. G. *Gibberella* 1982.
Kukkonen, I. *Anthracoidea* 1963, 1964.
Kulik, M. M. *Penicillium* 1968; *Phomopsis* 1984, 1985; *Pseudorobillarda* 1986.
Kurkela, T. *Linospora* 1984.
Kuter, G. A. *Phomopsis* 1985.
Laberry, R. *Uromyces* 1984.
Lai, Y.-q. *Microsphaera* 1982, 1983.
Lal, B. *Amerosporium* 1981; *Botryodiplodia* 1985.
Lal, S. *Sclerophthora* 1970.
Lamprecht, S. C. *Pleospora* 1984; *Stemphylium* 1984.
Langdon, R. F. N. *Cerebella* 1955; *Claviceps* 1954; *Sphacelotheca* 1978; *Sporisorium* 1978.
Langerfeld, E. *Synchytrium* 1984.
Lanquetin, P. *Aleurodiscus* 1985; *Vararia* 1975, 1976, 1977.
Latch, G. C. M. *Acremonium* 1984, 1985.
Latterell, F. M. *Hyalothyridium* 1984; *Marasmiellus* 1984; *Selenophoma* 1986.
Laundon, G. *Diplodia* 1984; *Microsphaeropsis* 1984.
Laundon, G. F. *Aecidium* 1967; *Botryosphaeria* 1973; *Kernkampella* 1975; *Maravalia* 1964; *Mycocentrospora* 1970; *Tranzschelia* 1975; Uredinales 1973.
LeClair, P. M. *Massaria* 1975; *Papulaspora* 1971.
Leach, C. M. *Epicoccum* 1972.
Leakey, C. L. A. *Dactuliophora* 1964.
Lemke, P. A. *Aleurodiscus* 1964, 1965.
Lenne, J. M. *Synchytrium* 1985.
Lentz, P. L. *Dicellomyces* 1976; *Marssonina* 1950; *Stereum* 1955.
Leonard, K. J. *Setosphaeria* 1974.
Leuchtmann, A. *Phaeosphaeria* 1984.
Levy, C. *Dactuliophora* 1986.
Lewis, B. G. *Mycocentrospora* 1980.
Lewis, R. M. *Diaporthe* 1959.
Liang, Z. R. *Phytophthora* 1984.
Lim, G. *Cercospora* 1980.
Lim, T. K. *Zimmermanniella* 1983.
Linderman, R. G. *Calonectria* 1972.
Lindquist, J. C. *Uropyxis* 1967.
Lindsey, J. P. Aphyllophorales 1978.
Linfield, C. *Tranzschelia* 1983.
Linnemann, G. Mucorales 1969.
Littlefield, L. J. Uredinales 1979.
Liu, B.-c. *Uncinula* 1984.
Liu, X.-j. *Cercosporidium* 1982.
Liu, Y. L. *Albugo* 1984.
Liyanage, A. de S. *Phytophthora* 1984.
Lock, N. Y. *Sirococcus* 1981.
Loeffler, W. *Dothidea* 1957.
Loerakker, W. M. *Laetisaria* 1982; *Leptosphaeria* 1981; *Phoma* 1981.
Lohman, M. L. *Nectria* 1943.
Lopes, M. C. Peronosporales 1982.
Lorenz, R. C. *Cystostereum* 1941.
Lorenzini, G. *Leptographium* 1977.
Loveless, A. R. *Sphacelia* 1964.
Lowe, J. L. *Fomes* 1957; *Poria* 1963, 1966.
Lozano, J. C. *Uromyces* 1984.
Lucas, L. T. *Marasmius* 1971.

242

Lucas, M. T. *Leptosphaeria* 1963, 1967, 1968; *Peronosporales* 1976, 1982.
Lundquist, J. E. *Lophodermium* 1984.
Lunghini, D. *Pleiochaeta* 1981.
Lunn, J. A. *Cunninghamella* 1983.
Luo, S.-b. *Uncinula* 1984.
Luttrell, E. S. *Ascomycotina* 1973; *Atkinsonella* 1984; *Balansia* 1984; *Bipolaris* 1977; *Catenophora* 1940; *Claviceps* 1984; *Cochliobolus* 1959, 1977; *Drechslera* 1977; *Epichloe* 1984; *Helminthosporium* 1963; *Leptosphaerulina* 1961; *Myriogenospora* 1977, 1982; *Pyrenophora* 1977; *Sphacelia* 1984.
Lyda, S. D. *Phymatotrichopsis* 1978.
Maas, J. L. *Dothiorella* 1984.
MacDonald, J. D. *Cylindrocarpon* 1981.
Maddock, S. E. *Pyrenopeziza* 1981.
Madelin, M. F. *Trichothecium* 1966.
Maffee, H. M. *Phomopsis* 1983.
Magasi, L. P. *Lophomerum* 1966.
Makela, K. *Leptosphaeria* 1975; *Mastigosporium* 1970; *Paraphaeosphaeria* 1975; *Phaeosphaeria* 1975; *Septoria* 1977.
Malloch, D. General 1981.
Manandhar, J. B. *Colletotrichum* 1986.
Manji, B. T. *Monilinia* 1982.
Manners, J. G. *Lachnellula* 1953.
Mantle, P. G. *Sphacelia* 1968.
Marasas, W. F. O. *Batcheloromyces* 1985; *Ceratocystis* 1980; *Fusarium* 1983, 1984; *Leptographium* 1981, 1983; *Neottiosporina* 1976; *Vizella* 1976.
Marlatt, R. B. *Cephaleuros* 1881.
Martin, B. *Rhizoctonia* 1987.
Martin, P. *Anthostomella* 1969; *Daldinia* 1969; *Hypoxylon* 1967, 1968, 1976; *Rosellinia* 1967; *Xylaria* 1970.
Martinez, A. T. *Rhizosphaera* 1983.
Mason, E. W. *Botryosporium* 1928.
Mathur, S. B. *Cochliobolus* 1973; *Curvularia* 1970; *Drechslera* 1973; *Myrothecium* 1973.
Matsushima, T. *Deuteromycotina* 1971, 1975.
Matthews, F. D. *Deightoniella* 1976.
McCain, A. H. *Phoma* 1985.
McCain, J. N. *Cumminsiella* 1982.
McGranahan, G. H. *Stegophora* 1984.
McKenzie, E. H. C. *Apiospora* 1976.
McKeown, B. M. *Ascochyta* 1985.
McNabb, R. F. R. *Exobasidium* 1962.
McPartland, J. M. *Phomopsis* 1983.
Mehrotra, B. S. *Blakeslea* 1968.
Mehta, S. C. *Neovossia* 1966.
Melendez, P. L. *Sphaeropsis* 1984.
Melnik, V. A. *Ascochyta* 1977.
Meloche, R. B. *Cladosporium* 1984.
Meng, Y. R. *Bremia* 1985.
Menzinger, W. *Botrytis* 1966.
Mer, G. S. *Pythium* 1984.
Meredith, D. S. *Mycosphaerella* 1970.
Merezhko, T. O. *Cytospora* 1984.
Messner, K. *Libertella* 1982.
Meyer, J. *Trichothecium* 1958.
Micales, J. A. *Cryphonectria* 1986; *Endothia* 1986.
Michaelides, J. *Deuteromycotina* 1979; *Myrothecium* 1980.
Michailides, T. J. *Tranzschelia* 1985.
Mihaljcevic, M. *Phomopsis* 1981, 1985.
Milholland, R. D. *Phomopsis* 1983.

Millar, C. S. *Lophodermella* 1986; *Lophodermium* 1978, 1980.
Miller, J. H. *Hypoxylon* 1961; *Myriangium* 1940.
Miller, S. G. *Rhytisma* 1987.
Millner, P. D. *Chaetomium* 1975.
Minter, D. W. *Arthrinium* 1985; *Cyclaneusma* 1983, 1986; *Elytroderma* 1979; *Hypoderma* 1983, 1986; *Leptostroma* 1980; *Lophodermium* 1978, 1980, 1981, 1982, 1983, 1986; *Nectria* 1980; *Ocotomyces* 1985; *Rhytisma* 1986.
Miranda, L. Rodrigues de *Geotrichum* 1977.
Mirza, F. *Cucurbitaria* 1968.
Misra, A. P. *Helminthosporium* 1974.
Mix, A. J. *Taphrina* 1949, 1954.
Mohanan, C. *Cylindrocladium* 1985.
Moline, H. E. *Microdochium* 1976.
Moller, W. J. *Eutypa* 1982, 1983; *Phomopsis* 1981.
Molnar, A. C. *Leptographium* 1965.
Monod, M. *Apiognomonia* 1983; *Gnomonia* 1983.
Monoson, H. L. *Uromyces* 1980.
Montemarini, C. *Endocalyx* 1962.
Moran, G. F. *Phytophthora* 1984.
Mordue, J. E. M. *Sclerotium* 1983; *Ustilaginales* 1984.
Morelet, M. *Venturia* 1985.
Moreno, G. *Lachnellula* 1985.
Morgan, D. J. *Botrytis* 1971.
Morgan-Jones, G. *Acremonium* 1982, 1987; *Cercospora* 1976, 1977; *Cryptocline* 1973; *Cytoplea* 1974; *Dinemasporium* 1971; *Discogloeum* 1971; *Epichloe* 1982; *Idriella* 1979; *Leptomelanconium* 1971; *Microsphaeropsis* 1974; *Monostichella* 1971; *Phaeoisariopsis* 1976, 1978; *Phoma* 1983, 1984, 1986, 1987; *Pseudorobillarda* 1972, 1973; *Pyrenochaeta* 1983; *Stachybotrys* 1976; *Stigmina* 1972.
Morgan-Jones, J. F. *Rhytisma* 1971.
Morquer, R. *Gliocladium* 1963.
Morris, E. F. *Phaeoisariopsis* 1968.
Morrison, D. J. *Armillaria* 1985.
Morrison, R. H. *Calonectria* 1972.
Morton, F. J. *Scopulariopsis* 1963.
Morton, H. L. *Phomopsis* 1983.
Moss, M. O. *Fusarium* 1984.
Mouchacca, J. *Microdochium* 1973.
Mouton, A. *Phymatotrichum* 1953.
Mueller, E. *Acantharia* 1965, 1984; *Ascocalyx* 1983; *Ascomycotina* 1954, 1962, 1973, 1975; *Atopospora* (1958)1959; *Biscogniauxia* 1986; *Ceratocystis* 1978; *Daldinia* 1986; *Dermatodothis* (1975)1976; *Diachora* 1986; *Diaporthopsis* 1955; *Didymella* 1952; *Herpotrichia* 1978, 1984; *Hypoxylon* 1986; *Idriella* 1983; *Lepteutypa* 1965; *Leptosphaeria* 1950, 1951; *Microdochium* 1980; *Monographella* 1977, 1984; *Nodulosphaeria* 1963; *Ophiobolus* 1952; *Ophiostoma* 1978; *Pleospora* 1951; *Seimatosporium* 1964.
Mukunya, D. M. *Phyllosticta* 1973.
Mulder, J. L. *Stenella* 1975, 1982.
Munk, A. *Ascomycotina* 1957.
Muntanola-Cvetkovic, M. *Phomopsis* 1981, 1985.
Murray, D. I. L. *Ceratobasidium* 1984; *Cryphonectria* 1985; *Endothia* 1985; *Rhizoctonia* 1982.
Muthumary, J. *Coryneum* 1986.
Muthyalu, G. *Pyrenopeziza* 1978.
Nag Raj, T. R. *Apostrasseria* 1983; *Catenophora*

1977; *Chalara* 1975; *Cryptosporella* 1981;
Cyclaneusma 1984; *Darkera* 1984;
Deuteromycotina 1979; *Dinemasporium* 1986;
Harknessia 1981; *Lepteutypa* 1985;
Lophophacidium 1984; *Monochaetia* 1985, 1986;
Pestalosphaeria 1985; *Pestalotia* 1985;
Pestalotiopsis 1986, 1985a, 1985b, 1985c;
Phacidiopycnis 1984; *Phacidium* 1984; *Phomopsis* 1974; *Potebniamyces* 1984; *Pseudophacidium* 1984; *Pseudorobillarda* 1972, 1973;
Seimatosporium 1986; *Seiridium* 1985;
Urohendersonia 1971.
Nagakubo, T. *Curvularia* 1985.
Nagasawa, E. *Ciborinia* 1984.
Naik, D. M. *Dactuliophora* 1986.
Nannfeldt, J. A. *Anthracoidea* 1979; *Exobasidium* 1981.
Narayanaswamy, T. *Waitea* 1984.
Narendra, D. V. *Physalospora* 1977.
Neergaard, P. *Alternaria* 1945; *Cochliobolus* 1973; *Drechslera* 1973.
Neergard, P. *Myrothecium* 1973.
Negrean, G. Peronosporales 1983.
Negru, A. *Septoria* 1973.
Nelson, P. E. *Fusarium* 1975, 1981, 1983, 1984; *Idriella* 1956.
Newhook, F. J. *Phytophthora* 1972, 1978.
Newsted, W. J. *Sclerotium* 1985.
Newton, A. C. *Puccinia* 1985.
Ngala, G. N. *Sarocladium* 1983.
Nguyen, T. H. *Myrothecium* 1973.
Nickerson, N. L. *Exobasidium* 1984.
Nielsen, J. *Ustilago* 1984, 1985.
Niemela, T. *Phellinus* 1984.
Nirenberg, H. *Fusarium* 1976, 1981, 1982.
Nirenberg, H. I. *Pseudocercosporella* 1981.
Nishigaki, H. *Phragmidium* 1980.
Noble, M. *Gloeotinia* 1954; *Septoria* 1970.
Nomura, Y. *Microsphaera* 1983; *Uncinula* 1983.
Noviello, C. *Phytophthora* 1983.
Nyland, G. *Tilletiopsis* 1950.
O'Brien, M. J. *Angiosorus* 1974.
O'Donnell, K. L. Mucorales 1979.
Oberwinkler, F. *Graphiola* 1982; *Herpobasidium* 1984; *Insolibasidium* 1984.
Occhinea, E. M. *Sphaerotheca* 1985.
Ofosu-Asiedu, A. *Pseudophaeolus* 1975.
Ogawa, J. M. *Monilinia* 1982; *Tranzschelia* 1985.
Ogoshi, A. *Erythricium* 1985; *Rhizoctonia* 1979, 1983, 1984, 1985; *Waitea* 1985.
Oguchi, T. *Lachnellula* 1980, 1981.
Okada, G. *Endocalyx* 1984.
Olchowecki, A. *Ceratocystis* 1974.
Old, K. M. *Cryphonectria* 1985; *Endothia* 1985; *Phytophthora* 1984.
Olive, L. S. *Dicellomyces* 1945; *Itersonilia* 1952.
Ondrej, M. *Cercospora* 1984; *Deightoniella* 1984; *Ovularia* 1972; *Pollaccia* 1984.
Oniki, M. *Erythricium* 1985; *Rhizoctonia* 1979, 1983, 1985; *Waitea* 1985.
Onions, A. H. S. *Paecilomyces* 1967; *Penicillium* 1984.
Ono, Y. *Chaconia, Goplana* 1983; *Maravalia* 1984; *Olivea* 1983.
Onofri, S. *Dactylaria* 1984; *Pleiochaeta* 1981.
Oorschot, C. A. N. van *Cordana* 1983; *Dactylaria* 1983, 1985; *Dichotomophthora* 1983.
Orlos, H. Aphyllophorales 1967.

Orton, C. R. *Phyllachora* 1944.
Otani, Y. *Phaeosphaeria* 1976.
Ou, S. H. *Magnaporthe* 1985; *Nakataea* 1985; *Pyricularia* 1985; *Sclerotium* 1985.
Ouellette, G. B. *Lophomerum* 1966.
Overholts, L. O. Aphyllophorales 1953.
Palmer, M. A. *Sphaeropsis* 1985.
Palti, J. Peronosporales 1984.
Pande, A. H. *Phyllachora* 1978.
Pandy, P. C. *Melampsoridium* 1972.
Papavizas, G. C. *Aphanomyces* 1974.
Parberry, D. G. *Phyllachora* 1967, 1971.
Park, R. F. *Hendersonia* 1984; *Mycosphaerella* 1982, 1984.
Parker, A. K. *Rhabdocline* 1969.
Parkinson, V. O. *Microdochium* 1980, 1981; *Monographella* 1981.
Parmelee, J. A. *Erysiphales* 1977; *Erysiphe* 1977; *Gymnosporangium* 1965, 1971; *Puccinia* 1981, 1986; *Strasseria* 1979.
Parmeter, J. R., Jr. *Rhizoctonia* 1970.
Partridge, A. D. *Leptographium* 1983.
Pascoe, I. G. *Vararia* 1984.
Paterson, R. R. *Penicillium* 1984.
Patil, M. S. *Diatrype* 1983.
Patil, S. D. *Diatrype* 1983.
Patton, R. F. *Mycosphaerella* 1983.
Pavgi, M. S. *Neovossia* 1973, 1979.
Payak, M. M. *Neovossia* 1966, 1968; *Sclerophthora* 1970.
Pearson, R. C. *Pseudopezicula* 1986.
Pegg, G. F. *Verticillium* 1974.
Pegler, D. N. Aphyllophorales 1973; *Crinipellis* 1978, 1983; *Inonotus* 1964; *Marasmiellus* 1977, 1983; *Marasmius* 1977, 1983; *Mycena* 1983.
Pennycook, S. R. *Botryosphaeria* 1985; *Colletotrichum* 1983; *Fusicoccum* 1985.
Penrose, L. J. *Monilinia* 1976.
Peredo, H. *Cyclaneusma* 1983.
Perez-Silva, E. *Daldinia* 1973; *Xylaria* 1975.
Peries, O. S. *Phytophthora* 1984.
Perrin, R. *Nectria* 1976, 1983.
Petch, T. *Myriangium* 1924.
Petersen, L. J. *Geotrichum* 1972.
Peterson, R. S. *Cronartium* 1973; *Phragmidium* 1962.
Petrak, F. *Discula* 1962, (1970)1971; *Dothichiza* 1957; *Macrophoma* 1927; *Ophiovalsa* 1965; *Sphaeropsis* 1927.
Petrini, L. *Biscogniauxia* 1985; *Daldinia* 1985; *Hypoxylon* 1985; *Nodulisporium* 1985; *Rosellinia* 1985; *Xylaria* 1985.
Petrini, L. E. *Biscogniauxia* 1986; *Daldinia* 1986; *Hypoxylon* 1984, 1986.
Petrini, O. *Biscogniauxia* 1985; *Cryptocline* 1984; *Daldinia* 1985; *Hypoxylon* 1985; *Leiosphaerella* 1983; *Nodulisporium* 1985; *Rosellinia* 1985; *Xylaria* 1985.
Petrov, M. *Phomopsis* 1981, 1985.
Pfister, D. H. Ascomycotina 1982.
Phaff, H. J. *Geotrichum* 1985.
Phitakpraiwan, P. *Cercospora* 1980.
Pinon, J. *Melampsora* 1973.
Pirozynski, K. A. *Darkera* 1975; *Herpotrichia* 1972; *Linocarpon* 1972; *Microdochium* 1972; *Seimatosporium* 1970; *Uncinula* 1965.
Pitt, J. I. *Aspergillus* 1985; *Eupenicillium* 1974; *Monascus* 1983; *Paecilomyces* 1985;

Penicillium 1974, 1979, 1985; *Talaromyces* 1979.
Pitt, J. I., eds. *Aspergillus* 1985; *Penicillium* 1985.
Plaats-Niterink, A. J. van der *Pythium* 1981.
Podger, F. D. *Phytophthora* 1972.
Poelt, J. *Ramularia* 1983.
Pollack, F. G. *Cercospora* 1987; *Deightoniella* 1976; *Microdochium* 1976; *Mycocentrospora* 1974; *Neohendersonia* 1974; *Phloeospora* 1974; *Septogloeum* 1974; *Sphaceloma* 1973.
Ponnappa, K. M. *Phomopsis* 1974.
Pound, G. S. *Albugo* 1963.
Pouzar, Z. *Biscogniauxia* 1979, 1986.
Powell, C. C. *Oidium* 1981.
Powell, J. M. *Cronartium* 1976.
Pramer, D. *Dactylaria* 1977.
Prasad, N. *Ravenelia* 1972.
Prasha, I. B. *Endocalyx* 1982.
Price, D. *Tranzschelia* 1983.
Price, T. V. *Microdochium* 1986.
Printz, H. *Cephaleuros* 1940.
Prior, C. *Gibberella* 1984.
Pugh, G. J. F. *Aureobasidium* 1971.
Punithalingam, E. *Ascochyta* 1979, 1981, 1985; *Ascochytulina* 1985; *Dendroseptoria* 1981; *Dimeriella* 1984; *Epicoccum* 1972; *Guignardia* 1974; *Leptodothiorella* 1974; *Macrophomina* 1982; *Massaria* 1969; *Neokellermania* 1981; *Phaeoseptoria* 1980, 1981; *Phomopsis* 1975, 1979, 1980, 1981; *Phyllosticta* 1982; *Septoria* 1965, 1981; *Sphaeropsis* 1969.
Punja, Z. K. *Athelia* 1982, 1985; *Sclerotium* 1982, 1985.
Punter, D. *Magnaporthe* 1984; *Sclerotium* 1984.
Pyykko, M. *Sclerotinia* 1980.
Quinn, J. A. *Oidium* 1981.
Raabe, R. D. *Armillaria* 1962.
Radulescu, E. *Septoria* 1973.
Rahman, M. A. *Ramichloridium* 1983; *Stenella* 1983; *Veronaea* 1983.
Rai, B. *Mycovellosiella* 1985.
Rai, M. K. *Phoma* 1982, 1983, 1984.
Raitviir, A. *Lachnellula* 1970.
Rajak, R. C. *Phoma* 1982, 1983, 1984.
Rajendren, R. B. *Muribasidiospora* 1968, 1970.
Ramachar, P. *Oidium* 1983; *Physopella* 1958.
Rambelli, A. *Pleiochaeta* 1981.
Ramirez, C. *Penicillium* 1982; *Rhizosphaera* 1983.
Rao, D. *Lacellinopsis* 1973.
Rao, K. N. *Kuehneola* 1985.
Rao, K. V. S. *Moesziomyces* 1983.
Rao, P. N. *Dichotomophthora* 1966.
Rao, R. *Ovularia* 1968.
Rao, V. *Myrothecium* 1983; *Pithomyces* 1986.
Rao, V. G. *Alternaria* 1969; *Calothyriopsis* 1980; *Peronospora* 1968; *Phyllosticta* 1964; *Physalospora* 1977.
Rao, V. P. *Claviceps* 1985.
Raper, K. B. *Aspergillus* 1965, 1978; *Emericella* 1978; *Penicillium* 1949.
Rappaz, F. *Eutypa* 1984.
Rathaiah, Y. *Pyricularia* 1980.
Rattan, S. S. *Scytinostroma* 1974.
Rawla, G. S. *Duosporium* 1977; *Ramulispora* 1973.
Rawlinson, C. J. *Pyrenopeziza* 1978, 1984.
Reddy, M. S. *Protomyces* 1975.
Redfern, D. B. *Plectophomella* 1981.
Redhead, S. A. *Cristulariella* 1975; *Mycopappus* 1985.
Reid, D. A. *Dicellomyces* 1976.
Reid, J. *Atropellis* 1966; *Ceratocystis* 1974; *Darkera* 1975; *Magnaporthe* 1984; *Phacidium* 1962; *Rhabdocline* 1969; *Sclerotium* 1984; *Therrya* 1961.
Reifschneider, F. J. B. *Sphaerotheca* 1985.
Reitsma, J. *Cylindrocladium* 1950.
Renfro, B. L. *Sclerophthora* 1970.
Ribeiro, O. K. *Phytophthora* 1978.
Richardson, M. J. *Sclerotium* 1970.
Rifai, M. A. *Trichoderma* 1968, 1969; *Trichothecium* 1966.
Rimpau, R. H. *Drepanopeziza* 1961; *Gloeosporidiella* 1961; *Marssonina* 1961.
Rishbeth, J. *Armillaria* 1982, 1986.
Rizwi, M. A. *Coryneum* 1980.
Roane, M. K. *Diaporthopsis* 1983; *Endothia* 1986.
Roberts, D. W. *Fusarium* 1983.
Robertson, G. I. *Pythium* 1980.
Robinson-Jeffrey, R. C. *Ophiostoma* 1968.
Rodriguez, R. *Sphaeropsis* 1984.
Roe, G. M. *Xylaria* 1984.
Roelfs, A. P., eds. Uredinales 1984.
Rogers, J. D. *Biscogniauxia* 1975; *Cryptosphaeria* 1984; *Diatrype* 1982, 1983, 1984; *Diatrypella* 1982, 1984; *Eutypa* 1982, 1984; *Eutypella* 1982, 1984; *Hypoxylon* 1972, 1975, 1979, 1986; *Nodulisporium* 1972; *Xylaria* 1983, 1984, 1985, 1986.
Rogerson, C. T. *Cochliobolus* 1959; *Nectria* 1984; *Nectriella* 1984.
Rohringer, R. *Ustilago* 1984.
Roll-Hansen, F. *Armillaria* 1985; *Melampsoridium* 1981.
Roll-Hansen, H. *Melampsoridium* 1981.
Rosenberger, D. A. *Phomopsis* 1982.
Rossi, A. E. *Hyalothyridium* 1984; *Marasmiellus* 1984; *Selenophoma* 1986.
Rossmann, A. Y. *Calonectria* 1983, 1979a, 1979b; *Nectria* 1979, 1983; *Nectriella* 1984.
Rostam, S. *Leveillula* 1984.
Rouch, J. *Gliocladium* 1963.
Roux, C. *Dothistroma* 1984; *Leptosphaerulina* 1986; *Lophodermium* 1984; *Pithomyces* 1986.
Roy, A. *Phellinus* 1979.
Rykard, D. M. *Atkinsonella* 1984; *Balansia* 1984; *Claviceps* 1984; *Epichloe* 1984; *Myriogenospora* 1982; *Sphacelia* 1984.
Ryvarden, L. Aphyllophorales 1973, 1975, 1976, 1978, 1980, 1981, 1984, 1986.
Sabet, K. A. *Marasmiellus* 1970.
Saccas, A. M. *Rosellinia* 1956.
Saho, H. *Peridermium* 1981.
Sakai, R. *Rhizoctonia* 1979; *Waitea* 1985.
Salogga, D. S. *Discula* 1983.
Sampson, M. G. *Urocystis* 1985.
Samra, A. S. *Marasmiellus* 1970.
Samson, R. A. *Aspergillus* 1979, 1984, 1985; *Cunninghamella* 1969; *Eupenicillium* 1983; *Microdochium* 1973; *Moniliophthora* 1978; *Paecilomyces* 1974; *Penicillium* 1976, 1983, 1985; *Talaromyces* 1971.
Samuels, G. J. *Acremonium* 1976, 1984; *Botryosphaeria* 1985, 1986; *Ceratocystis* 1978; *Cylindrocarpon* 1978; *Fusicoccum* 1985, 1986; *Herpotrichia* 1978; *Microdochium* 1983;

Monographella 1983, 1984; Nectria 1973, 1976, 1977, 1978, 1979, 1982, 1984, 1985; Nectriella 1979, 1984; Ophiostoma 1978; Volutella 1977.
Samuelson, D. A. Corynelia 1985.
Sanders, P. L. Rhizoctonia 1980.
Santesson, R. Cephaleuros 1952.
Sarbhoy, A. K. Aspergillus 1985; Chaetomella 1976, 1984.
Sasaki, K. Phomatospora 1982; Plagiosphaera 1982; Seiridium 1974; Tubakia 1979.
Sato, S. Aecidium 1985; Ceraceopsora 1984; Nyssopsora 1984; Pileolaria 1980, 1984; Thekopsora 1976, 1979.
Sato, T. Aecidium 1985; Ceraceopsora 1984; Nyssopsora 1984; Pileolaria 1984.
Savile, D. B. O. Arcticomyces 1959; Chrysomyxa 1950; Entyloma 1947; Exobasidium 1959; Puccinia 1973, 1979, 1981, 1985.
Savulescu, O. Bremia 1962.
Savulescu, T. Uredinales 1953; Ustilaginales 1957.
Scala, F. Phytophthora 1983.
Scharpf, R. F. Cytospora 1983.
Schauz, K. Tilletia 1985.
Scheffer, R. J. Ceratocystis 1984; Ophiostoma 1984.
Scheinpflug, H. Didymosphaeria 1958.
Schenck, S. Dactylaria 1977.
Schieber, E. Hemileia 1984.
Schipper, M. A. A. Rhizopus 1984.
Schlaepfer-Bernard, E. Ascocalyx 1968; Godronia 1968.
Schlesser, P. E. Uromyces 1980.
Schmitthenner, A. F. Phomopsis 1985.
Schneider, R. Kabatina 1966, 1976; Pyrenochaeta 1979, 1984.
Schoen, J. F. Sclerotium 1983.
Schoknecht, J. D. Pesotum 1973.
Schol-Schwarz, M. B. Cerebella 1959; Epicoccum 1959; Phialophora 1970.
Schoulties, C. L. Calonectria 1983, 1986.
Schuepp, H. Leptotrochila 1959; Pseudopeziza 1959.
Schumacher, T. Ciboria 1978; Gloeotinia 1979.
Schwarz, M. R. Strasseria 1983.
Schwarz, R. Pyrenochaeta 1979.
Scott, D. B. Magnaporthe 1983.
Scott, W. W. Aphanomyces 1961.
Seaver, F. J. Ascomycotina 1928, 1951; Phyllosticta 1922.
Seeler, E. V., Jr. Thyronectria 1940.
Seemuller, E. Fusarium 1968.
Seifert, K. Gliocladium 1985; Nectria 1985; Tubercularia 1985.
Sekar, G. Herpotrichia 1980.
Seneca, E. D. Marasmius 1971.
Senula, A. Cryptosporiopsis 1985.
Seshadri, V. S. Phyllachora 1978.
Seth, H. K. Chaetomium (1970) 1972.
Shahin, E. A. Dothiorella 1980.
Shahjahan, A. K. M. Mycovellosiella 1981.
Sharma, A. K. Uncinula 1985.
Sharma, J. K. Cylindrocladium 1985.
Sharma, M. P. Endocalyx 1982; Lophodermium 1982, 1983; Sclerotinia 1983.
Sharma, N. D. Phomopsis 1979.
Sharma, R. Lophodermium 1983.
Shaw, C. G. Hypodermella 1966; Peronosclerospora 1978, 1980; Peronospora 1959; Sclerospora 1978.
Shaw, C. G. III Armillaria 1985.
Shaw, D. E. Verrucispora 1967.
Shearer, C. A. Nakataea 1979.
Sherwood, M. A. Coccomyces 1980.
Sherwood, R. T. Rhizoctonia 1980.
Sherwood-Pike, M. Rhabdocline 1986.
Sherwood-Pike, M. A. Ascomycotina 1985.
Shipton, W. A. Cunninghamella 1983.
Shoemaker, R. A. Bipolaris 1959; Botryosphaeria 1964; Drechslera 1959, 1962; Entodesmium 1984; Lepteutypa 1965; Leptosphaeria 1984; Lophophacidium 1984; Massaria 1975; Nodulosphaeria 1984; Ophiobolus 1976; Paraphaeosphaeria 1985; Pestalosphaeria 1981; Pyrenophora 1966; Seimatosporium 1964, 1970.
Siegel, M. R. Acremonium 1985.
Siepmann, R. Mucorales 1976.
Sigler, L. Briosia 1976; Deuteromycotina 1980; Geotrichum 1976; Oospora 1976.
Simmons, E. G. Alternaria 1967, 1981, 1982, 1986; Embellisia 1971, 1983; Pleospora 1969, 1985, 1986; Stemphylium 1967, 1969, 1985; Ulocladium 1967.
Simpson, J. Pythium 1984.
Simpson, J. A. Pestalosphaeria 1981.
Sinclair, J. B. Colletotrichum 1986; Dactuliophora 1986; Phakopsora 1982.
Sinclair, J. B., eds. Macrophomina 1977.
Singer, R. Crinipellis 1976; Marasmiellus 1973; Marasmius 1976.
Singh, B. Botryosphaeria 1986; Fusicoccum 1986.
Singh, R. A. Neovossia 1973, 1979.
Singh, S. Melampsoridium 1972.
Sisterna, M. N. Helminthosporium 1984.
Sivanesan, A. Acantharia 1984; Ascomycotina 1984; Bipolaris 1985; Cercospora 1985; Cochliobolus 1985; Curvularia 1985; Exserohilum 1984; Herpotrichia 1971; Marasmius 1986; Microcyclus 1970, 1975; Microdochium 1981; Microsphaera 1971; Monographella 1981; Mycosphaerella 1985; Myriogenospora 1986; Venturia 1977, 1985; Vizella 1973.
Sivasithamparam, K. Phialophora 1975.
Skidmore, D. I. Bremia 1985.
Skirgiello, A. Aphyllophorales 1967.
Skolko, A. J. Alternaria 1944.
Skotland, C. B. Eutypa 1982.
Smalley, E. B. Stegophora 1984.
Smedegard-Petersen, V. Pyrenophora 1971, 1978.
Smerlis, E. Godronia 1969.
Smiley, R. W. Leptosphaeria 1984.
Smit, W. A. Batcheloromyces 1983.
Smith, A. M. Leptosphaeria 1972.
Smith, G. Paecilomyces 1957; Scopulariopsis 1963.
Smith, H. C. Verticillium 1965.
Smith, J. D. Nectriella 1984.
Smith, J. E., eds. Fusarium 1984.
Smith, M. T. Geotrichum 1977.
Smith, R. B. Potebniamyces 1981.
Smith, R. S., Jr. Phoma 1985.
Snell, W. H. General 1971.
Snider, R. D. Taphrina 1974.
Snow, G. A. Cronartium 1977.
Snyder, W. C. Phoma 1985.
Sobers, E. K. Calonectria 1972.
Solheim, W. G. Peronospora 1973.

Somal, B. S. *Curvularia* 1976.
Sonoda, R. M. *Monilinia* 1982.
Sontirat, P. *Cercospora* 1980.
Sowell, G., Jr. *Itersonilia* 1960.
Sparrow, F. K. General 1973; *Olpidium* 1960; *Synchytrium* 1973.
Spaulding, P. *Helicobasidium* 1961.
Spear, R. N. *Mycosphaerella* 1983.
Spencer, D. M., ed. Erysiphales 1978; Peronosporales 1981.
Spielman, L. J. *Cytospora* 1985; *Valsa* 1985.
Spiers, A. G. *Drepanopeziza* 1983; *Marssonina* 1983.
Spooner, B. M. *Chalara* 1984; *Encoelia* 1985; *Gibberella* 1984; *Melanopsichium* 1985.
Sprague, R. Deuteromycotina 1950; *Septoria* 1950.
Srivastava, G. *Amerosporium* 1981.
Srivastava, R. C. *Botryodiplodia* 1985.
Stafleu, F. A. General 1976-1985.
Staley, J. M. *Lophodermium* 1975, 1978.
Stalpers, J. A. Aphyllophorales 1980; *Laetisaria* 1982; *Moniliophthora* 1978; *Rhizopus* 1984.
Stamps, D. J. *Phytophthora* 1978.
States, J. S. *Aspergillus* 1982.
Stavely, J. R. *Erysiphe* 1966.
Stephan, B. R. *Lophodermium* 1973.
Stevens, R. B., ed. General 1981.
Stevenson, J. A. *Chaetoseptoria* 1946; *Sphaceloma* 1971.
Steyaert, R. L. *Ganoderma* 1962, 1980; *Pestalotia* 1949; *Pestalotiopsis* 1961.
Stillwell, M. A. *Amylostereum* 1966.
Stipes, R. J. *Cryphonectria* 1986; *Endothia* 1982, 1986.
Stockwell, V. O. *Tilletia* 1986.
Stolk, A. C. *Chaetomella* 1963; *Eupenicillium* 1983; *Penicillium* 1976, 1983; *Talaromyces* 1971.
Stone, J. *Phaeocryptopus* 1985.
Stone, J. K. *Rhabdocline* 1986.
Stover, P. R. *Cylindrocarpon* 1974.
Stowell, E. A. *Entomosporium* 1966, 1967.
Streets, R. B. *Phymatotrichopsis* 1973.
Stroo, H. F. *Entomosporium* 1985.
Subhedar, A. W. *Calothyriopsis* 1980.
Subramanian, C. V. *Discosia* 1974; *Herpotrichia* 1980; *Monographella* 1978; *Nectria* 1984; *Thyronectria* 1984.
Suggs, E. G. *Setosphaeria* 1974.
Sukapure, R. K. *Septoria* 1963.
Sundstrom, K.-R. *Exobasidium* 1964.
Sussman, A. S., eds. General 1973.
Sutherland, J. R. *Sirococcus* 1981.
Suto, Y. *Hypoderma* 1983; *Phomatospora* 1983.
Sutton, B. C. *Chaetomella* 1976; *Chaetoseptoria* 1964; *Coryneum* 1975, 1980, 1986; *Dendrophoma* 1965; Deuteromycotina 1973, 1977, 1980; *Dilophospora* 1974; *Dinemasporium* 1969; General 1983; *Harknessia* 1971; *Libertella* 1982; *Lidophia* 1974; *Melanconium* 1964; *Microdochium* 1972, 1976, 1986; *Microsphaeropsis* 1971, 1974, 1980; *Monochaetia* 1969; *Monochaetiellopsis* 1977; *Mycoleptodiscus* 1976, 1985; *Mycovellosiella* 1981; *Neohendersonia* 1974, 1975; *Neottiosporina* 1974, 1976, 1985; *Pestalotia* 1969; *Phaeocytostroma* 1964; *Phloeospora* 1974; *Phoma* 1964; *Plectophomella* 1981; *Pyrenopeziza* 1978; *Septogloeum* 1974,

1984; *Sphaceloma* 1973; *Stegonsporium* 1981; *Stenocarpella* 1964; *Stigmina* 1972; *Trimmatostroma* 1978; *Tubakia* 1973; *Wojnowicia* 1975; *Zyxiphora* 1980.
Swart, H. J. *Corynespora* 1985; *Lepteutypa* 1973; *Ophiodothella* 1982; *Pestalotia* 1974; *Seimatosporium* 1974; *Vizella* 1971, 1975.
Swinburne, T. R. *Nectria* 1977.
Sydow, H. *Macrophoma* 1927; *Sphaeropsis* 1927; Uredinales 1904-1924.
Sydow, P. Uredinales 1904-1924.
Taber, R. A. *Annellophora* 1985.
Taga, M. *Curvularia* 1985.
Talbot, P. H. B. *Amylostereum* 1977; Aphyllophorales 1973; *Ceratobasidium* 1965, 1967, 1970, 1971, 1980; *Stereum* 1954; *Thanatephorus* 1965, 1967, 1970; *Waitea* 1962.
Talde, U. K. *Nectria* 1979.
Tanaka, S. *Rhizoctonia* 1985; *Waitea* 1985.
Tanda, S. *Microsphaera* 1983; *Uncinula* 1983.
Tandon, M. P. *Amerosporium* 1981.
Tariq, V.-N. *Sclerotinia* 1985.
Tarran, J. *Monilinia* 1976.
Taylor, G. S. *Cryptosporiopsis* 1983.
Taylor, J. *Dothichiza* 1959.
Teetro-Barsch, G. H. *Fusarium* 1983.
Tehon, L. R. *Lophodermium* 1935.
Terra, P. *Vararia* 1976.
Terrell, E. E. *Ustilago* 1982.
Teterevnikova-Babayan, D. N. *Septoria* 1976, 1985.
Thakur, R. P. *Claviceps* 1985; *Moesziomyces* 1983.
Thaung, M. M. *Cercospora* 1984; *Newinia* 1973.
Thind, K. S. *Daldinia* 1978; *Duosporium* 1961; *Hypoxylon* 1976; *Rosellinia* 1975.
Thirumalachar, M. J. *Angiosorus* 1974; *Balansia* 1961, 1973; *Hemileia* 1947; *Melanotaenium* 1953; *Septoria* 1963.
Thom, C. *Penicillium* 1949.
Thomas, M. D. *Microdochium* 1984.
Thompson, A. H. *Pleospora* 1984; *Stemphylium* 1984.
Thyr, B. D. *Hypodermella* 1966.
Tiffany, L. H. *Diaporthe* 1959; *Diatrype* 1965.
Tilak, S. T. *Botryosphaeria* 1978; *Nectria* 1979.
Tokeshi, H. *Cercospora* 1978.
Tomilin, B. A. *Mycosphaerella* 1979.
Tommerup, I. C. *Apiospora* 1976.
Torkelson, A.-E. *Cenangium* 1977.
Tortolero, O. *Brasiliomyces* 1984.
Toussaint, T. A. *Fusarium* 1975, 1981, 1983, 1984.
Traquair, J. A. *Cladosporium* 1984.
Tredick, J. *Geotrichum* 1985.
Trigaux, G. *Encoelia* 1985.
Trione, E. J. *Tilletia* 1988, 1986.
Trujillo, E. E. *Selenophoma* 1986.
Tsao, P. H., eds. *Phytophthora* 1983.
Tsuda, M. *Curvularia* 1985; *Duosporium* 1982; *Magnaporthe* 1978; *Pseudocochliobolus* 1985.
Tu, C. C. *Athelia* 1978; *Rhizoctonia* 1975, 1978.
Tubaki, K. *Endocalyx* 1984; *Tilletiopsis* 1952; *Tubakia* 1971.
Tuite, J. General 1969.
Tullis, E. C. *Neovossia* 1952.
Tulloch, M. *Epicoccum* 1972; *Myrothecium* 1972.
Tyagi, R. N. S. *Kernkampella* 1974; *Ravenelia* 1972.
Tylutki, E. E. *Rhizina* 1979.
Tzavella-Klonari, K. *Macrophoma* 1979.
Udagawa, S. *Chaetomium* 1960, 1969; *Eupenicillium*

1973; *Pyricularia* 1978.
Uecker, F. A. *Dothiorella* 1984; *Pseudorobillarda* 1986.
Ueyama, A. *Curvularia* 1985; *Duosporium* 1982; *Magnaporthe* 1978; *Pseudocochliobolus* 1985.
Ui, T. *Rhizoctonia* 1979, 1983.
Ullasa, B. A. *Balansia* 1969.
Ullrich, R. C. *Armillaria* 1978, 1979, 1980.
Ulvinen, T. *Urocystis* 1980.
Upadhyay, H. P. *Ceratocystis* 1981.
Urban, Z. *Cytospora* 1958; *Leucostoma* 1958; *Valsa* 1958.
Vajna, L. *Trichoderma* 1983.
Valder, P. G. *Helicobasidium* 1958.
Vallavieille, C. de *Phytophthora* 1984.
Vaneb, S. *Ramularia* 1974.
Vanev, S. G. *Mycoleptodiscus* 1983.
Vanky, K. *Moesziomyces* 1977, 1986; Ustilaginales 1985.
Vann, S. R. *Annellophora* 1985.
Vassiljevsky, N. I. Deuteromycotina 1950.
Vasudeva, R. S. *Cercospora* 1963.
Vegh, I. *Marssonina* 1983.
Velastegui, J. *Marssonina* 1983.
Venkata, Rao A. *Waitea* 1984.
Verhoeff, K. *Botrytis* 1980.
Verma, B. L. *Pythium* 1984.
Viala, G. *Gliocladium* 1963.
Vital, A. F. *Phyllosticta* 1952.
Vloten, H. van *Cenangium* 1953.
Voss, E. G., et al., General 1983.
Vries, G. A. de *Cladosporium* 1952.
Vukojevic, J. *Phomopsis* 1985.
Walker, J. *Cryphonectria* 1985; *Didymella* 1983; *Dilophospora* 1974, 1980; *Endothia* 1985; *Gaeumannomyces* 1972, 1975, 1980; *Leptosphaeria* 1972; *Lidophia* 1974, 1980; *Linocarpon* 1980; *Linospora* 1980; *Ophiobolus* 1980; *Plagiosphaera* 1980; *Wettsteinina* 1984.
Wall, C. J. *Mycocentrospora* 1980.
Wall, E. *Aulographina* 1984.
Waller, J. M. *Marasmius* 1986; *Myriogenospora* 1986.
Wang, C.-g. *Sphaeropsis* 1985.
Wang, Y. X. *Albugo* 1984.
Wang, Y.-c. *Gymnosporangium* 1984.
Waraitch, K. S. *Hypoxylon* 1976.
Warcup, J. H. *Ceratobasidium* 1967, 1971, 1980; *Thanatephorus* 1967; *Waitea* 1962.
Wargo, P. M. *Armillaria* 1985.
Warmelo, K. T. van *Stegonsporium* 1981.
Warren, T. B. *Marasmius* 1971.
Washington, W. S. *Vararia* 1984.
Waterhouse, G. M. *Peronosclerospora* 1980; Peronosporales 1973; *Phytophthora* 1963, 1970, 1978; *Pseudoperonospora* 1981; *Pythium* 1967, 1968.
Watkins, G. M. *Eremothecium* 1981.
Watling, R. *Armillaria* 1982, 1983, 1985.
Watson, A. J. *Nectria* 1943.
Watson, A. K. *Urocystis* 1985.
Webber, J. F. *Phomopsis* 1984.
Weber, G. *Tilletia* 1985.
Webster, J. *Dinemasporium* 1955; *Leptosphaeria* 1967; *Phomatospora* 1955; *Septogloeum* 1984; *Trichoderma* 1968.
Webster, R. K. *Magnaporthe* 1972.
Wehmeyer, L. E. *Clathrospora* 1961; *Diaporthe* 1933; *Diaporthopsis* 1933; *Melanconis* 1941; *Pleospora* 1961.
Wei, A. J. *Agaricodochium* 1983.
Weidemann, G. J. *Phyllosticta* 1982.
Weiler, R. *Pezicula* 1984.
Weitzman, I. *Cunninghamella* 1984.
Welacky, T. W. *Botryosporium* 1983.
Welden, A. L. *Stereum* 1971; *Vararia* 1965.
Wellman, A. M. *Rhizopus* 1973.
Wells, H. *Chaetomium* 1973.
Weresub, L. K. *Butlerelfia* 1980; *Oidium* 1973; *Papulaspora* 1971.
Westhuizen, G. C. A. van der *Colletotrichum* 1983, 1984, 1985; *Phyllosticta* 1980; *Pseudophaeolus* 1973; *Sclerospora* 1977.
Whalley, A. J. S. *Biscogniauxia* 1985; *Hypoxylon* 1973, 1983, 1984; *Nodulisporium* 1975.
Wheeler, B. E. J. *Septoria* 1965.
White, G. P. *Mycopappus* 1985.
White, J. F. *Phoma* 1983, 1984, 1986; *Pyrenochaeta* 1983.
White, J. F., Jr. *Acremonium* 1985, 1987; *Phoma* 1987.
White, L. T. *Scytinostroma* 1951.
Whitehead, M. D. *Melanotaenium* 1953; *Neovossia* 1979.
Whitney, H. S. *Darkera* 1975; *Rhizoctonia* 1970.
Wiley, B. J. *Aspergillus* 1973.
Wilhelm, S. *Idriella* 1956; *Phoma* 1985.
Willetts, H. J. *Botrytis* 1984; *Monilinia* 1969, 1977, 1984; *Sclerotinia* 1973, 1975, 1979, 1980; *Sclerotium* 1972.
Williams, P. H. *Albugo* 1963.
Wilson, M. *Gloeotinia* 1954; Uredinales 1966.
Wingfield, M. J. *Ceratocystis* 1980; *Leptographium* 1981, 1983, 1985; *Thyronectria* 1983.
Witcher, W. *Monochaetia* 1971.
Wolf, F. A. *Cephaleuros* 1930.
Wollenweber, H. W. *Ascochyta* 1936; *Diplodia* 1943.
Wong, A.-L. *Monilinia* 1976, 1977; *Sclerotinia* 1973, 1975, 1979, 1980.
Wong, P. T. W. *Verrucalvus* 1984.
Woo, J. Y. *Leptographium* 1983.
Woodhams, J. E. *Phyllosticta* 1982.
Woodhouse, W. W., Jr. *Marasmius* 1971.
Wright, E. F. *Ceratocystis* 1961.
Wyk, P. S. van *Batcheloromyces* 1985; *Vizella* 1976.
Yaegashi, H. *Pyricularia* 1978.
Yamazaki, M. *Tilletiopsis* 1985.
Yarrow, D. *Geotrichum* 1977.
Yarwood, C. E. Erysiphales 1973; *Oidium* 1973, 1978.
Yen, J.-M. *Cercospora* 1980, 1982; *Pseudocercospora* 1979.
Yerkes, W. D., Jr. *Peronospora* 1959.
Yokoyama, T. *Monochaetia* 1975; *Tubakia* 1971.
Yu, Y. N. *Phytophthora* 1984.
Yu, Y.-n. *Microsphaera* 1982, 1983.
Zaayen, A. van *Verticillium* 1982.
Zachos, D. G. *Macrophoma* 1979.
Zambettakis, C. *Botryodiplodia* 1954; *Diplodia* 1954; *Melanotaenium* 1972; *Microdiplodia* 1954; *Thecaphora* 1975; *Tolyposporium* 1973; *Ustilago* 1973.
Zaracovitis, C. *Oidium* 1965.
Zavrel, H. *Ovularia* 1972.

Zeller, S. M. *Atropellis* 1930.
Zentmeyer, G. A. *Hemileia* 1984; *Phytophthora* 1980.
Zhang, J.-z. *Nectria* 1984.
Zhang, Z. Y. *Albugo* 1984.
Zheng, R.-y. *Brasiliomyces* 1984; Erysiphales 1985; *Erysiphe* 1981; *Sawadaea* 1980; *Uncinula* 1977; *Uncinuliella* 1979, 1982.
Zhuang, W. Y. *Phytophthora* 1984.

Zhuang, W.-Y. *Pseudopezicula* 1986.
Ziegler, P. *Gnomonia* 1983.
Ziller, W. G. *Hypoderma* 1978; *Lirula* 1978; *Lophodermium* 1978; *Ploioderma* 1978; Uredinales 1974.
Zogg, H. *Tilletia* 1972, 1983.
Zsuffa, L. *Melampsora* 1983.
Zucconi, L. *Dactylaria* 1984.
Zundel, G. L. Ustilaginales 1953.

INDEX TO GENERA

Acantharia 17
Acremonium 17
Acroconidiella 18
Actinopelte 18
Aecidium 18
Agaricodochium 18
Albugo 19
Aleurocorticium 19
Aleurodiscus 19
Alternaria 20
Amerosporium 21
Amylostereum 21
Angiosorus 21
Anguillospora 21
Angusia 21
Anisogramma 21
Annellophora 22
Anthostomella 22
Anthracoidea 22
Aphanomyces 23
Aphyllophorales 7
Apiognomonia 23
Apiospora 23
Apiosporina 23
Apostrasseria 24
Arcticomyces 24
Armillaria 24
Arthrinium 25
Arthrobotrys 26
Ascocalyx 26
Ascochyta 26
Ascochytulina 27
Ascomycotina 9
Ascotricha 28
Ashbya 28
Aspergillus 28
Asperisporium 29
Asteromella 29
Athelia 29
Atkinsonella 30
Atopospora 30
Atropellis 30
Aulographina 31
Aureobasidium 31
Balansia 31
Basidiophora 32
Batcheloromyces 32
Bifusella 32
Bipolaris 32
Biscogniauxia 34
Blakeslea 34
Blumeriella 35
Bondarzewia 35
Bothrodiscus 35
Botryodiplodia 35
Botryosphaeria 35
Botryosporium 36
Botryotinia 37
Botrytis 37
Brachybasidium 38
Brasiliomyces 38
Bremia 38
Bremiella 38
Briosia 39
Brunchorstia 39
Butlerelfia 39
Byssochlamys 39
Caliciopsis 39

Calonectria 39
Calothyriopsis 40
Catenophora 41
Cenangium 41
Centrospora 41
Cephaleuros 41
Cephalosporium 42
Ceraceopsora 42
Ceraceosorus 42
Ceratobasidium 42
Ceratocystis 43
Cercoseptoria 44
Cercospora 45
Cercosporella 47
Cercosporidium 47
Cerebella 48
Cerotelium 48
Ceuthospora 48
Chaconia 48
Chaetochalara 49
Chaetomella 49
Chaetomium 49
Chaetoseptoria 50
Chalara 50
Cheilaria 51
Choanephora 51
Chondroplea 51
Chondrostereum 51
Chrysocelis 51
Chrysomyxa 51
Ciboria 51
Ciborinia 51
Cladosporium 52
Clathrospora 53
Claviceps 53
Climacocystis 53
Clypeoporthe 53
Coccomyces 54
Cochliobolus 54
Coleosporium 55
Coleroa 55
Colletotrichum 56
Coniella 57
Coniothyrium 57
Cordana 57
Corniculariella 57
Corticium 57
Corynelia 57
Corynespora 58
Coryneum 58
Crinipellis 59
Cristulariella 59
Cronartium 59
Crumenula 60
Cryphonectria 60
Cryptocline 60
Cryptodiaporthe 61
Cryptosphaeria 61
Cryptosporella 61
Cryptosporiopsis 61
Cryptostictis 62
Cryptostroma 62
Cucurbitaria 62
Cumminsiella 62
Cunninghamella 62
Curvularia 63
Cyclaneusma 64
Cylindrocarpon 64

Cylindrocladiella 65
Cylindrocladium 65
Cylindrosporium 66
Cymadothea 66
Cystostereum 66
Cystotheca 66
Cytoplea 66
Cytospora 66
Dactuliophora 67
Dactylaria 68
Daldinia 69
Darkera 69
Davisomycella 69
Deightoniella 70
Dendrophoma 70
Dendroseptoria 70
Dendryphion 71
Dermatodothis 71
Dermea 71
Deuteromycotina 11
Deuterophoma 71
Diachora 71
Diaporthe 71
Diaporthopsis 72
Diatractium 72
Diatrype 73
Diatrypella 73
Dibotryon 73
Dicellomyces 74
Dichomitus 74
Dichotomophthora 74
Dicyma 74
Didymella 74
Didymosphaeria 75
Didymosporina 75
Dilophia 76
Dilophospora 76
Dimeriella 76
Dinemasporium 76
Diplocarpon 77
Diplodia 77
Diplodina 77
Diplorhinotrichum 77
Discella 77
Discochora 77
Discogloeum 78
Discosia 78
Discosporium 78
Discostroma 78
Discula 78
Dothichiza 79
Dothidea 79
Dothiora 79
Dothiorella 79
Dothistroma 80
Drechslera 80
Drepanopeziza 81
Duosporium 81
Echinodontium 82
Elsinoe 82
Elytroderma 82
Embellisia 83
Emericella 83
Encoelia 83
Endocalyx 83
Endocronartium 84
Endothia 84
Endothiella 85

Entodesmium 85
Entomosporium 85
Entyloma 86
Ephelis 86
Epichloe 86
Epicoccum 86
Eremothecium 87
Erysiphales 12
Erysiphe 87
Erythricium 88
Eupenicillium 89
Europhium 89
Eurotium 89
Eutypa 89
Eutypella 90
Exobasidium 90
Exserohilum 90
Fabraea 91
Fistulina 91
Fomes 91
Fomitopsis 91
Foveostroma 92
Fulvia 92
Fusarium 92
Fusicladium 94
Fusicoccum 94
Gaeumannomyces 95
Ganoderma 95
Gelatinosporium 96
General 5
Geniculosporium 96
Geotrichum 96
Gerlachia 96
Gibberella 97
Gibellina 97
Gilbertella 97
Gliocladium 98
Gloeocercospora 98
Gloeosporidiella 98
Gloeosporium 98
Gloeotinia 99
Glomerella 99
Gnomonia 99
Godronia 99
Goplana 100
Graphiola 100
Graphium 100
Greeneria 100
Gremmeniella 100
Grifola 100
Grovesinia 101
Guignardia 101
Gymnoconia 101
Gymnosporangium 101
Haplobasidion 102
Harknessia 103
Helicobasidium 103
Helminthosporium 103
Hemileia 104
Hendersonia 104
Hendersonina 105
Hendersonula 105
Herpobasidium 105
Herpotrichia 105
Heterobasidion 106
Heteropatella 106
Hormotheca 106
Hyalodendron 106
Hyalothyridium 106
Hymenella 107

Hymenula 107
Hypnotheca 107
Hypocrea 107
Hypoderma 107
Hypodermella 107
Hypodermina 108
Hypoxylon 108
Idriella 110
Inonotus 110
Insolibasidium 110
Isthmiella 111
Itersonilia 111
Kabatiella 111
Kabatina 111
Kernkampella 112
Khuskia 112
Kordyana 112
Kriegeria 112
Kuehneola 112
Lacellina 113
Lacellinopsis 113
Lachnellula 113
Laeticorticium 114
Laetiporus 114
Laetisaria 114
Lasiodiplodia 114
Laurobasidium 115
Lecanosticta 115
Leiosphaerella 115
Lepteutypa 115
Leptodothiorella 116
Leptographium 116
Leptomelanconium 117
Leptosphaeria 117
Leptosphaerulina 119
Leptostroma 119
Leptotrochila 119
Leucostoma 120
Leveillula 120
Libertella 120
Lidophia 121
Limonomyces 121
Linocarpon 121
Linospora 121
Lirula 122
Lophodermella 122
Lophodermium 122
Lophomerum 124
Lophophacidium 124
Macrophoma 124
Macrophomina 125
Magnaporthe 125
Marasmiellus 126
Marasmius 127
Maravalia 127
Marssonina 128
Massaria 129
Mastigosporium 129
Melampsora 129
Melampsoridium 130
Melanconis 131
Melanconium 131
Melanodothis 131
Melanopsichium 131
Melanotaenium 132
Meloderma 132
Memnoniella 132
Meria 132
Metacoleroa 132
Microascus 132

Microcyclus 132
Microdiplodia 133
Microdochium 133
Micronectriella 134
Micropera 134
Microsphaera 134
Microsphaeropsis 136
Moesziomyces 136
Monascus 136
Monilia 137
Monilinia 137
Moniliophthora 138
Monochaetia 139
Monochaetiella 139
Monochaetiellopsis 140
Monographella 140
Monostichella 141
Mucorales 13
Muribasidiospora 141
Mycena 141
Mycocentrospora 141
Mycoleptodiscus 142
Mycopappus 143
Mycosphaerella 143
Mycovellosiella 145
Myriangium 145
Myriogenospora 145
Myrothecium 146
Myxofusicoccum 146
Naemacyclus 146
Nakataea 146
Nectria 147
Nectriella 149
Nematospora 150
Neocosmospora 150
Neohendersonia 150
Neokellermania 150
Neopeckia 151
Neopycnodothis 151
Neottiosporina 151
Neovossia 151
Newinia 152
Nigrospora 152
Nodulisporium 152
Nodulosphaeria 153
Nummularia 153
Nyssopsora 153
Ocotomyces 153
Oidium 153
Olivea 155
Olpidium 155
Omphalia 155
Oospora 155
Ophiobolus 155
Ophiodothella 156
Ophiostoma 156
Ophiovalsa 156
Ovularia 157
Ovulariopsis 157
Paecilomyces 157
Papulaspora 158
Paracercospora 158
Paraphaeosphaeria 158
Passalora 158
Pellicularia 159
Penicillium 159
Periconia 160
Periconiella 160
Peridermium 160
Peronoplasmopara 160

Peronosclerospora 160
Peronospora 161
Peronosporales 13
Pesotum 162
Pestalosphaeria 162
Pestalotia 162
Pestalotiopsis 163
Pestalozzina 164
Pezicula 164
Pezizella 164
Phacidiopycnis 165
Phacidium 165
Phaeocryptopus 166
Phaeocytostroma 166
Phaeoisariopsis 166
Phaeolus 167
Phaeoramularia 167
Phaeoseptoria 167
Phaeosphaeria 167
Phakopsora 168
Phellinus 168
Phialophora 169
Phloeospora 169
Phloeosporella 170
Phoma 170
Phomatospora 172
Phomopsis 172
Phragmidium 174
Phyllachora 175
Phyllactinia 175
Phyllosticta 176
Phyllostictina 177
Phymatotrichopsis 177
Phymatotrichum 177
Physalospora 177
Physoderma 178
Physopella 178
Phytophthora 178
Pileolaria 180
Pithomyces 181
Plagiosphaera 181
Plasmodiophora 181
Plasmopara 182
Plectophomella 182
Pleiochaeta 182
Pleospora 182
Ploioderma 183
Podosphaera 183
Pollaccia 183
Polymyxa 184
Polyporus 184
Polyscytalum 184
Polythrincium 184
Poria 184
Postia 185
Potebniamyces 185
Pragmopara 185
Prospodium 186
Protomyces 186
Pseudocercospora 186
Pseudocercosporella 186
Pseudocochliobolus 187
Pseudoperonospora 187
Pseudopezicula 187
Pseudopeziza 187
Pseudophacidium 188
Pseudophaeolus 188
Pseudorobillarda 188
Pseudoseptoria 189
Puccinia 189

Pucciniastrum 191
Pucciniosira 192
Pycnostysanus 192
Pyrenochaeta 192
Pyrenopeziza 192
Pyrenophora 193
Pyricularia 194
Pythium 194
Ramichloridium 195
Ramularia 195
Ramulispora 196
Ravenelia 196
Rhabdocline 196
Rhizina 197
Rhizoctonia 197
Rhizopus 198
Rhizosphaera 199
Rhynchosphaeria 199
Rhynchosporium 199
Rhytisma 199
Rigidoporus 200
Roestelia 200
Rosellinia 200
Sarocladium 201
Sawadaea 201
Schizothyrium 201
Sciniatosporium 201
Scirrhia 201
Scleroderris 202
Sclerophoma 202
Sclerophthora 202
Sclerospora 202
Sclerotinia 202
Sclerotium 203
Scopella 204
Scopulariopsis 204
Scytinostroma 205
Seimatosporium 205
Seiridium 206
Selenophoma 206
Septocyta 206
Septofusidium 205
Septogloeum 206
Septoria 207
Setosphaeria 209
Sirococcus 209
Soleella 209
Sorosporium 209
Sphacelia 209
Sphaceloma 210
Sphacelotheca 210
Sphaeropsis 210
Sphaerostilbe 211
Sphaerotheca 211
Sphaerulina 212
Spilocaea 212
Spongospora 212
Sporisorium 212
Sporonema 213
Sporothrix 213
Stachybotrys 213
Stagonospora 214
Stegonsporium 214
Stegophora 214
Stemphylium 214
Stenella 215
Stenocarpella 215
Stereum 216
Stigmella 216
Stigmina 216

Strasseria 217
Stromatinia 217
Sydowia 217
Synchytrium 218
Talaromyces 218
Taphrina 218
Thanatephorus 219
Thecaphora 219
Thekopsora 220
Therrya 220
Thielaviopsis 220
Thyronectria 220
Tiarosporella 221
Tilletia 221
Tilletiopsis 221
Tolyposporium 222
Trachysphaera 222
Trachyspora 222
Tranzschelia 222
Trechispora 223
Trichocladium 223
Trichoderma 223
Trichometasphaeria 224
Trichoscyphella 224
Trichothecium 224
Trimmatostroma 224
Tubakia 224
Tubercularia 225
Typhula 225
Ulocladium 225
Uncinula 225
Uncinuliella 226
Uredinales 14
Uredo 226
Urocystis 226
Urohendersonia 227
Uromyces 227
Urophlyctis 228
Uropyxis 228
Ustilaginales 15
Ustilaginoidea 228
Ustilago 228
Ustulina 229
Valsa 229
Vararia 230
Venturia 230
Veronaea 231
Verrucalvus 231
Verrucispora 231
Verticicladiella 232
Verticillium 232
Vizella 232
Volutella 233
Waitea 233
Wettsteinina 234
Wojnowicia 234
Xylaria 234
Xylobolus 235
Zimmermanniella 235
Zythia 236
Zythiostroma 236
Zyxiphora 236